JUDGE ADVOCATES IN COMBAT

Army Lawyers in Military Operations from Vietnam to Haiti

by

Frederic L. Borch

OFFICE OF THE JUDGE ADVOCATE GENERAL
and
CENTER OF MILITARY HISTORY
UNITED STATES ARMY
WASHINGTON, D.C., 2001

Library of Congress Cataloging-in-Publication Data

Borch, Frederic L., 1954–
Judge advocates in combat : Army lawyers in military operations from Vietnam to Haiti /by Frederic L. Borch
p. cm.
Includes bibliographical references and index.
1. Judge advocates—United States—History. 2. United States. Army—Lawyers—History. I. Title.

KF7307 .B67 2001 2001031309
355.3'4—dc21

First Printing—CMH Pub 70–77

For sale by the Superintendent of Documents, U.S. Government Printing Office
Washington, D.C. 20402

Foreword

As the U.S. Army has evolved in the past half-century, the Judge Advocate General's Corps has been an important part of its maturing ability to provide effective military force to meet a broad range of challenges. Since the opening days of American involvement in Vietnam, the U.S. Army has been working to meet national security objectives under close public scrutiny in complex, demanding situations. Those conditions call for commanders to make full use of all available staff input, and the special training of the Staff Judge Advocate has often made the lawyer one of the most important sources of insight. This volume recounts numerous instances when new challenges would not have been met so effectively had that specialized staff work not been available.

At one level this is the chronicle of judge advocates at work in the theater of active operations. It provides valuable information on the organization, tasks, and performance of legal offices in a wide array of activities. The author uses the term "combat" to evoke the theater of active operations—justifiable shorthand, but calling too little attention to the operations other than war covered very ably in the last chapter. Throughout, the reader is introduced to Army lawyers who met unexpected requirements while working under tough, demanding conditions.

At another level, this is the history of the evolution of "operational law"—the concept that put those Army lawyers at the right hand of commanders during the deployments of the 1990s so that everything from Status of Forces Agreements to application of the principles of the Law of Land Warfare would be integrated into the planning and execution of operations such as JUST CAUSE and DESERT STORM as well as the many "peacekeeping" operations and deployments in support of civil authorities. This operational focus of judge advocate staff support—in addition to traditional legal support—has enhanced mission success in the politically charged and militarily ambiguous operations that have become common in our era.

Commanders and staff officers should read this book to see how the Army lawyer's role has evolved. Judge advocates should read it because it offers a shortcut to knowledge that ordinarily is gained only through experience. Those interested in the Army's history should read it because it provides details published in no other source. It provides operational context so that the military lawyer's contribution can be easily understood against the larger backdrop, and it describes the lawyer's work in ways that will be accessible to all.

Washington, D.C.
29 May 2001

GORDON R. SULLIVAN
General, USA (Ret.)

The Author

Col. Frederic L. Borch III is a career Army judge advocate. He has an A.B. in history from Davidson College, a J.D. from the University of North Carolina at Chapel Hill, an LL.M. in international and comparative law from the University of Brussels, Belgium, and an LL.M. in military law from the Judge Advocate General's School in Charlottesville, Virginia.

Since entering the Judge Advocate General's Corps in 1980, Colonel Borch has served in a variety of assignments in the United States and overseas, including tours at the Infantry Center at Fort Benning, XVIII Airborne Corps at Fort Bragg, 21st Support Command in Germany, and the 4th Battalion, 325th Airborne Infantry Combat Team, in Italy. His area of expertise is military justice, and he taught criminal law at the Judge Advocate General's School in Charlottesville and served as a member of the Department of Defense Joint Service Committee on Military Justice while assigned to the Office of the Judge Advocate General in the Pentagon. Colonel Borch also supervised the high-profile prosecution of drill sergeants charged with sexual misconduct at Aberdeen Proving Ground and subsequently served as the Deputy Chief, Government Appellate Division, U.S. Army Legal Services Agency. He also served as Staff Judge Advocate, U.S. Army Signal Center, Fort Gordon, Georgia. Colonel Borch is presently attending the Naval War College, Newport, Rhode Island. He is the co-author of two books: *The Purple Heart: A History of America's Oldest Military Decoration* and *The Soldier's Medal: A History of the U.S. Army's Highest Award for Non-Combat Valor.*

Preface

This is a narrative history of Army lawyers in military operations from 1959—when the first judge advocate reported for duty in Vietnam—to 1996—when the last Army lawyers participating in United Nations operations in Haiti returned home to the United States. Its principal theme is the evolution of the role of judge advocates in military operations and how this development has enhanced commanders' ability to succeed. As this role changed dramatically during this period, *Judge Advocates in Combat* explores how soldier-lawyers have evolved from their Vietnam-era responsibility simply to provide traditional legal services—military justice, claims, legal assistance, administrative law—to today's practice of "operational law" in which Army lawyers provide a broad range of legal services that directly affect the conduct of an operation. This new judge advocate role, and the accompanying emergence of operational law—a process that came to full bloom in the 1990s—has increased commanders' ability to achieve mission success in a variety of environments, from conventional combat to operations other than war.

The book explores this theme by examining what individuals did as judge advocates in Vietnam, Grenada, Panama, the Persian Gulf, Somalia, Haiti, and selected other deployments. When people read about those who served at home and abroad as soldier-lawyers, they want answers to at least three questions: Who was there? What did they do? How did that enhance the commanders' ability to accomplish the assigned mission?

In answering the first two questions, *Judge Advocates in Combat* identifies the men and women who deployed in a particular military operation, and it looks at selected courts-martial, military personnel and foreign claims, legal assistance, administrative and contract law issues, and international and operational law matters handled by those judge advocates. Examining who was there and what they did is important, because it captures for posterity the contributions of judge advocates of an earlier era. Viewed from this perspective, *Judge Advocates in Combat* is a contemporary branch history. But, in light of the principal theme—the evolution of the Army lawyer's role from that of a special staff officer providing traditional legal support to the current role in which judge advocates are integrated into operations at all levels—

Judge Advocates in Combat answers the third question by also focusing on those events where Army lawyers blazed new paths, enhancing mission success in nontraditional ways.

During the Vietnam era, lawyers who took on nontraditional roles did so on an individual basis; there was no institutional recognition that such matters were appropriate issues for judge advocates. In 1960, for example, during a coup d'etat led by disaffected South Vietnamese paratroopers, Army judge advocate Lt. Col. Paul J. Durbin left the safety of his home to observe the rebels in action. As a result, Durbin was able to see—and explain—to an American adviser accompanying the coup leader that "advising" this Vietnamese paratroop colonel did not include participating in a rebellion against the Saigon government. Lt. Col. George C. Eblen, who followed Durbin as the lone Army judge advocate in Vietnam, decided to begin monitoring war crimes committed by the Viet Cong against Americans. Eblen's decision to tape record all interviews of U.S. personnel claiming mistreatment resulted in a command policy that a military lawyer participate in all future debriefings involving war crimes. Again, like Durbin, Colonel Eblen stepped outside his traditional role.

Similarly, then-Col. George S. Prugh, the staff judge advocate for the Military Assistance Command, Vietnam (MACV), from 1964 to 1966, spearheaded a number of unique efforts ranging from compiling and translating all existing Vietnamese laws to establishing a legal advisory program that monitored the real-world operation of South Vietnam's criminal justice system. Of particular significance was Prugh's successful effort in persuading the South Vietnamese military that its conflict with the Viet Cong and North Vietnamese was no longer an internal civil disorder. This was a significant achievement for, once its military leaders had accepted the international nature of the conflict, the South Vietnamese government also acceded to this view—and agreed that the provisions of the 1949 Geneva Convention on Prisoners of War would be applied.

Persuading the South Vietnamese armed forces to change their position concerning the conflict—and therefore their view of the status and treatment of Viet Cong and North Vietnamese prisoners—was not a judge advocate responsibility, and Colonel Prugh had not been tasked with resolving this matter. Recognizing, however, that the increasing number of Americans captured by the Viet Cong and North Vietnamese would have significantly enhanced chances to survive if South Vietnam applied the Geneva Prisoners of War Convention to enemy soldiers in its custody, Prugh and his staff spearheaded the efforts to bring about this change.

After Prugh's departure from Vietnam, Col. Edward W. Haughney, his successor as MACV staff judge advocate, continued using the law to support the mission in related ways. Thus, while the MACV provost marshal was primarily responsible for advising the Vietnamese on prisoner of war issues, Haughney and his staff promulgated the first procedural framework for classifying combat captives, using so-called Article 5 tribunals. They also took the initiative in establishing a records system identifying and listing all prisoners of war.

The individual initiatives of Colonels Durbin, Eblen, Prugh, and Haughney illustrated how judge advocates could provide support on a broad range of legal and nonlegal issues associated with operations at the Military Assistance Advisory Group and the Military Assistance Command, Vietnam. Their efforts also demonstrated that Army lawyers could properly focus on more than the traditional peacetime issues of military justice, claims, administrative law, and legal assistance.

As an institution, however, the Judge Advocate General's Corps was slow to recognize that the role of the judge advocate should go beyond traditional peacetime legal support. Consequently, while a number of Army lawyers assigned to Vietnam after Durbin, Eblen, Prugh, and Haughney did provide support beyond the traditional judge advocate niche, the corps institutionally held fast to its traditional view of the proper role of the judge advocate. That is, while appreciating that individual initiatives could enhance mission success, the corps as an organization continued to envision the role of the judge advocate in combat operations as one of providing essentially the same legal services as those offered in a peacetime garrison setting. Accordingly, with the enactment of sweeping changes to the Uniform Code of Military Justice in 1968, most military lawyers continued to concentrate on courts-martial in their day-to-day work.

The Army's experience in Vietnam did, however, plant the seeds for an end to the almost exclusive focus of judge advocates on military justice and peacetime legal issues. The murders at My Lai, and the investigations and courts-martial that followed, all culminated in a 1974 Department of Defense directive tasking Army judge advocates with a new mission: ensuring that all U.S. military operations complied strictly with the Law of War. Accomplishing this new responsibility now required Army lawyers regularly to immerse themselves in many aspects of operational planning and execution—and thus to assume a role that earlier judge advocates did not see as a part of their duties.

A number of perceptive judge advocates realized that this new legal mission inexorably meant judge advocate integration into operations at all levels, and they initiated efforts to move the Corps toward this end.

These efforts, however, were both fragmented and slow. The real catalyst for change occurred in late 1983, when American forces launched Operation URGENT FURY. This operation was a "wake-up call": the wide range of nontraditional legal issues confronted by judge advocates in Grenada propelled the Corps' leadership to move toward a formal recognition that a contingency-oriented Army required judge advocates adept at handling more than traditional peacetime legal missions. It was essential that judge advocates now be schooled in a new role and a new legal discipline: operational law—a compendium of domestic, foreign, and international law applicable to U.S. forces engaged in military operations at home and abroad.

Beginning in 1986, the Corps reconfigured its assets and training to define and support this new judge advocate role. By 1989, when U.S. forces successfully removed Panamanian dictator Manuel Noriega from power during Operation JUST CAUSE, judge advocates were fully prepared to advise commanders on a broad range of legal and nonlegal issues. Thus, Lt. Col. James J. Smith, the 82d Airborne Division staff judge advocate, parachuted into Panama with the lead elements of the assault command post so that he would be able to provide accurate and timely support from the outset of the operation. Less than a year later, when U.S. forces deployed to the Persian Gulf in Operations DESERT SHIELD and DESERT STORM, the evolution in the role of the judge advocate was virtually complete. Commanders at all levels now saw their judge advocates as important force-multipliers. They were first-class attorneys who prosecuted and defended courts-martial, adjudicated claims, and provided legal assistance. But their new role meant that these same lawyers also contributed to mission success in countless other ways—from drafting rules of engagement and providing advice on targeting, using combat contracting to purchase special fabric for force protection, and assisting division intelligence (G–2) personnel in gathering war crimes evidence, to constructing bunkers and fighting positions, investigating friendly fire incidents, and drafting war trophy policies. By the time judge advocates deployed to Somalia in Operation RESTORE HOPE and to Haiti in Operation UPHOLD DEMOCRACY, they were enhancing mission success in still other ways—organizing "cash for guns" programs, overseeing the operation of detainee centers, and even advising on critical political-military matters ordinarily considered the exclusive domain of professional diplomats.

In describing this transformation in the role of the judge advocate in the Army—and in answering "Who was there? What did they do? How did this enhance mission success?"—*Judge Advocates in Combat* examines major and minor operations, both at home and overseas. Separate chapters are devoted to Vietnam, Grenada, Panama, the Persian Gulf, Somalia, and Haiti. A final catch-all chapter looks quickly at judge advocate participation in eleven operations other than war: the U.S. intervention in the Dominican Republic in 1965 and 1966, Cuban refugee resettlement efforts in the United States in the early 1980s, activities of the Multinational Force and Observers in the Sinai in the 1980s and 1990s, Hurricane Hugo relief work in the U.S. Virgin Islands in 1989, disaster relief operations in Western Samoa in 1990, humanitarian operations in northern Iraq in 1991, relief operations in Bangladesh after a cyclone struck that country in 1991, migrant camp activities at Guantanamo Bay Naval Base from 1991 to 1994, the Los Angeles civil disturbance operations of 1992, relief efforts in south Florida in the aftermath of Hurricane Andrew in 1992, and humanitarian aid efforts in Rwanda in 1994. Some deployments receive a fuller treatment than others, but the goal of all is to record accurately the story of Army lawyers participating in the full spectrum of military operations.

Having explained what the book *is* about, it is just as important to state what it *is not* about. It is not a history of the Judge Advocate General's Corps. It is not a history of wartime legal issues, nor a social history of the Army as seen from the judge advocate perspective. It also is not a collection of "lessons learned" or a history of operational law, although the origin of the Judge Advocate General's Corps' institutional recognition of operational law will be obvious to the reader. Consequently, *Judge Advocates in Combat* does not address such topics as the Corps' training base, personnel acquisition, or legal philosophy or the role of the Office of the Judge Advocate General on the Army Staff—except where those matters illustrate the evolution of the role of the judge advocate between 1959 and 1996.

The story of judge advocates in military operations from Vietnam to Haiti is a rich and varied one, and this book offers some interpretations about the participation of Army lawyers in such deployments. Ultimately, however, conclusions about the impact of judge advocates on Army operations—and the continuing evolution of the role of the judge advocate—are best left to each reader.

In December 1995, then–Brig. Gens. John D. Altenburg, Jr., and Walter B. Huffman and Maj. Gen. Michael J. Nardotti, Jr., then the Judge Advocate General, decided to capture the history of judge advocates in recent contingency operations. But for their vision, *Judge Advocates in Combat: Army Lawyers in Military Operations from Vietnam to Haiti* would not exist. Moreover, this project would not have been brought to a successful conclusion without the continuous encouragement and support that followed from Generals Altenburg and Huffman after their promotions to be the Assistant Judge Advocate General and the Judge Advocate General, respectively.

The project got under way in February 1996 with an initial conference involving Chief Historian Dr. Jeffrey J. Clarke and a group of professional historians and book production experts at the U.S. Army Center of Military History. The original concept was a 250-page manuscript that would begin with the experiences of Army lawyers in Vietnam and end with a discussion of the role played by judge advocates in Bosnia. What began as a one-year project ended five years later with this book. Its contents reflect the contributions of more than two hundred individuals who consented to be interviewed and material gleaned from hundreds of official documents and secondary sources.

Many people deserve credit and my thanks, but four deserve special recognition: Dr. Joel D. Meyerson, Col. David E. Graham, Maj. Gen. John D. Altenburg, Jr., and Maj. Gen. (Ret.) George S. Prugh. Dr. Meyerson, a truly outstanding historian at the Center of Military History and an expert on the U.S. Army's experience in Vietnam, spent many hours discussing—and editing—the chapter on lawyering in Vietnam. But even after I had stopped researching and writing on Vietnam, Dr. Meyerson continued discussing the book with me and made invaluable suggestions for improvement. At the Pentagon, Colonel Graham, serving as Chief of the International and Operational Law Division at the Office of the Judge Advocate General, put his own important work aside to read and reread each page of the manuscript. Colonel Graham also spent many hours reworking the Vietnam portions of the book, for which I am most grateful. General Altenburg's participation in this project was pivotal. He and I spent many hours talking about the evolution of the role of the judge advocate. He saw more clearly than anyone else the radical metamorphosis that the role of judge advocates had undergone in the last twenty years and the importance to the Corps and the Army that someone write about this changed role. *Judge Advocates in Combat* would not exist without his frequent guidance and unwavering and direct support when it was needed most. Finally, Maj. Gen. George S. Prugh, the Judge Advocate General from 1971 to 1974 and the author of *Law at War*, the first historical work ever published on judge advocate operations in a combat environment, deserves special thanks. General Prugh wrote letters, sent books and articles, shared his private papers, and spent hours on the telephone with me. His perspective was invaluable.

Those involved in providing research, editing, proofing, page layout, and production of *Judge Advocates in Combat* also deserve special mention. Mr. James Boyd aided my research in military personnel records at the National Archives. Ms. Susan Carroll spent hours editing the manuscript and made many helpful suggestions that improved the text. She also prepared the index. Mrs. Rosemary Land ably keyed successive iterations of changes. At the Center of Military History's Production Services Division, Ms. Catherine Heerin, Mrs. Diane S. Arms, and Ms. Diane Donovan proofread and checked printer's proofs. Under the supervision of Mr. Steve Hardyman, Ms. Sherry Dowdy drew the maps and Ms. Beth MacKenzie prepared the layout. All took a personal interest in ensuring that this book was a first-class production in every respect.

Most of the judge advocates who took part in the military operations discussed in these pages have read portions of the manuscript relating to their participation. Additionally, I have checked and rechecked facts. Nonetheless, there are sure to be errors, and these are my responsibility alone.

Newport, R.I.
29 May 2001

FREDERIC L. BORCH III
Colonel, Judge Advocate
General's Corps

Contents

Charts

Maps

Illustrations

Illustrations courtesy of the following sources: p. 8, George C. Eblen; p. 14, George S. Prugh; p. 32, John J. Douglass; p. 35, J. Stevens Berry; p. 43 *(top)*, Leonard G. Crowley; p. 43 *(bottom)*, Ned E. Felder; p. 50, Jerome W. Scanlon; p. 70, Jack T. Tomarchio; p. 100, James J. Smith; p. 131, Raymond C. Ruppert; p. 136, Mark P. Brewer; p. 150, Patricia A. Ham; p. 182, William B. Hagan; p. 185, Charles E. Trant; p. 223, Victor L. Horton; p. 254, Catherine M. With; p. 259, Mark S. Ackerman; pp. 34 *(top and bottom)*, and 144, Frederic L. Borch; pp. 204 and 225, Richard E. Gordon. All other illustrations from Department of the Army files.

Judge Advocates in Combat

1

VIETNAM

"Will you go to Vietnam?" I was asked in late 1958.
I said: "Where is that? And what will I do?"[1]

—*Col. Paul J. Durbin*
First Judge Advocate in Vietnam

The American involvement in Vietnam began at the end of World War II. Believing that Ho Chi Minh and the Viet Minh would establish a Communist state if the French were forced out of Indochina, the United States went to the active aid of the French. For the next thirty years, Vietnam was the centerpiece of U.S. containment policy in Southeast Asia and the battleground for America's longest war. Before its involvement ended in 1975, some 3.5 million members of the Army, Navy, Air Force, Marine Corps, and Coast Guard would serve in Vietnam, and roughly 58,000 would lose their lives there.

The U.S. Army's presence in Vietnam began in August 1950, when President Harry S. Truman established the U.S. Military Assistance Advisory Group (MAAG), Indochina. Initially, the advisory group funneled American equipment to the French and advised only on the use of this materiel. With the departure of the French and the creation of the Republic of Vietnam in 1955, however, American soldiers assigned to the renamed Military Assistance Advisory Group, Vietnam, began advising South Vietnamese Army units on tactics, training, and logistics—any matter that would improve combat effectiveness. *(Map 1)*

By mid-1960, the MAAG numbered nearly 700 U.S. Army, Navy, Air Force, and Marine Corps personnel, all of whom advised their counterparts in the roughly 150,000-man Republic of Vietnam armed forces. Since the Army of the Republic of Vietnam, with a strength of about 140,000, made up the bulk of the South Vietnamese military, U.S.

P'u-erh
CHINA
Ha Giang
Lao Cai
Nan-ning
CHINA
Tuyen Quang
Lang Son
Phong Saly
Thai Nguyen
Mong Cai
BURMA
Dien Bien Phu
HANOI
Muong Luong
Nam Tha
Haiphong
Samneua
Luang Prabang
NORTH
VIETNAM
Ban Ban
Muong Sen
LAOS
GULF
OF
TONKIN
HAINAN
Muang Nan
Vinh
Paksane
VIENTIANE
Nong Khai
Udon Thani
(Udorn)
Dong Hoi
Muang Nakhon
Phanom
Khammouane
Vinh Linh
DEMARCATION LINE
Tchepone
THAILAND
Quang Tri
Savannakhet
Hue
Khon Kaen
Da Nang
Saravane
Muang Nakhon Sawan
Pakse
Nakhon Ratchasima
(Khorat)
Attopeu
Kontum
Qui Nhon
Pleiku
BANGKOK
Sisophon
Stung Treng
CAMBODIA
Ban Me Thuot
Pursat
Kratie
Mekong River
Snuol
Da Lat
Loc Ninh
GULF
OF
THAILAND
SOUTH
VIETNAM
PHNOM PENH
Svay Rieng
SAIGON
Sihanoukville
DAO PHU
QUOC
Can Tho
SOUTH
CHINA
SEA
INDOCHINA
1960
0
150
Miles
CON SON

Map 1

Army personnel were the largest advisory component, and the chief of the Military Assistance Advisory Group was a senior Army general officer.

With the Korean War in mind, the primary mission of the South Vietnamese military forces was to resist an external attack from the north.[2] This meant delaying any North Vietnamese invasion until the arrival of American reinforcements. Consequently, advisers helped organize the South Vietnamese Army into standard infantry divisions compatible with this conventional task. In time, the advisers busied themselves with every aspect of the new army, from administrative procedures, personnel management, logistics, and intelligence to unit training, mobilization, war planning, and leadership.[3]

After President John F. Kennedy took office in January 1961, the United States assumed an increasingly greater role in South Vietnam in response to a growing internal Communist insurgency. Soon U.S. Army Special Forces teams and Army helicopter units arrived there. Advisers, who previously had been placed at the division level, were now permanently assigned to infantry battalions and certain lower-echelon combat units.[4] Many began to see active combat. In February 1962 the Joint Chiefs of Staff created the United States Military Assistance Command, Vietnam (MACV), as the senior American military headquarters in Vietnam. Although the MAAG was not formally dissolved until May 1964, MACV now directed the ever-expanding American involvement in that country. By 1962 U.S. forces had increased to 11,000 men, and the MACV commander, a four-star Army general, worked diligently to combat the growing strength of the guerrillas who, aided by the North Vietnamese, were everywhere undermining the government of President Ngo Dinh Diem.

The Military Assistance Advisory Group, 1959–1962

The mission of the MAAG staff judge advocate was to render legal aid and advice to the members of the advisory element and to act as legal adviser to the Director of Military Justice, the Judge Advocate General equivalent in the South Vietnamese armed forces.[5] As the U.S. embassy in Saigon had no lawyer among its personnel, the MAAG staff judge advocate also provided legal advice to the ambassador and his staff.

The first military lawyer assigned for duty in Vietnam, Lt. Col. Paul J. Durbin, arrived in June 1959. Other than a part-time Vietnamese secretary assisting with typing, he had no staff—nor had he received much guidance from the Judge Advocate General's Office in Washington, D.C., or JAGO as it was known. Overseas communication

Lt. Col. Paul J. Durbin, deputy staff judge advocate of U.S. Army, Pacific, left, *and Pfc. Matthew Guarino, a legal clerk in the 25th Infantry Division, during a military exercise in Thailand, June 1963. The first military lawyer in Vietnam, Durbin served as MAAG staff judge advocate, June 1959–July 1961.*

was difficult, and judge advocates in those days were unaccustomed to requesting technical assistance from the Pentagon, much less coordinating with it on a routine basis. Colonel Durbin was thus on his own.[6]

At the time an international agreement provided MAAG officers with diplomatic status, which carried with it complete criminal and civil immunity from Vietnamese law. Enlisted soldiers enjoyed diplomatic status equivalent to that of clerical personnel assigned to the U.S. embassy. The government of South Vietnam thus had neither criminal nor civil jurisdiction over those soldiers assigned to the advisory group, and criminal jurisdiction over MAAG personnel was exercised exclusively under the Uniform Code of Military Justice, an arrangement that continued throughout the duration of the conflict.[7]

Colonel Durbin and the Army lawyers who immediately followed him thus provided a full range of traditional judge advocate legal services to members of the advisory component, ranging from wills, powers of attorney, and tax assistance to advice on domestic relations, civil suits, and the filing of claims for damaged property. But no courts-martial were convened prior to Colonel Durbin's arrival or during his tenure as staff judge advocate. The small size of the advisory element and the quality of people assigned meant that there was little crime that could not be handled under Article 15 of the Uniform Code.

In order to deal with claims, Durbin established a claims office for Vietnamese whose property was damaged by MAAG members, mostly in traffic accidents involving military vehicles. He discovered, however, that the concept of filing a claim against the U.S. government was completely foreign to the Vietnamese; they did not make claims against their own government, and thus did not readily pursue claims against the United States.[8]

A few months after settling his family in Saigon, Colonel Durbin found himself in the midst of an attempted coup against the Diem government. On 11 November 1960, three battalions of South Vietnamese paratroopers surrounded the presidential palace and demanded reforms.[9] That morning, Durbin awoke in the dark to the sound of automatic weapons fire. He finally ventured out that afternoon, heading toward the presidential palace. En route, he noticed a jeep pass with a Vietnamese paratrooper colonel, accompanied by an American Army captain. Immediately he flagged down the jeep, asked the American officer if "he was advising on the coup," and subsequently "advised" him to return to his quarters.[10] But the whole incident forced Durbin to consider how one might handle such unanticipated legal issues should they reoccur in the future. As a result, on 28 June 1961, he produced written guidance for MAAG personnel in the "event of a breakdown of internal law and order within South Vietnam," which later became part of the legal annex to MAAG Vietnam Operations Plan 61–61, which addressed such contingencies.[11]

Looking for other ways to enhance MAAG success, and believing that South Vietnam's military justice system would work better if modeled after American, rather than French, military law, Durbin met regularly with the head of the Vietnamese Directorate of Military Justice in order to draft a new criminal code for Saigon's armed forces. The 1928 *U.S. Manual for Courts-Martial*, he thought, would be "ideal for the Vietnamese Army, as it was much more simple than the 1951 *Manual*—not necessarily better—just simpler." His work on a new Vietnamese Code of Military Justice was never finished, however, and the project was abandoned after his departure in July 1961.[12]

Durbin was replaced by Lt. Col. George C. Eblen. Fluent in French, Eblen was well suited to work with Vietnamese government officials, many of whom were French-educated. But to accommodate a growing

Lt. Col. George C. Eblen, left, *with Lt. Gen. Le Van Ty, Chief of Staff, Army of the Republic of Vietnam, before leaving on an inspection tour of Vietnam, 10 February 1962*

workload, he was soon joined by an Air Force and a Navy lawyer as well as by two more Army judge advocates.[13] Together they continued to provide traditional judge advocate legal services to MAAG members. Significantly, however, they quickly moved beyond the delivery of such services and began to investigate alleged violations of the Law of War. Several Special Forces advisers captured by the Viet Cong had escaped, and Eblen interviewed them, tape recording their allegations of mistreatment. His work prompted a MAAG policy requiring that military lawyers participate in all interviews or debriefings involving alleged war crimes, and by mid-1962 such incidents had become so common that Eblen tasked his Air Force judge advocate with creating case files indexing allegations of mistreatment by subject matter and perpetrator.[14] His interest in monitoring war crimes later became the basis for the first MACV directive requiring the reporting and investigation of all such incidents, one of the first major command decisions that clearly reflected the impact that judge advocates could have on traditionally "nonlegal" operational concerns.

In the area of traditional military justice, Colonel Eblen decided in early 1962 that the advisory group's increased size, and the related increase in criminal misconduct, made it desirable to convene summary and special courts-martial in Vietnam.[15] No general courts-martial were convened during his tenure as staff judge advocate, however, as the advisory group was not a general court-martial convening authority. Consequently, when a general court was appropriate, charges were preferred and an Article 32 investigation held in Vietnam. For referral, the accused and the entire case packet were then sent to Schofield Barracks in Hawaii or to Clark Air Force Base or Subic Bay Naval Base in the Philippines, depending on the accused's branch of service.[16]

MACV and U.S. Army Support Group, Vietnam, 1962–1965

The creation of MACV as a unified command in February 1962 and the establishment one month later of the U.S. Army Support Group, Vietnam, as the Army component under MACV headquarters heralded a greater commitment of men and materiel to Vietnam, including lawyers. In August 1962, Colonel Eblen was replaced by Lt. Col. George F. Westerman, and a year later, in 1963, Westerman was replaced by Lt. Col. Richard L. Jones. All Army attorneys were assigned to the advisory group but served both MAAG and MACV headquarters, advising the commands on nonjudicial proceedings under Article 15 of the Uniform Code and assisting with a few summary and special courts-martial. In any event, no general courts-martial were conducted.

The full-time claims judge advocate at MACV headquarters was fully engaged. For example, Maj. William Myers, who arrived in December 1963, was an experienced military attorney and handled all monetary claims filed in Vietnam and payable under the Personnel Claims Act, the Military Claims Act, or the Foreign Claims Act. The most serious were those filed by the Vietnamese under the Foreign Claims Act, generally for property damage or personal injury suffered in traffic accidents involving MACV vehicles. These claims were settled promptly to promote better relations between U.S. forces and the Vietnamese.[17]

Complementing the lawyer buildup at the advisory group and MACV was the addition of an Army attorney to the U.S. Army Support Group. Capt. Arthur H. Taylor arrived in September 1962 and acted as a one-man legal adviser to the brigadier general in command. Taylor's conditions were less than ideal. His office was a tent open to the local weather, in which desktops were quickly covered with insects, paper clips rusted so quickly that they could be used only once, and the frayed electrical wire strung about the makeshift office caused the canvas cloth to catch fire. Security was also a concern. Shortly after arriving, Taylor learned that a Viet Cong attack was imminent, but could find no spare personal weapons for his use. He had his brother in the United States quickly send him a .45-caliber semiautomatic pistol.[18]

Although the support group headquarters was located at Tan Son Nhut airport in Saigon, Taylor frequently journeyed by helicopter and plane as far north as Da Nang and as far west as Bangkok to provide legal advice to the command and its soldiers. Most of his work concerned military justice and legal assistance. One of his most time-consuming tasks, however, was updating commanders on amendments to

Article 15 of the Uniform Code of Military Justice. Congress had amended the article in 1962 by increasing a commander's power to punish nonjudicially, thus providing a better alternative to trial by court-martial for minor offenses. Ultimately, Taylor had an airplane assigned to him for travel throughout Vietnam to see that these changes were properly implemented.[19]

Capt. Charles Baldree replaced Taylor in 1963 and was in turn replaced by Capt. Alfred A. McNamee a year later. A former infantry officer before entering the Judge Advocate General's Corps in 1963, McNamee also found himself serving as a doorgunner on helicopter missions and doing other odd jobs, such as developing hostile-fire pay policies for the command. For this and other "nonlegal" staff work, McNamee received the Legion of Merit.[20] Once again, a judge advocate had stepped out of the "traditional," restricted role of the military lawyer and become a more integrated part of the overall command.

With the formal disestablishment of the Military Assistance Advisory Group in May 1964, Lt. Col. Robert J. DeMund became the first MACV staff judge advocate. He was followed in November by Col. George S. Prugh. Prugh, a graduate of the Army War College, was the first judge advocate colonel to serve in Vietnam as a lawyer. A combat veteran of the World War II Pacific theater, he instinctively appreciated many of the difficulties encountered by American soldiers in Vietnam. But having also been a judge advocate since 1949, with three previous tours in the Pentagon and overseas lawyering in Germany and Korea, he would prove adept at handling legal policy questions at a high level. In fact, because of his broad experience, he was also to serve as legal adviser to the U.S. embassy, the U.S. Information Service, and the U.S. Agency for International Development. Prugh was also the last judge advocate officially to have his family with him; the bombing of the Brink Hotel on Christmas Eve 1964, and subsequent guerrilla attacks on U.S. forces at Pleiku and Qui Nhon, resulted in the return of all dependents to the United States in February 1965.[21]

A Break with Tradition

The staff judge advocate's operation at the MACV was so small that there was minimal formal organization. However, Colonel Prugh assigned his three-person staff specific responsibilities. He tasked Lt. Col. George R. Robinson with the claims mission; his Navy lawyer, the sole legal assistance officer, with administrative law and international affairs; and his Air Force judge advocate with military justice and discipline operations. All four men, however, provided legal assistance and command legal advice as necessary. Again, these responsibilities reflected the traditional legal services provided by judge advocates since World War II.

As the senior American legal officer in Vietnam, however, Colonel Prugh made a decision that would prove a benchmark for change in the way the Judge Advocate General's Corps, and the Army as a whole, historically viewed the role of military attorneys. Shortly after his arrival, he identified three major areas that were, at the time, deemed beyond the scope of "traditional" judge advocate responsibilities. The first involved the status and treatment of captured enemy personnel, the second concerned the investigation and reporting of war crimes, and the third dealt with assisting the South Vietnamese with resource control. Each would take the Army lawyers into uncharted waters.

By the end of 1964, more than 24,000 American soldiers were in Vietnam, with many participating in combat operations. A few were, inevitably, captured by the enemy. Although some survived, Colonel Prugh learned that both sides—Viet Cong and South Vietnamese—often killed enemy soldiers wounded or captured on the battlefield. The fratricidal nature of the war explained these killings, at least in part. But some guerrillas were executed by the South Vietnamese simply because they viewed the guerrillas as "Communist rebel combat captives" who deserved summary treatment as illegitimate insurgents. In short, the Saigon government refused to treat Viet Cong captives as prisoners of war (POWs), maintaining that the Geneva Conventions addressed only armed conflicts between states and not civil insurrections such as the one taking place in South Vietnam. In fact, those guerrillas who did survive capture in the field were generally imprisoned in provincial and national jails along with political prisoners and common criminals.[22] In sum, the South Vietnamese government viewed the enemy as criminals and treated them accordingly. The Viet Cong were usually even harsher in their treatment of captives, executing South Vietnamese soldiers falling into their hands as a matter of routine. Initially, captured U.S. advisers were spared, but when the government of South Vietnam publicly executed a number of enemy agents, the Viet Cong killed several captured U.S. advisers in retribution.[23]

Prugh and his staff quickly concluded that the Viet Cong might reciprocate with better treatment of U.S. captives if South Vietnam were to reverse its position regarding the status of Viet Cong prisoners. A unilateral decision by the Saigon government to acknowledge the applicability of the Geneva Prisoners of War Convention might also, he felt, "ameliorate domestic and international criticism of the war."[24] Prugh and his staff thus worked to convince Col. Nguyen Monh Bich, the

Director of Military Justice, that it was in South Vietnam's best interest to construct prison camps for enemy captives and to ensure their humane treatment during imprisonment. In fact, the more enemy POWs there were in custody, the more likely that an exchange of South Vietnamese and American POWs could be effected.

In December 1964, Colonels Prugh and Bich visited Vietnamese confinement facilities throughout South Vietnam. By American standards, conditions were exceptionally poor—overcrowding, insufficient food, and a shortage of qualified security personnel prevailed. In Da Nang, for example, Prugh saw that one jail, built by the French to house 250 individuals, contained 750 people. Not only were far too many confined in the facility, but combat captives were mingled with prostitutes, thieves, and other criminals, as well as juveniles.[25] But persuading the South Vietnamese to reverse course proved agonizingly slow. Both the MACV commander and the U.S. ambassador strongly supported Prugh's position, but to no immediate avail. Not until mid-1966 did the South Vietnamese set up suitable POW facilities, and the number of such prisoners rose to nearly 36,000 by the end of 1971. On the other hand, the Viet Cong and North Vietnamese never acknowledged the applicability of the Geneva Convention, and their treatment of American and South Vietnamese captives continued to be brutal. Nevertheless, the humane treatment eventually afforded Viet Cong and North Vietnamese Army prisoners exerted constant pressure on the enemy to reciprocate, and more American soldiers and airmen did begin to survive capture.[26] Colonel Prugh's decision to take on this issue again served as an example of the positive results that could be achieved when a judge advocate chose to act beyond the scope of his traditional legal responsibilities.

The second issue of critical importance to Colonel Prugh involved war crimes investigations. When he arrived in 1964, Prugh discovered that MACV had no official policy on how violations of the Law of War should be investigated or on who should conduct such investigations. Believing that the command not only needed "uniform procedures for the collection . . . of evidence relative to war crimes incidents," but that it also must "designate the agencies responsible for the conduct of [such] investigations," Prugh authored MACV Directive 20–4, *Inspections and Investigations of War Crimes*.[27] In preparing the directive in early 1965, he produced a document that defined the different types of war crimes and prohibited acts and that required their reporting to the MACV staff judge advocate. Prugh's original directive governed only investigations of war crimes committed *against* U.S. forces. Subsequently, however, MACV lawyers revised it to include war crimes committed both by and against U.S. military and civilian personnel. By mid-1965, MACV judge advocates were advising, assisting, and reviewing all war crimes investigations in Vietnam. Again, this was a significant responsibility not previously assumed by Army lawyers, and it remained a major mission for MACV lawyers until the end of the war.[28]

The third area identified by Colonel Prugh concerned resource control in South Vietnam. Believing that the defeat of the enemy was impossible without a "plan of national pacification in the form of the blockade of all enemy sources of supply," the Saigon government had issued nearly one hundred legal decrees controlling the distribution of resources.[29] These were to be implemented by a decentralized administrative structure of province and district chiefs as well as many military commanders who had been assigned area responsibilities, with violators being tried by military courts. Materiel critical to the enemy effort—food, medicine, transport, and other items—was to be strictly controlled by monitoring its use and by storing excess supply in government-controlled buildings. In the absence of an effective civil court system in South Vietnam, MACV judge advocates not surprisingly became a focal point for advice on enforcement of resource control regulations. Effective advising, however, meant collecting, translating, indexing, interpreting, mimeographing, and distributing all relevant government decrees and directives. It also meant learning the mechanics of resource control so that practical guidance could be provided to U.S. advisers in the field.[30]

Intervention

The arrival of the 173d Airborne Brigade in May 1965 marked the end of relatively small-scale U.S. Army involvement in Vietnam and the beginning of direct intervention. The unit was soon followed by a host of others, Army and Marine Corps alike, as well as support units and air units of all types and kinds. More soldiers also meant more lawyers and major changes in judge advocate operations, particularly in the area of military justice. In fact, due to the rising number of U.S. soldiers, it was on this subject that the Judge Advocate General's Corps would increasingly focus the bulk of its attention and resources. Legal advisory missions and other "unconventional" endeavors would grow comparatively less significant to those toiling on the battlefield.

The decision to intervene with ground troops quickly established battlefield patterns that would see the United States through the next four years of war. In entering upon this course of escalation, no source

Col. George S. Prugh, right, *MACV staff judge advocate, presents the Air Medal to Capt. John T. Sherwood, Jr., April 1966. Prugh established a unique advisory program in which Army lawyers like Sherwood advised their Vietnamese counterparts on ways to improve their legal facilities and programs.*

of military pressure was overlooked. The bombing of North Vietnam, begun in February 1965, was one part of an evolving American war strategy. In the south, support for Saigon's pacification effort in the countryside continued, but the main focus was on ground combat against the enemy's main forces, both North Vietnamese and Viet Cong, wherever they could be found. And it was in furthering this mission—managing an ever-expanding ground war by maneuver elements of the U.S. Army and Marine Corps—that General William C. Westmoreland and MACV headquarters held center stage.[31]

Westmoreland exercised operational control over U.S. ground forces through three corps-size commands: I Field Force and II Field Force for U.S. Army units and III Marine Amphibious Force for the marines. Each was mated with a South Vietnamese regional command. The field forces were the senior Army tactical commands, and they reported directly to Westmoreland in Saigon. But, in addition to exercising operational control over U.S. units (and any Australian, South Korean, or other allied forces subordinate to them), the two field forces were "to maintain close liaison with MACV's senior advisers with Vietnamese troops" and coordinate with Vietnamese Army corps commanders in their areas of operation.[32]

All Army units arriving in Vietnam were assigned to U.S. Army, Vietnam (USARV), the service component, which exercised command, less operational control of combat forces, and was headed by the senior Army lieutenant general in Vietnam. Established in July 1965, the USARV command grew rapidly—a burgeoning establishment of logistical, engineer, signal, medical, military police, and aviation units. The numbers tell the story: of the Army's eighteen divisions, seven were in Vietnam by the end of 1967.[33] These divisions were the 1st Cavalry Division (Airmobile); the 1st, 4th, 9th, 23d, and 25th Infantry Divisions; and the 101st Airborne Division. The 23d Infantry (Americal) Division was formed in Vietnam as an amalgamation of the 11th, 196th, and 198th Light Infantry Brigades. At the peak of the buildup in early 1969, there were 543,000 U.S. troops from all the services in Vietnam, including recently deployed units such as the 3d Brigade, 82d Airborne Division, and the 1st Brigade, 5th Infantry Division (Mechanized). Joining these soldiers were some 1,100 U.S. civilian employees of the Department of Defense and about 9,000 U.S. civilian employees of U.S. contractors.[34]

The typical U.S. Army division or separate brigade had a designated area of operations, usually covering several Vietnamese provinces within one of the four Vietnamese corps areas, in which subordinate elements sought out the enemy's forces. The 1st Cavalry Division in 1969 illustrates how a typical division operated. Its main headquarters, the location of the commanding general and his principal staff, was north of Saigon at Phuoc Vinh, protected by a battalion-size "palace guard." The division rear headquarters was at Bien Hoa, the location of most of its logistical and administrative support. The 1st Cavalry's three brigades were dispersed, with their respective headquarters at three different base camps located 50 to 100 miles from each other. Battalions in these brigades were located at still other bases, usually settled in with artillery, and the battalions themselves were often dispersed into two or three smaller bases. In sum, the 1st Cavalry Division was spread among a dozen or more base camps and firebases. While the division and brigade bases were fairly permanent in location, the firebases were not, opening or closing depending on the division's mission. Helicopters linked the firebases with base camps, ferrying troops, supplies, and equipment to and from them. Platoon- and company-size elements left their firebases—either on foot or by air—to conduct operations. Most combat operations in Vietnam were no larger than company size. Many were run at night.[35]

From 1965 to 1969, the number of Army lawyers in Vietnam mirrored the ground combat buildup. In early 1965 there were only four Army lawyers in Vietnam—three at the Military Assistance Command and one at the Support Command; by 1969 more than 135 U.S. Army attorneys were "in country."[16] From 1965 to 1969, lawyers served at the headquarters of Military Assistance Command, Vietnam; U.S. Army, Vietnam; I and II Field Forces; and every division and separate brigade, as well as at a number of large support organizations such as transportation and engineer commands. *Map 2* shows the locations of all units with judge advocate support between 1965 and 1969. *Chart 1* illustrates the legal organization of U.S. Army units in Vietnam during the same period.

Lawyering at MACV

In early 1965, the MACV staff judge advocate's office provided legal support in the areas of claims, legal assistance, military justice, international law, and administrative law and also advised the Vietnamese Director of Military Justice and his staff. The reorganization of the American command structure that year resulted in the disappearance of certain of the traditional lawyering tasks. By late 1966, for example, the MACV staff judge advocate had transferred responsibility for claims adjudication to U.S. Army, Vietnam. Additionally, the command no longer convened courts-martial; this mission fell to USARV judge advocates and those assigned to its subordinate units. Consequently, by 1967 the MACV legal office had a slimmed-down organizational structure: a Civil Law and Military Affairs Division, a Criminal and International Law Division, and an Advisory Division.

In the Civil Law and Military Affairs Division, MACV judge advocates advised on currency control and black marketeering, ruled on civilian contractor employees' privileges and access to U.S. military installations and facilities by U.S. civilians, and made determinations of unacceptability for employment under U.S. government contracts. This same division also advised on real estate matters, such as compensating owners for land appropriated for use as military bases or facilities and negotiating commercial leases of property (there were more than 1,300 such leases in Saigon alone by 1970). Finally, it advised the Central Purchasing Agency, Vietnam, on importing, distributing, and selling all post exchange items in Vietnam.

MACV's Criminal and International Law Division furnished "advice and guidance" to subordinate commands on disciplinary and criminal matters. In the area of international law, the division main-

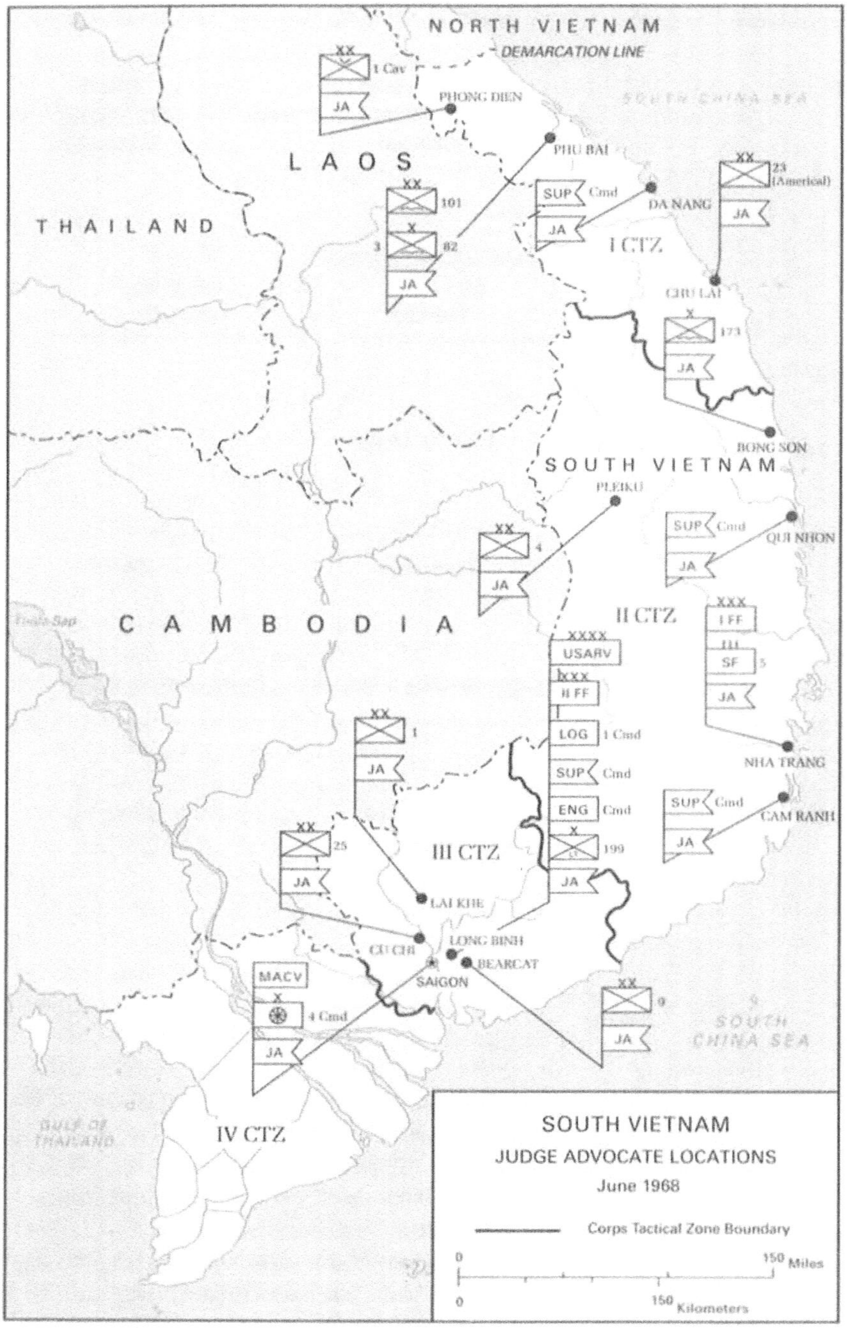
NORTH VIETNAM
DEMARCATION LINE
SOUTH CHINA SEA
1 Cav
JA
PHONG DIEN
LAOS
PHU BAI
THAILAND
101
3
82
SUP Cmd
JA
DA NANG
23 (Americal)
I CTZ
CHU LAI
173
BONG SON
SOUTH VIETNAM
PLEIKU
4
SUP Cmd
QUI NHON
CAMBODIA
II CTZ
USARV
I FF
SF 5
II FF
LOG 1 Cmd
SUP Cmd
ENG Cmd
199
NHA TRANG
CAM RANH
III CTZ
25
LAI KHE
CU CHI
LONG BINH
BEARCAT
SAIGON
MACV
4 Cmd
9
SOUTH CHINA SEA
GULF OF THAILAND
IV CTZ
SOUTH VIETNAM
JUDGE ADVOCATE LOCATIONS
June 1968
Corps Tactical Zone Boundary
0
150 Miles
0
150 Kilometers

Map 2

CHART 1—LEGAL ORGANIZATION OF U.S. ARMY UNITS IN VIETNAM

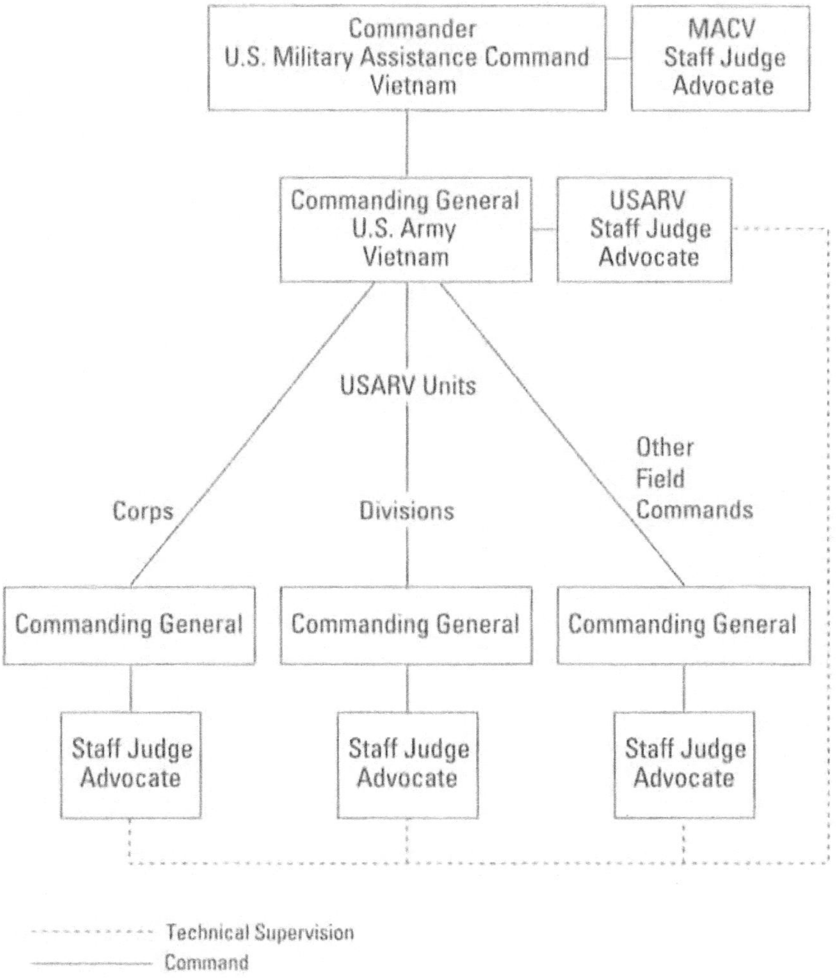

tained files of war crimes investigations and issued opinions on the 1949 Geneva Conventions and the Law of War.[37] The Advisory Division coordinated with the Vietnamese Directorate of Military Justice, participated in legal society and educational programs in Saigon, and monitored the activities of its judge advocate field advisers. These lawyers worked in all four Vietnamese corps areas on legal issues ranging from desertion control, resource control, and security operations to obtaining transportation for Vietnamese judge advocates, providing storage for records of trials, and obtaining material for local prisons.[38] Several Vietnamese attorney-advisers and interpreter-translators served in the office, as well.

MACV's multiservice composition meant that one or more Air Force and/or Navy judge advocates were always assigned to the MACV legal staff, acting as liaisons with their respective services in addition to performing legal tasks. An Army colonel always served as the MACV staff judge advocate (SJA), however, as Army personnel constituted the largest MACV component. The number of Army attorneys at MACV headquarters ranged from a low of three in 1965 to a high of nine in 1967. In early 1967, for example, eight Army attorneys worked for the staff judge advocate, Col. Edward W. Haughney, along with three Navy and five Air Force lawyers, one of whom was a colonel who served as Haughney's deputy staff judge advocate. Supporting these American attorneys were seven Vietnamese lawyers and some fifty Vietnamese clerks and translators.[39] *Chart 2* shows the organization of the MACV Office of the Staff Judge Advocate in 1967. After that date, the number of Army lawyers at MACV headquarters steadily declined.

By 1967 the MACV staff judge advocate's office was formulating legal policy in three major areas: prisoners of war and war crimes, discipline and criminal law, and claims. Agreed-upon policies were promulgated in MACV directives, and over the next few years MACV lawyers wrote and periodically updated more than twenty regulations.

On the subject of prisoners of war and war crimes, MACV continued to develop legal policy based on the 1949 Geneva Convention for the Protection of War Victims (GPW) and U.S. policy. As previously noted, by August 1965 the South Vietnamese had acceded to the American view that the ongoing hostilities constituted an armed international conflict, that North Vietnam was a belligerent, and that the Viet Cong were agents of the government of North Vietnam. Shortly thereafter, the MACV commander directed that all suspected guerrillas captured by U.S. combat units be treated initially as POWs and that U.S. units be responsible for prisoners from the time of capture until their release to Vietnamese authorities. The decision was made, as well, that prisoners would be detained by U.S. units only long enough to be interrogated for tactical intelligence. Thereafter, they were sent to a combined U.S.-Vietnamese center for classification and further processing by the South Vietnamese. However, although Vietnamese authorities took custody of all prisoners, Article 12 of the GPW Convention required that the United States ensure that Vietnamese treatment of these captives complied with the convention. MACV lawyers helped to implement a program that ensured that Vietnamese POW camps complied with international law.

CHART 2—ORGANIZATION OF THE OFFICE OF THE STAFF JUDGE ADVOCATE, MILITARY ASSISTANCE COMMAND, VIETNAM

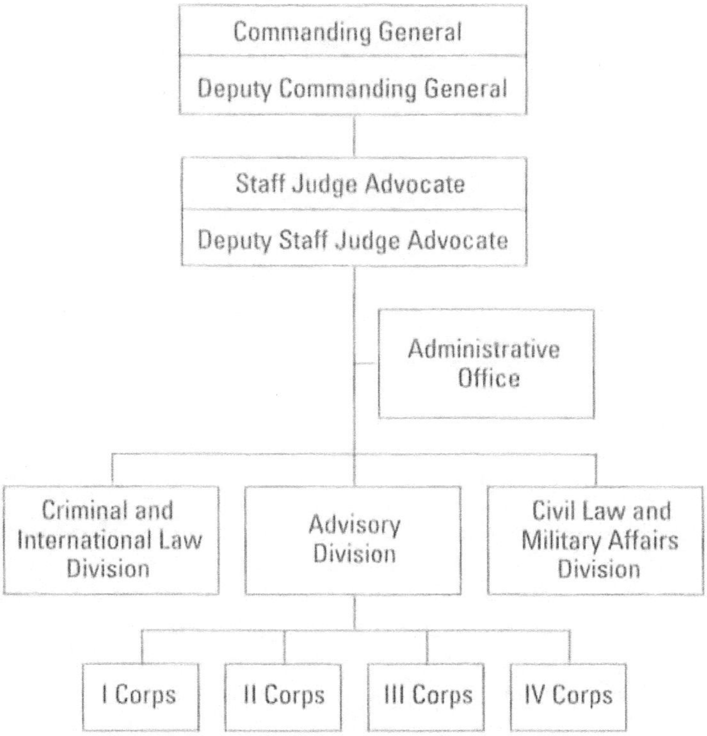

Note: Judge advocate advisers in I, II, and IV Corps were assigned to the MACV corps advisory headquarters in each of these military regions. The judge advocate adviser for III Corps was a member of the MACV Staff Judge Advocate's Office in Saigon.

MACV judge advocates assumed the lead on several other prisoner issues. Most noteworthy was the work done during Colonel Haughney's tenure as the MACV staff judge advocate from July 1966 to July 1967. Haughney and his staff promulgated the first procedural framework for classifying combat captives using so-called Article 5 tribunals. Under that article of the GPW Convention, a "competent tribunal" was used to determine if a person was entitled to POW status. MACV Directive 20–5, *Prisoners of War—Determination of Eligibility*, first issued in September 1966 and updated in March 1968, established and provided authority for a procedural framework for Article 5 tribunals. The directive explained that "the responsibility for determining the status of persons captured by U.S. forces rests with the United States" and that no combat captive or detainee could be transferred to the Vietnamese until "his status as a prisoner of war or non–prisoner of war" was determined. Consequently, a tribunal of three or more officers, including at least one lawyer familiar with the GPW Convention, would hold a formal hearing to decide each doubtful case.

No Article 5 tribunal was required for persons who "obviously" were prisoners of war, such as North Vietnamese Army or Viet Cong regulars captured while fighting on the battlefield. A tribunal was needed only for a detained person whose legal status was in doubt. This was often the case in Vietnam, however, as rarely did the Viet Cong wear a recognizable uniform, and only occasionally did the guerrillas carry their arms openly. Additionally, some combat captives were compelled to act for the Viet Cong out of fear of harm to themselves or their families. Despite these complications, however, the tribunal could still find that such a person merited POW status. Or it could determine that an individual was a "civil defendant" subject to Veitnamese courts or an innocent civilian who should be released. Detailed guidance on conducting an Article 5 tribunal was contained in Annex A of MACV Directive 20–5, including the rights of the detainee and counsel, voting procedures, powers of the tribunal, and posthearing procedures. The MACV staff judge advocate reviewed all tribunal decisions "to insure there were no irregularities in the proceedings."[40] However, the issue regarding the treatment of those regarded as political prisoners by these tribunals remained unaddressed.

In the area of war crimes investigations, MACV attorneys continued the work started by their predecessors, setting out detailed written guidance on investigating and reporting war crimes. Significantly, the command decided, as a matter of policy, that the MACV staff judge advocate—as opposed to the provost marshal or any subordinate headquarters legal officer—would oversee all war crimes matters. Thus, by mid-1968 an updated MACV Directive 20–4, *Inspections and Investigations, War Crimes*, required the reporting of all war crimes committed by or against U.S.forces. All investigations were to be coordinated with MACV lawyers, with technical assistance furnished by qualified criminal investigators.

By the time American troop strength peaked in 1969, MACV Directive 20–4 and other MACV directives contained a significant body of law that defined, prohibited, and provided for the investigation of war crimes. During this time, the most grievous of breaches of the Geneva Conventions were those committed by the enemy. However, American soldiers also committed war crimes, and from 1965 to 1973

Col. Edward J. Haughney
MACV Staff Judge Advocate,
1966–1967

Col. Irvin M. Kent
II Field Force Staff Judge Advocate,
1968–1969

there were 241 cases in which Americans were alleged to have committed such offenses, of which 160 were found to be unsubstantiated. Thirty-six war crime incidents, however, resulted in trials by courts-martial on charges ranging from premeditated murder, rape, and assault with intent to commit murder or rape to involuntary manslaughter, negligent homicide, and the mutilation of enemy dead. Sixteen trials involving thirty men resulted in findings of not guilty or dismissal after arraignment, while twenty cases involving thirty-one soldiers resulted in conviction.[41]

In the area of discipline and criminal law, MACV developed criminal law policy in two major areas. First, it implemented a coherent program for dealing with misconduct committed by MACV members, as well as by U.S. civilians connected with the war effort. Second, MACV judge advocates worked with other U.S. government agencies in Vietnam in suppressing black-marketeering and similar practices.[42]

A particularly challenging legal policy issue involved criminal activity by U.S. civilians. Such misconduct fell into three categories: disorderly conduct, abuse of military privileges, and black-market activities and currency manipulation. In April 1966, at the request of the U.S. ambassador, the MACV staff judge advocate prepared a staff study dealing with the ambassador's authority over U.S. civilians in

Vietnam. That study concluded that the ambassador could issue police regulations for all U.S. citizens in Vietnam, provided that the regulations did not conflict with U.S. or Vietnamese laws.[43] The study also concluded that armed forces police could be used to enforce these regulations and that civilians who violated Vietnamese or American laws could be punished through administrative measures, such as the withdrawal of military privileges and the loss of employment.

The increase in serious crimes committed by U.S. civilians, however, soon made criminal prosecutions appropriate. But who would prosecute? Although some American laws applied extraterritorially, only two practical possibilities existed: the U.S. military or Vietnamese civilian authorities. While American military authorities could exercise control over uniformed personnel using the Uniform Code of Military Justice or MACV directives, their authority over civilians in Vietnam was tenuous at best. Although Article 2 of the Uniform Code did permit the courts-martial of civilians "accompanying an armed force in the field," that provision applied only "in time of war," and it was unclear as to whether the fighting in Vietnam legally constituted a "war." Additionally, even if such was the case, criminal jurisdiction over civilians extended only to those civilians accompanying U.S. forces "in the field." Consequently, while civilian employees of government contractors engaged on military projects, war correspondents with troops on combat missions, and merchant sailors unloading cargo in U.S. Army ports might be subject to military criminal jurisdiction, the more than 6,000 U.S. civilian employees of private contractors, independent businessmen, and tourists in Vietnam were not subject to the Uniform Code under any circumstances. Furthermore, the Vietnamese were either unable or unwilling to prosecute Americans.

As a result, the MACV staff judge advocate devised a two-pronged approach toward civilian misconduct. First, administrative sanctions were meted out to punish and deter civilian wrongdoing. Withdrawing the privilege of a civilian to use the post exchange and commissary or denying him entry onto military bases, along with notification to his employer that this official action had been taken, generally resulted in the civilian offender's having his employment terminated.[44] Beyond the loss of employment, however, nothing could be done. Moreover, if an employee refused to leave Vietnam following the termination of his employment, there was little American authorities could do other than ask the Vietnamese to deport him.[45] Consequently, the use of administrative sanctions to punish civilian misconduct was complemented with a second MACV policy that authorized, when absolutely necessary, the military prosecution of civilians accompanying U.S. forces. With the approval of Ambassador Henry Cabot Lodge, a few such civilian cases were prosecuted by U.S. Army, Vietnam, and 1st Logistical Command. However, this practice was terminated in 1970 when the U.S. Court of Military Appeals declared that civilians were not subject to military criminal jurisdiction in Vietnam.[46]

In an effort to curb American criminal activity in Vietnam, MACV judge advocates like Col. Lawrence H. Williams worked with the Irregular Practices Committee. Formed in August 1967 and consisting of three U.S. embassy representatives and the MACV staff judge advocate, the committee had no operational resources. Rather, it coordinated the work of those elements of the U.S. mission—such as the Military Assistance Command—that had resources to suppress black-marketeering, currency manipulation, and other illegal activities adversely affecting the Vietnamese economy.[47]

Setting uniform criteria for reporting, investigating, processing, and supervising claims in Vietnam was the last major area in which MACV judge advocates formulated legal policy. The buildup of troops and materiel from 1965 to 1969 resulted in an increase in claims for compensation, and MACV lawyers designed and implemented an indemnification program to compensate individuals for losses resulting from U.S. government activity. This effort promoted two important policy goals. First, it was hoped that fair and timely restitution would show the Vietnamese that the government was interested in justice and the welfare of its citizens. Second, an effective claims program was viewed as supporting the war against the guerrillas.[48]

Lt. Col. George R. Robinson, MACV claims judge advocate from November 1964 to November 1965, was chiefly responsible for implementing a fast and fair claims service during the early months of the U.S. buildup. In early 1965 Robinson headed the revision of MACV Directive 25–1, *Claims*, which governed the payment of claims for noncombat damage. When reissued in May 1965, the new directive was easier for nonlawyer unit claims officers to understand and included trilingual (English, Vietnamese, and Chinese) claims forms and a sample letter of condolence, in both English and Vietnamese, for use in making a solatia payment. Such a payment or gift reflected institutional compassion for a serious personal injury or death incurred, and MACV headquarters encouraged unit claims officers to utilize this custom. As a result, a solatia payment, accompanied by the letter found in MACV Directive 25–1, would routinely be made by a unit's claims officer in appropriate situations.[49]

The more difficult policy issue was the payment of combat-related claims. Traditionally, the host country is responsible for such claims, but, initially at least, the Republic of Vietnam had no program to compensate its citizens for injuries or damage suffered in combat situations. For example, in August 1965 a U.S. Air Force B–57 bomber returning from a combat mission crashed in the city of Nha Trang, killing a number of civilians and destroying a great deal of property. Viet Cong radio broadcasts accused the United States of criminal recklessness, and the incident generated adverse feelings toward Americans. Colonel Robinson flew immediately to Nha Trang with two other members of the MACV staff judge advocate's office and began accepting claims from Vietnamese civilians. While Robinson was processing claims, however, the Pentagon issued a statement that no compensation for this disaster could be paid, as damage resulting directly or indirectly from combat was not permitted under the Foreign Claims Act. Colonels Robinson and Prugh, however, convinced MACV headquarters that payments to claimants would gain the goodwill of the people, and Defense Department contingency funds were subsequently used to pay them.

Similar situations resulted in the Military Assistance Command's recommendation that the Foreign Claims Act be amended to allow payment of certain claims indirectly related to the combat activities of U.S. forces, and Congress made such a change to the law in 1968. Subsequently, claims such as those Colonel Robinson had handled in Nha Trang in August 1965 could be honored directly.[50]

Believing that "a successful counterinsurgency program" required respect for law and order, the MACV staff judge advocate established an Advisory Branch in July 1965.[51] Using the law and lawyers to further the allied mission in Vietnam was a unique approach, and by late 1965 the work done by the Advisory Branch accounted for roughly 40 percent of the MACV staff judge advocate's total workload.[52]

At the Saigon level, the advisory effort was aimed at the Directorate of Military Justice and other Vietnamese government agencies and focused on improving such matters as budgeting, desertion control, tables of organization and equipment, and the administration of the courts and prison systems. Outside Saigon, the Advisory Division's field advisers, located in each of the four corps areas, were the eyes and ears of the MACV staff judge advocate, monitoring military discipline in South Vietnamese units, the effectiveness of resource control, and the functioning of South Vietnamese military courts and prisons.[53] Inevitably the pervasive administrative corruption in South Vietnam, coupled with the inherent weakness of Saigon's now troubled civil and military justice system, greatly limited the effectiveness of such endeavors, no matter how strongly pressed. But whatever their ultimate success, the fact that Army lawyers were attempting to address such problems institutionally rather than on an individual basis represented a major change in itself.

Lawyering at USARV

The mission of the U.S. Army, Vietnam, staff judge advocate was to provide legal services for the USARV commander, deputy commander, and staff, as well as for all major subordinate commanders. He also "exercised staff supervision over all judge advocate activities in the U.S. Army, Vietnam."[54] When organized in 1965, the USARV staff judge advocate office had five military lawyers—one colonel, two majors, and two captains. It expanded rapidly, however, and between 1966 and 1969 no fewer than ten lawyers occupied the headquarters office. Initially, the operation was divided into two sections. A Military Affairs Division, with Legal Assistance, Claims, and International Affairs Branches, handled all noncriminal legal matters. A Military Justice Division, with Trial, Inferior Courts, and Review Branches, provided all criminal law support. This two-part framework had been the norm for staff judge advocate operations since World War II. However, when Col. John Jay Douglass became the USARV staff judge advocate in July 1968, he determined that this traditional way of providing legal services was no longer suitable and restructured the office into four divisions: Civil Law, Claims, Military Justice, and Legal Assistance.[55] *(Charts 3 and 4)*

From 1965 to 1969, approximately sixty judge advocates served at U.S. Army, Vietnam, headquarters. Staff judge advocates during these years were Cols. Dean R. Dort (1965–66), Hal H. Bookout (1966–67), William B. Carne (1967–68), John J. Douglass (1968–69), and Wilton B. Persons, Jr. (1969–70). Colonel Persons later served as the Judge Advocate General from 1975 to 1979. Other notables were Maj. John L. Fugh and Capt. Kenneth D. Gray. Fugh later served as the Judge Advocate General from 1991 to 1993, and Gray as the Assistant Judge Advocate General from 1993 to 1997.

The USARV staff judge advocate Civil Law Division prepared opinions and advised on the interpretation and application of laws, regulations, and directives. Subjects handled by the division included issues involving the status of USARV military and civilian personnel (except criminal matters), military security, operations, logistics, and civil affairs. Lawyers in this division reviewed, for legal sufficiency, investigations concerning post exchanges, clubs and messes, security violations, and postal losses; reports of survey; elimination boards; and collateral investigations involving aircraft accidents. The division also

CHART 3—ORGANIZATION OF STAFF JUDGE ADVOCATE'S OFFICE
U.S. ARMY, VIETNAM, DECEMBER 1967

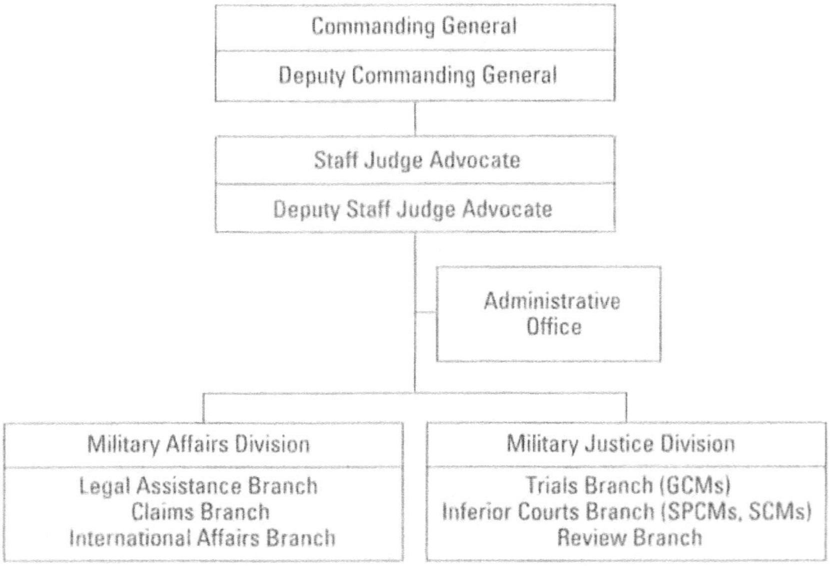

arranged for the travel of soldiers from Vietnam to the United States when they were needed as witnesses in legal proceedings, issued legal opinions on international law, monitored Geneva Convention lectures to USARV troops, provided counsel for respondents at administrative elimination boards, and advised on procurement law matters.

The Army exercised single-service responsibility for processing claims in favor of or against U.S. forces in Vietnam. As MACV had ceased processing claims by 1966, USARV judge advocates had sole responsibility for administering a claims program. The number of claims for damaged or destroyed possessions, equipment, and clothing grew rapidly as the war intensified. Similarly, the buildup of American forces in Vietnam brought with it increased claims by Vietnamese nationals for personal injury and property damage. The impact of heavy military truck traffic on a people accustomed to the bicycle, small car, and animal-drawn wagons also led to many claims.[56] By the end of 1969, the number of claims filed and the resulting backlog were significant.

The USARV commander had the authority to create two foreign claims commissions with a monetary jurisdiction of up to $15,000 each and twelve single-man commissions with a monetary jurisdiction of up to $1,000 each. An award in excess of $5,000 was subject to approval

Chart 4—Organization of Staff Judge Advocate's Office
U.S. Army, Vietnam, September 1968

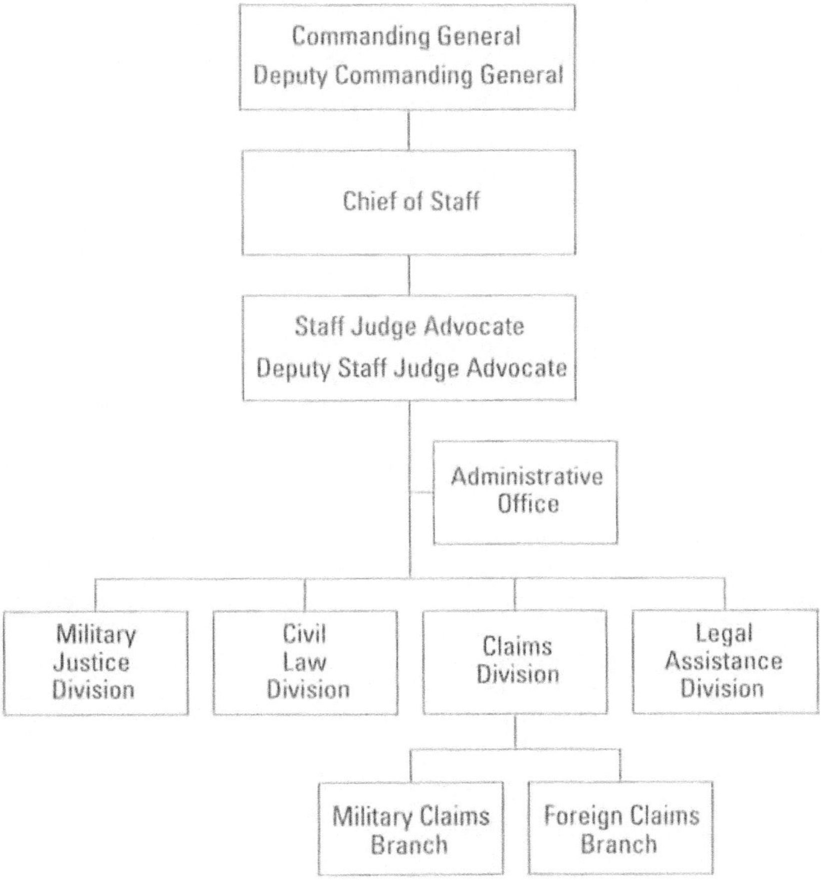

by the appointing authority, and the USARV staff judge advocate was delegated by the USARV commander to act for him in claims matters.[57] The line between combat and noncombat claims was often difficult to draw, but since in almost every case innocent victims required relief, the Vietnamese and Americans worked together so that compensation was available, regardless of cause.[58]

Whatever its importance, claims work continued to be overshadowed by the more traditional labor of military justice, specifically prosecuting and defending during military courts-martial. The Military Justice Act in 1968 required more trained lawyers to serve as defense counsel and as judges, greatly increasing the burden on trial lawyers, while implementing the changes to the code necessitated a new *Manual for Courts-Martial*, which was published just in time for the 1 September 1969 effective date of the new act. The gross numbers tell the story. USARV and its subordinate units conducted roughly 25,000 courts-martial between 1965 and 1969. Of these, 9,922 courts-martial were tried in 1969 alone, at the peak of the U.S. buildup, of which 377 were general courts, 7,314 were special courts, and 2,231 were summary courts. Similarly, a large number of Article 15s were administered between 1965 and 1969—66,702 in 1969 alone. And while there were few illegal drug prosecutions in 1966, a constantly rising drug use rate among U.S. troops translated into more and more criminal prosecutions. By 1969 roughly 20 percent of the special courts tried in Vietnam were for drug-related offenses.[59]

Perhaps the best known criminal incident occurring in Vietnam between 1965 and 1969 was the killing of Vietnamese civilians by soldiers in My Lai in 1968. On 16 March 1968, members of Company C, 1st Battalion, 20th Infantry, an element of the Americal Division, murdered approximately 350 Vietnamese civilians at the small village of My Lai in southern I Corps. There was no official knowledge of the atrocity until April 1969, when a veteran who had heard of the killings wrote to General Westmoreland, then Army Chief of Staff, describing his suspicions and requesting an inquiry. The Army's Criminal Investigation Division determined that 1st Lt. William L. Calley and twelve men under his command were primarily responsible for the killings. In September 1969 Calley was charged with the murder of 109 Vietnamese civilians, and in November of that same year a second soldier, S.Sgt. David Mitchell, was charged with multiple counts of murder and assault with intent to commit murder. Eleven other soldiers were also charged with murder.[60] Of the thirteen men charged, only Calley was convicted. Proceedings against six of the accused were dismissed for insufficient evidence. The others were tried by court-martial and found not guilty.

Of the twelve Americal Division officers accused of covering up the atrocity, only Calley's company commander, Capt. Ernest L. Medina, and his brigade commander, Col. Oran K. Henderson, came to trial. Both were court-martialed, and both were acquitted. Charges against Maj. Gen. Samuel W. Koster, the division commander, for failing to report the killings to MACV headquarters were also dismissed. Secretary of the Army Stanley R. Resor, however, punished Koster administratively by demoting him from major general to his permanent grade of brigadier general and revoking his Distinguished Service Medal.[61]

On 29 March 1971, Calley was found guilty of premeditated murder by a general court-martial convened at Fort Benning, Georgia, and was sentenced to life imprisonment. On 20 August 1971, the commanding general, Third U.S. Army, took action as the general court-martial convening authority. He approved the findings of premeditated murder against Calley, but reduced his sentence to twenty years' confinement. In April 1974, after both the Army Court of Military Review and the U.S. Court of Military Appeals had rejected Calley's appeals and had affirmed the findings and his sentence, the newly appointed Secretary of the Army, Howard H. Callaway, reduced his sentence further, to ten years. With time served, this made Calley eligible for parole after six months and, after serving a short time in jail at Fort Leavenworth, Kansas, he was paroled in November 1974.[62]

While the war crimes committed at My Lai caused much consternation and soul-searching among Americans generally, the ramifications of this tragedy on the Army were just as far-reaching. The Peers Inquiry, so-named because its senior member was Lt. Gen. William R. Peers, thoroughly investigated the murders. In addition to identifying thirty individuals involved in the My Lai killings or in the subsequent cover-up at the Americal Division, the Peers Inquiry also examined the causes of the incident. For Army lawyers, the most significant Peers Report finding was the determination that inadequate training in the Law of War was a contributory cause of the killings. Particularly damning was the report's finding that Law of War training in Calley's unit was deficient with regard to the proper treatment of civilians and the responsibility for reporting war crimes.

Almost immediately, senior members of the Judge Advocate General's Corps began examining ways to correct this deficiency. In May 1970 the regulation governing Law of War training was revised so that soldiers received more thorough instruction in The Hague and Geneva Conventions. Significantly, the revised regulation also required that such instruction be presented by judge advocates, "together with officers with command experience, preferably in combat." This ensured that the training had a firm grounding in real-world experience, while also demonstrating that Law of War instruction was a command responsibility.

Of even greater importance, however, was the initiative taken by retired Col. Waldemar A. Solf. In 1972, while serving as the chief of the International Affairs Division at the Office of the Judge Advocate General, Solf recommended that the Army propose to the Defense Department that it create a DOD-level Law of War Program. This idea was wholeheartedly endorsed by General Prugh, who was then serving as the Judge Advocate General. As a result of Solf's recommendation, DOD Directive 5100.77, promulgated by the secretary of defense on 5 November 1974, not only established a uniform Law of War Program for the armed forces but also made the Army Judge Advocate General's Corps the lead organization in its implementation.[63]

In establishing the Law of War Program, DOD mandated that not only must extensive Law of War training be provided to armed forces personnel, but that judge advocates must be involved in both the development and review of operations plans (OPLANS) in order to ensure that these plans complied with Law of War requirements. This action was of particular significance as it represented the first institutionally mandated involvement of military attorneys in the operational planning process. It was this event that served both as the initial step in modifying the historic mindset of the Army regarding the "appropriate" role to be played by attorneys within the military and as the precursor for the later development of operational law.

Lawyering in the Field

Each major American combat and support unit had its own legal staff. Initially, the Army's table of organization and equipment authorized five lawyers for a division: one lieutenant colonel, two majors, and two captains. A division deployed in Vietnam, however, was often overstrength by one or more judge advocates. Additionally, non–judge advocate attorneys often supplemented a staff judge advocate operation, particularly after the passage of the Military Justice Act in 1968, when more lawyers were needed. For example, although the 1st Cavalry Division was authorized only five attorneys, its staff judge advocate office had fifteen in 1969, roughly half of whom were not members of the Judge Advocate General's Corps.[64]

From 1965 to 1969 more than 350 judge advocates served at units in the field, outside of MACV and USARV. Their work focused almost exclusively on the traditional legal tasks of the judge advocate, particularly that of military justice. The experiences of judge advocates in the 101st Airborne Division serve to illustrate. The 1st Brigade, 101st Airborne Division, and its sole judge advocate, Capt. Frank R. Stone, arrived in Vietnam in July 1965. The division's remaining elements did not deploy until December 1967. Although its table of organization and equipment authorized five judge advocates, the division had seven lawyers by 1968, headed by Lt. Col. Victor A. DeFiori as the staff judge advocate and Maj. Steven R. Norman as the deputy staff judge advocate. In accordance with doctrine, DeFiori and most of his lawyers were

Judge Advocate General Maj. Gen. Kenneth J. Hodson, center, *visits Vietnam in November 1968. With him are,* left to right, *Col. Thomas H. Reese, Staff Judge Advocate, 1st Logistical Command; Col. John J. Douglass, Staff Judge Advocate, U.S. Army, Vietnam; Lt. Col. Robert Clark, Chief, Personnel, Plans, and Training Office, Office of the Judge Advocate General; and Col. Paul Tobin, Military Judge, U.S. Army Trial Judiciary, Vietnam.*

located at the division rear headquarters at Bien Hoa, outside of Saigon. In December 1969, however, the newly arrived staff judge advocate, Lt. Col. George C. Ryker, moved most of his lawyers to the division main headquarters at Camp Eagle in I Corps. Ryker's rationale was that he and his attorneys could provide better legal support at this location, as Maj. Gen. Melvin Zais, the division commander, and his principal staff were there. In addition to Ryker, his deputy, and five judge advocates, five more lawyers, both enlisted men and officers, served with the division.[65] They worked and lived in wooden huts. During the dry season, ceiling fans provided some relief from the 100-degree days, but during monsoon weather, from November to February, almost everyone used an electric blanket or sheet to keep both dry and warm.[66]

Military justice practice in the 101st Airborne was typical of that for a deployed division, with the majority of the offenses being absence without leave, disobedience of orders, and assaults. These were prosecuted at general, special, or summary courts, depending on the severity of the offense. Marijuana use generally was handled under Article 15 of the Uniform Code. Special courts were usually tried by a panel; a military judge was used only if the case turned on a particular legal issue. Initially, at least, confinement of soldiers both before and after trial was a significant problem. Camp Eagle was more than 300 miles from Long Binh Jail, the confinement facility for all U.S. Army troops in Vietnam, and it took nearly a week to send two guards on a C–130 aircraft to deliver or return a jailed soldier. Consequently, in December 1969 the division began sending its pretrial and posttrial confinees to the Marine Corps Brig in Da Nang. Overall, military justice functioned fairly well, although basic reference materials were often lacking. For example, the division had only one copy of the newly published *Manual for Courts-Martial*. Its owner, the new deputy staff judge advocate, Maj. Thomas R. Cuthbert, had received it while attending the new Special Court Judge's Course prior to coming to Vietnam and guarded the book closely until more copies arrived six months later.[67]

The amendments to the Uniform Code contained in the Military Justice Act of 1968 became effective on 1 September 1969. Some commanders, however, opposed relinquishing control over special courts-martial, even after lawyers began serving as defense counsel. For example, in convening special courts, the division's aviation group and artillery commanders continued using nonlawyers as prosecutors, believing that a line officer, rather than a judge advocate, would better represent the command's interest. These commanders accepted that felony-level general courts required judge advocates, but did not appreciate the intrusion of lawyers into their special courts, which they saw as tools of discipline rather than as instruments of justice. However, as nonlawyer trial counsels often did not do well against legally trained defense counsels, even the most reluctant special court-martial convening authorities eventually accepted the presence of judge advocates at these courts. By mid-1970, when USARV regulations required all jurisdictions in Vietnam to attempt to secure a military judge in all special courts-martial, control over special court proceedings passed irrevocably to military lawyers.[68]

Legal assistance for division soldiers was provided primarily by enlisted lawyers. For example, Pfc. Howard R. Andrews, an Alabama lawyer who had been serving in one of the division's field artillery battalions, joined the legal assistance shop at Camp Eagle. While there, Andrews applied for and received a commission in the Judge Advocate General's Corps, and Maj. Gen. John M. Wright, Zais' successor, personally administered his oath of office on the day Andrews was pro-

Working at the USARV staff judge advocate's office at Long Binh in 1968, above, *was different from practicing law,* below, *at the 101st Airborne Division that same year.*

Defense counsel in the "Green Beret Affair," in which Special Forces members were accused of murdering a South Vietnamese double agent, 1969. Left to right, *Capt. J. William Hart, XXIV Corps; Capt. Myron D. Stutzman, USARV; and Capt. J. Stevens Berry, II Field Force.*

moted from private, first class, to captain. After becoming a judge advocate, Captain Andrews was transferred to the 25th Infantry Division and was killed in a helicopter crash only a few months later.[69]

As the number-two lawyer in the division, Major Cuthbert did "a little bit of everything," but "because he could speak artillery" by virtue of his prior service as a line officer with the 1st Cavalry Division, his major responsibility became that of reviewing friendly fire investigations. Although such investigations could have been conducted pursuant to Army regulations, Generals Zais and Wright directed that all friendly fire incidents be investigated under paragraph 32*b* of the 1969 *Manual for Courts-Martial*. This provision required a commander with immediate jurisdiction over a wrongdoer to "make or cause to be made, a preliminary inquiry into the charges or the suspected offenses." As a result, an experienced major in the division was directed to interview witnesses and collect other evidence essential to determining fault in a particular friendly fire incident. After the investigation was complete, Major Cuthbert would conduct a review. This meant examining regulations on fire control and applying the principles of causation and negligence. After receiving Cuthbert's review and pursuing further discussion with principal staff officers in the division, usually the adjutant and operations officer, the division commander would take appropriate action. If the investigation led to a finding of misconduct, the individual at fault usually received an Article 15 as punishment.[70]

Lawyering in the Final Years, 1970–1975

Although American offensive operations continued after 1970, President Richard Nixon had decided to withdraw U.S. forces from Vietnam. He called his strategy Vietnamization, and its intent was to create a strong South Vietnamese military capable of carrying the burden of fighting. Under this new policy, all American operations were designed to buy time for the South Vietnamese in order that they might improve and modernize their forces. Primary targets for U.S. forces were enemy bases in South Vietnam and across its borders. The denial of these bases as staging areas for enemy operations seemed the best way of reducing the long-term threat to South Vietnam.

As a result, while American troops began withdrawing, with most units leaving in 1970 and 1971, aggressive operations continued. One of the largest of these began on 1 May 1970, as units of the 1st Cavalry Division, 25th Infantry Division, and 11th Armored Cavalry Regiment pushed into Cambodia. The Americans discovered large, well-stocked storage sites, training camps, and hospitals, all recently occupied. But most enemy units had retreated beyond the self-imposed limit of the U.S. advance. Despite mixed success in Cambodia, the South Vietnamese, with U.S. aviation support, moved across the border into Laos in February 1971. The aim was to sever the Ho Chi Minh Trail, the enemy supply line into South Vietnam. The result, however, was near-disaster for the South Vietnamese, whose operational weakness at all levels of their army became embarrassingly obvious.[71]

From 1970 to 1972 the number of Army lawyers at MACV headquarters ranged from three to five, with an Army colonel continuing to serve as the staff judge advocate. While providing the same types of legal services as their predecessors, they again became more involved with advising the Vietnamese. In Saigon, these efforts focused on the organization and budget of the Directorate of Military Justice. The Americans also collected, translated, and indexed Vietnamese laws and decrees, prepared staff studies, and participated as members of various MACV and joint MACV-Vietnamese committees.[72] Everywhere the policy of Vietnamization lent to their work a sense of urgency that had been absent since 1965.

Officers of the Staff Judge Advocate's Office, 25th Infantry Division, in Vietnam, March 1970. Front row, left to right: *Maj. Burnett H. Radosh, Staff Judge Advocate; Capt. James A. Burger; Capt. Howard R. Andrews, Jr., the only Army lawyer killed in Vietnam; Capt. Thomas P. DeBerry; Capt. Leary G. Skinner; Capt. Robert B. Warren;* back row, left to right: *Maj. Richard K. Dahlinger, Deputy Staff Judge Advocate; 1st Lt. J. Michael Evans, an Infantry lawyer; Capt. Louis R. Davis; Capt. Michael J. Kemp; CW2 Daniel D. Daniel; Capt. John L. Caden.*

In the field, judge advocates' advisory activities varied widely. Some judge advocates worked with their Vietnamese counterparts on a daily basis and devoted most of their time to Vietnamese military justice procedures, the operation of Vietnamese provincial jails and military prisons, the Vietnamese claims program, desertion control, resource control, and security programs. As the judge advocate field advisers were collocated with senior U.S. advisers, they sometimes also functioned as command judge advocates. As always, a field adviser's success depended on many nontraditional factors, ranging from his own personality and ability to establish rapport with his Vietnamese colleagues to the support provided him by the local U.S. commander. More than anything else, however, a field adviser had to be innovative, identifying problems and discovering practical solutions. Sometimes, the most pressing problems were nonlegal, such as arranging transportation for Vietnamese legal officers, providing storage for trial records, or obtaining materials and equipment to improve the Vietnamese military courts and prisons.[73] *(Map 3)*

Although most advisory efforts during this period focused on programs already in existence, a new challenge was that of working with the Vietnamese military prison system. In the American Army, confinement facilities are the responsibility of the Military Police Corps; in the Vietnamese armed forces, prisons were administered by the Military Justice Corps. Consequently, U.S. Army lawyers at the MACV staff judge advocate's office in Saigon served as advisers to the military prisons, a role for which they had little preparation. Most of the work consisted of periodic visits to the prisons in each corps area, monitoring progress, providing administrative assistance (to obtain building supplies, for example), and coordinating advisory programs with the field advisers. As a practical matter, the MACV staff judge advocate also augmented its advisory staff with a U.S. Military Police Corps officer, who focused on pretrial confinement facilities under Vietnamese control.[74] But, as elsewhere in the advisory effort, the effectiveness of the efforts of judge advocates depended on the willingness of the Vietnamese to accept American advice, and on many key issues the two cultures remained far apart.

Until December 1972, when U.S. Army, Vietnam, merged with the Military Assistance Command, USARV judge advocates also provided the same range of legal services—military justice, administrative law, legal assistance, and claims—offered by their predecessors. The number of military lawyers at USARV headquarters from 1970 to 1972 ranged from eight to twelve. The changing makeup of the corps was reflected in the assignment of a husband and wife "JAG team" to Vietnam, Capt. Nancy W. Keough at U.S. Army Area Command and Capt. James E. Keough at U.S. Army Procurement Agency. Although not the first, Nancy Keough was one of the few female judge advocates to serve in Vietnam.[75]

After 1970 USARV headquarters lawyers handled all courts-martial conducted for U.S. support troops in Vietnam. With more than 40,000 personnel in this category, they found themselves with the largest general court-martial jurisdiction in Vietnam. These same attorneys also provided guidance and assistance to thirteen subordinate general court-martial jurisdictions and approximately a hundred special court-martial convening authorities. This large number of special court jurisdictions resulted from Article 23 of the Uniform Code, a provision that permitted the commanding officer of a detached battalion to convene a special court. In Vietnam, this resulted in some divisions having as many as fifteen special court-martial convening authorities.

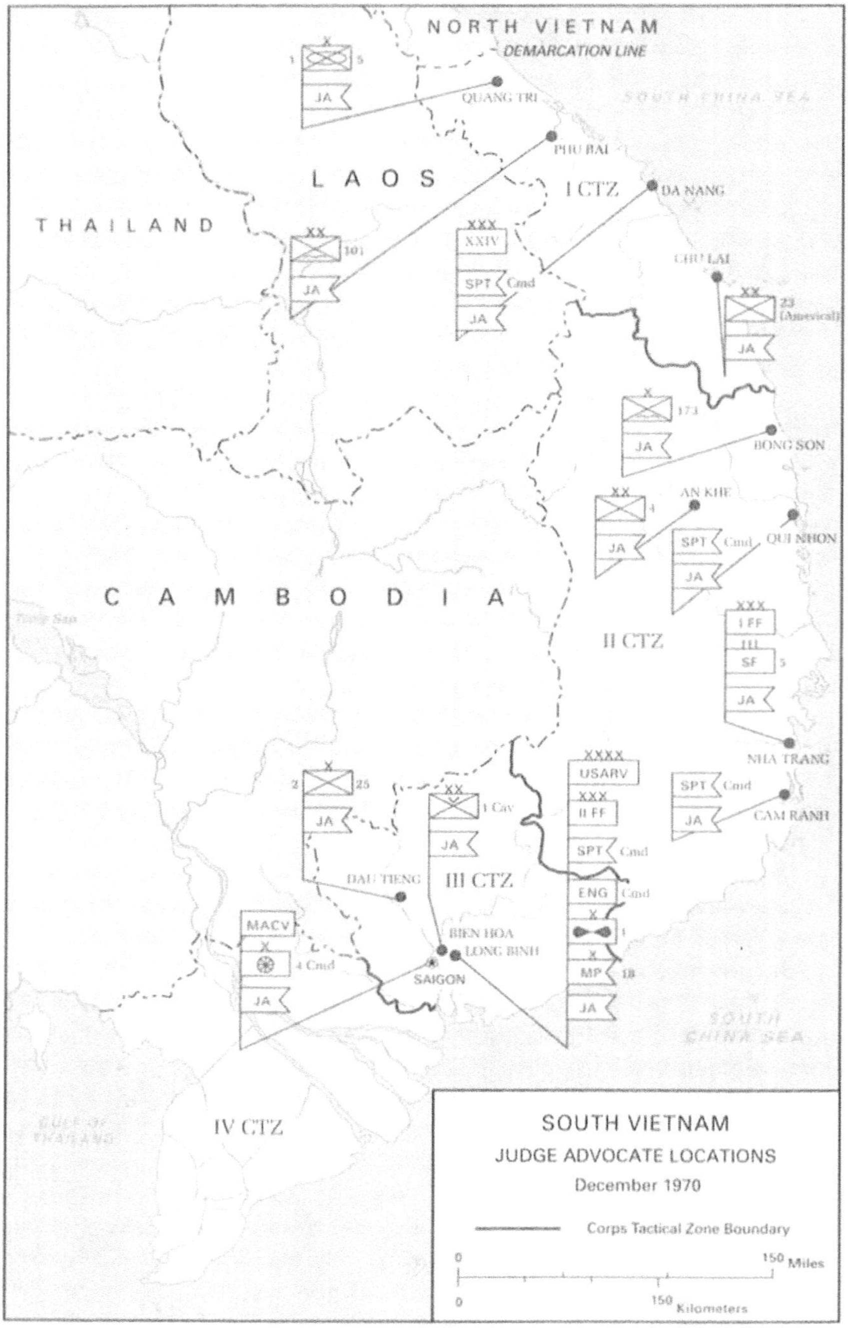
NORTH VIETNAM
DEMARCATION LINE
SOUTH CHINA SEA
QUANG TRI
PHU BAI
LAOS
I CTZ
DA NANG
THAILAND
XXIV
SPT Cmd
JA
101
CHU LAI
23 (Americal)
173
BONG SON
AN KHE
QUI NHON
CAMBODIA
II CTZ
I FF
SF
NHA TRANG
USARV
II FF
CAM RANH
25
1 Cav
DAU TIENG
III CTZ
ENG Cmd
MACV
4 Cmd
BIEN HOA
LONG BINH
SAIGON
MP
18
SOUTH CHINA SEA
IV CTZ
SOUTH VIETNAM
JUDGE ADVOCATE LOCATIONS
December 1970
Corps Tactical Zone Boundary
150 Miles
150 Kilometers

Map 3

With the implementation of the Military Justice Act of 1968 and the resulting lawyer participation at special courts-martial, the many special court-martial convening authorities made managing legal activity difficult. Lawyers, court reporters, and legal clerks who previously had limited roles in the operation of special courts now discovered that prosecuting, defending, transcribing, and processing these courts-martial had increased their work by more than twentyfold in just one year.[76] Consequently Colonel Persons, the USARV staff judge advocate, urged field force and division staff judge advocates to convince their commanding generals to consolidate their special courts at the brigade level. Most did, but some did not. As a result, uniformity in military justice matters was not always achieved.[77]

While thousands of courts-martial were successfully prosecuted in Vietnam, the military justice system was severely challenged. The breakdown of order and discipline in the Army, beginning in the late 1960s, created extraordinary institutional turbulence in Vietnam and elsewhere, raising questions about the Uniform Code and the effectiveness of the military justice system in a combat environment. Drug addiction, racial strife, mutinous behavior on the battlefield indicated that the Army, like the nation, was mired in a crisis of confidence in its mission; fewer and fewer soldiers, especially young draftees, were willing to risk their lives in an unpopular war.

Army leaders looked to the military justice system as a weapon in the fight against rampant drug use. In 1970 Army authorities in Vietnam arrested 11,058 soldiers for illegal drug possession, sale, or use, of which 1,146 cases involved either opium or heroin. Many of these arrests resulted in courts-martial.[78]

Racial tension also played a part in the decline of discipline in Vietnam. Although blacks and whites were united by common needs during combat, the story was different in rear areas, where race relations were often poor. Some black soldiers viewed the military as a racist institution and saw Vietnam as a white man's war.[79] This belief, combined with their experience of discrimination in the United States, made some black soldiers suspicious of the mostly white officer and noncommissioned officer corps. They resented the attempts of Army leaders to prohibit, as contrary to good order and discipline, expression of racial pride, such as black bootlace jewelry and neck chains, "Afro" haircuts, and "dapping," a racial salute involving a series of mirrored, uniform motions. Sometimes, racial unrest escalated into violence. Although most brawls involved only a few soldiers, there were some major confrontations. In 1968 more than 200 black prisoners rioted at Long Binh Jail, and in 1970 a race riot exploded at Camp Baxter in Da Nang.[80] Years later, one judge advocate observed that the unpopularity of the war, the perception that black soldiers were disproportionately represented in the combat arms, and racial dissent in the United States were major contributing factors in the deterioration of discipline and the complementary challenges to authority.[81]

The breakdown in discipline was also reflected in "combat refusals," the official term for disobedience of orders to fight. Although most refusals involved individuals, on at least two occasions company-size units resisted lawful orders. In September 1969 a company of the 196th Light Infantry Brigade refused to recover bodies from a downed helicopter, and in April 1970 CBS Evening News reported the reluctance of a company in the 1st Cavalry Division to advance down a dangerous trail.[82]

The most serious mutinous activity, however, was not combat refusal, but the killing or attempted killing of officers and noncommissioned officers. Called fragging, slang derived from the fragmentary grenade, it was carried out by soldiers against unpopular or overly aggressive leaders. Because most fraggings, or "assaults with explosives" as they were officially called, resulted in injury rather than death, the Army concluded that "in the majority of cases the intent is to intimidate or to scare." Nonetheless, with 209 reported fraggings in Vietnam in 1970, some resulting in death, and with similar attacks continuing over the next two years, Army leaders again looked to the military justice system for a solution.[83] At the same time, some commentators began to voice concerns about the adequacy of the military justice system. They blamed it, at least in part, for failing to maintain good order and discipline during combat operations.[84]

Claims remained a significant part of USARV legal operations. Since those payable to Americans under the Military Personnel Claims Act were handled by unit claims officers, almost all work done by lawyers at the USARV Foreign Claims Division involved claims filed by Vietnamese or other foreign nationals. However, as U.S. law still prevented paying compensation for combat-related damage, and as the Vietnamese government was responsible for paying all claims arising from the combat activities of American forces, USARV lawyers theoretically adjudicated only noncombat claims. In practice, however, Vietnamese claimants always found it easier to approach American officials on such matters and construed their particular cases accordingly. Thus, with 70 to 90 percent of the total processing time in a foreign claim spent in the investigation stage, USARV claims officials often found themselves devoting considerable time to claims that eventually were found to be the result of combat.[85] The time spent on such claims and the inability to pay them posed a constant problem.

By January 1970 the USARV Foreign Claims Division operated two three-man foreign claims commissions with an approval authority for claims up to $15,000. Located in downtown Saigon, one commission processed only those claims arising out of an April 1969 explosion at the Da Nang ammunition supply point. Extensive damage to civilian property from the explosion resulted in some 9,000 claims being filed by November 1971. Some were fraudulent and others were untimely, but all had to be processed.[86] The other three-person commission processed the routine workload received from the field at a rate of about 225 claims per month; all cases that could not be settled by one of twelve one-man commissions in an amount of $1,000 or less were forwarded to this commission. The unusual case that exceeded the jurisdiction of this three-man commission would be forwarded to the Pentagon for a decision by the assistant secretary of the Army (financial management).[87]

During these final years of lawyering in Vietnam, USARV claims judge advocates looked for solutions to three major questions. First, should compensation be paid for combat-related damage or loss based on the reckless and wanton conduct of U.S. forces? Second, who should have claims responsibility upon the complete withdrawal of U.S. forces from Vietnam? Finally, what should be done about increasingly violent Vietnamese-American confrontations over claims for damage or loss?

Under U.S. law, appropriated monies could not be used to compensate for combat-related damage or loss of life. Due to the nature of the war in Vietnam, however, this prohibition resulted in inequities. The battlefield was anywhere and everywhere, with no identifiable front lines. Innocent civilians could not avoid the war or its suffering. Recognizing that compensation for losses relating to the combat activities of U.S. forces could not be paid under the Foreign Claims Act, but believing that this position did not best serve U.S. interests, MACV determined that its assistance-in-kind funds could be used to pay for some combat-related damage. As a result, the USARV Foreign Claims Division processed Vietnamese claims springing indirectly from combat, provided the loss or damage was caused by the reckless or wanton conduct of U.S. forces. While injuries resulting from a firefight between U.S. troops and guerrilla forces were not compensable, loss of life or damage to property caused by a soldier on patrol who indiscriminately fired his weapon into a village was. Paying these claims demonstrated that the Americans took responsibility for their behavior, showed the Vietnamese people that the law could confer a benefit, and, it was hoped, fostered popular respect for the law in Vietnam.[88]

Lt. Col. James D. Clause, far left, *MACV SJA Advisory Division, and Maj. Leonard G. Crowley,* far right, *USARV SJA Foreign Claims Division, receive the Vietnamese Justice Medal (Second Class) from Minister of Justice Le Van Thu,* second from left, *1970. The decoration was Vietnam's highest judicial honor for foreigners.* Bottom, *Capt. Ned E. Felder, the only judge advocate to serve two consecutive twelve-month tours in Vietnam, receives the Bronze Star Medal from Brig. Gen. John S. Lekson, Chief of Staff, II Field Force, Vietnam, February 1968.*

Who should have claims responsibility upon complete withdrawal of U.S. forces from Vietnam? As early as October 1971, Maj. Ralph G. Miranda, chief of the Foreign Claims Division, had recommended to the USARV staff judge advocate that a plan be formulated for processing foreign claims submitted after U.S. forces departed Vietnam. Miranda anticipated that Vietnamese nationals would continue filing claims then handled by the USARV Foreign Claims Division. He also believed that when departing U.S. forces returned leased real properties prior to the expiration of the leases, Vietnamese landlords would file substantial claims against the United States. Anticipated as well was the fact that, as U.S. troop strength decreased and various support agencies terminated operations, the need for local national employees would diminish and claims for termination pay would result.[89] Finally, claims would also arise out of contracts with Vietnamese businesses for goods or services. After coordination with the Military Assistance Command and the Air Force and Navy, it was decided that the Army would continue to process foreign claims at U.S. Army, Pacific, the theater-level component headquarters in Hawaii. Subsequently, foreign claims were accepted at the Defense Attaché Office in Saigon and by the U.S. consular staff throughout South Vietnam and forwarded to Army headquarters in Hawaii.[90]

The third claims issue of personal interest to claims judge advocates was that of the actions that could be taken "to cool off potentially explosive situations" involving claims for loss or damage. After 1970, as the Vietnamese saw American units departing and as the backlog of claims cases increased, a general officer reported that "they visualize that the only means of getting a prompt and adequate settlement is via the confrontation approach."[91] On one occasion, several hundred Vietnamese claimants blocked the entrance to a U.S. military compound in the XXIV Corps area, refusing to leave until their claims were paid. The disturbance was quelled only after the chief of the USARV Foreign Claims Division flew from Saigon to Da Nang, met personally with the village and hamlet chiefs, and assured them that "we would do all within our power to settle the problem as soon as possible."[92]

Matters were often not that simple. In May 1970 a 2 1/2-ton Marine Corps truck struck and killed a young Vietnamese boy. Almost at once, a crowd of more than a hundred Vietnamese surrounded the truck containing the marines. The local commander requested that Capt. Donald A. Deline, the new Da Nang claims officer, go to the accident scene at once. Arriving with some claims forms in his old International Harvester truck, Deline discovered that concertina wire had been placed around the Marine Corps truck. The dead child was lying on an altar in front of the truck, and the boy's mother and others were praying loudly. South Vietnamese soldiers were also on the scene, and they, together with the local mayor, informed the captain that they wanted money. Late that evening, at 11 P.M., a Marine Corps officer appeared on the scene. After making a small solatia payment to the victim's family, he and Deline started to leave the house in which the discussions had taken place. Although armed with a .45-caliber pistol, Captain Deline was held down in his chair, and the Marine officer was escorted out. For the next two to three very tense hours, Deline and the marines in the truck remained captive.

Early the next day, at 2 A.M., a Marine Corps colonel arrived by jeep with $3,000 to $5,000 in Vietnamese piasters of his own money. The crowd permitted the colonel and the marines in the truck to depart, leaving Deline at the scene. The Marine Corps colonel returned his men to their barracks and sent two military policemen for Deline. At the time, the captain was undecided as to whether he and the police should use force and "push our way out or not." Finally, they did force their way out of the house, and although the Vietnamese were yelling angrily and striking the three Americans, Deline and the two military policemen managed to escape.[93] Such was the initiation of the young JAG officer to his tour in Vietnam.

During this period, one of the most celebrated investigations reviewed by USARV lawyers did not involve a claim or a war crime. Rather, it concerned the attack by enemy sappers on Fire Support Base MARY ANN, an Americal Division outpost in I Corps. In March 1971 a group of fifty to sixty well-prepared enemy penetrated MARY ANN's perimeter and, tossing grenades and satchel charges into the tactical operations center, killed or wounded virtually all of the officers. An investigation concluded that the failure of the officers in charge to post guards or follow other proper defensive procedures was grossly negligent and contributed directly to the heavy American casualties—thirty dead and eighty-two wounded.[94] Maj. William K. Suter, newly assigned to the USARV staff judge advocate office, was tasked with reviewing the MARY ANN investigation, fixing responsibility for the incident, and recommending an appropriate course of action. After digesting the classified report's eleven volumes, Suter briefed Lt. Gen. William J. McCaffrey, the USARV deputy commander. He recommended no courts-martial, but urged reprimands, administrative elimination actions, and adverse efficiency reports. General McCaffrey approved all of these recommendations.[95]

Until the last combat units departed Vietnam in 1972, judge advocates lawyered actively with these units. The experiences of military attorneys at the 1st Cavalry Division illustrate the lawyering in the final years of U.S. presence in Vietnam. By 1970 fifteen lawyers were providing legal services in the division. One of the attorneys trying courts-martial was Capt. Royce C. Lamberth. Lamberth had served briefly as a judge advocate at XVIII Airborne Corps at Fort Bragg, North Carolina, before arriving at Phuoc Vinh in November 1969, where he immediately assumed a heavy courts-martial caseload, serving as both prosecutor and defense counsel. While the general courts-martial were tried at division headquarters, the inferior courts-martial were often tried at the brigade bases, as the commanders did not want witnesses "leaving the field." Consequently, Captain Lamberth, accompanied by the military judge and his opposing counsel, routinely flew in a small unarmed observation helicopter to these bases for the trials. Proceedings were typically held in a tent.[96]

During his year in Vietnam, Captain Lamberth tried more than 200 cases. The most memorable involved defending a team of six Rangers accused of mutilating the bodies of enemy soldiers. The Rangers had ambushed some North Vietnamese soldiers bicycling down the "Jolley Trail," a major infiltration route into South Vietnam. One or more of the Rangers later boasted over a few beers that, after killing the enemy soldiers, they had "cut open the bodies from throat to groin and stuffed them with rice" from the 100-pound burlap bags strapped to the enemy bicycles. This "calling card" was intended to strike fear into any enemy who later happened upon the dead men. The Rangers, however, soon regretted their braggadocio as their alleged mutilation of the dead was reported by others as a war crime.

A lieutenant colonel with the MACV inspector general's office arrived at the 1st Cavalry Division to interview the six Rangers. Each man told the same story: they had ambushed and killed the enemy, but no mutilation of the dead had occurred; that had just been bragging. After reducing their statements to writing, the investigator asked the six Rangers to submit to a polygraph. They balked. All asked for a lawyer, and Captain Lamberth was assigned to represent all six men. With his clients facing courts-martial, Lamberth filed a motion requesting that Maj. Gen. Elvy B. Roberts, the division commander, "produce" the bodies of the dead North Vietnamese. He argued that only if the bodies were produced would the six Americans "be able to establish their innocence." After a late-night staff meeting that included the chief of staff and the G–3 (operations), the commanding general decided that it would be consistent with planned operations in the area to send an aerial rifle platoon to search for the bodies of the North Vietnamese. Lt. Col. Ronald M. Holdaway insisted that the defense counsel go on the mission to ensure that there would be no later claim of a cover-up. As a result, Lamberth learned he would be departing by helicopter at first light.

Air Force jets and Cobra helicopter gunships "prepped" the insertion site for the Huey utility helicopter, or "Slick," carrying Lamberth and the six Rangers. Then, about 100 feet above the bomb crater where the insertion was to occur, the engine stalled and the helicopter crashed. Believing that they had been shot down, Lamberth, the only officer aboard other than the warrant officer pilot, and the Rangers "fired like hell" from their perimeter into the jungle. When no fire was returned, the men realized that mechanical failure had caused the crash. They radioed for a Sky Crane helicopter to recover the crashed aircraft and for a new Slick to retrieve them. Meanwhile, Lamberth and the Rangers proceeded on foot down the Jolley Trail, eventually finding the bicycles, burlap bags containing rice, and much blood. One soldier came across an enemy bunker, which was blown up with hand grenades; others located a bridge along the trail, which was also destroyed. But no bodies were found. Lamberth and his clients returned without further incident, and in the absence of corroborative evidence no courts-martial charges were preferred.

Last Army Lawyers, 1972–1975

The continued withdrawal of U.S. forces meant decreasing mobility, firepower, intelligence, and air support. When the North Vietnamese Army launched its Easter offensive in March 1972, total U.S. military strength in theater was about 95,000, of which only 6,000 were combat troops. Responsibility for countering the enemy invasion thus fell almost completely on the South Vietnamese Army, strongly supported by U.S. advisers and American airpower. In the end Hanoi's conventional troops were thrown back, but only after heavy losses had been suffered by both sides. Subsequently, the United States, North and South Vietnam, and the Viet Cong signed an armistice that went into effect early the following year, promising a cease-fire and national reconciliation. Almost immediately, the U.S. military command stood down, all remaining American troops in Vietnam departed, and direct American military involvement there came to an end. U.S. advisers, who had provided the backbone of the South Vietnamese command structure, were also withdrawn, never to return.

For Saigon, the U.S. withdrawal proved calamitous. Far from ending the fighting, the armistice left South Vietnam competing with the enemy for territory. Soon the inherent weakness of the Saigon regime and its military forces became increasingly apparent. At the same time a war-weary U.S. Congress steadily reduced American military aid, forcing Saigon to reduce its tempo of operations in order to husband its diminishing resources. The end was not long in coming. In January 1975 North Vietnamese military forces seized Phuoc Long Province in III Corps and, when the United States failed to respond, continued their offensive. When President Nguyen Van Thieu withdrew his forces to defend Saigon to the south, the action provoked panic among both troops and civilians, making a coordinated defense impossible. Some South Vietnamese units fought well, but most disintegrated. Saigon fell to the enemy on 30 April 1975, and the remaining American technicians, embassy personnel, and others were hastily evacuated.

The agreement that had ended American participation in the war, known popularly as the Paris Peace Accords, had been signed on 27 January 1973. Initially, it provided for a Four-Party Joint Military Commission for sixty days to oversee a mutual troop withdrawal, serve as a forum for communication among the four parties, and assist in the implementation and verification of the agreement. Additionally, the commission was to arrange the return of prisoners of war and gather information about those missing in action.[97] The four parties quickly agreed to create the main commission in Saigon, supported by seven regional commissions throughout the country. Military representatives from each of the four parties were appointed to each of these commissions. Having determined that Army judge advocates should participate in the work of the Joint Military Commission, Col. Joseph Tenhet, the USARV/MACV staff judge advocate, selected Maj. Paul P. Dommer, the incumbent chief of the Advisory Division, as the legal adviser to the U.S. delegation to the central Four-Party Joint Military Commission in Saigon.[98] Junior judge advocates from Tenhet's office were detailed as legal advisers to the regional commissions.[99]

Capt. Vahan Moushegian, Jr., was one of those selected to serve as a regional Joint Military Commission legal adviser. Assigned to the USARV staff judge advocate office, Moushegian joined the Joint Military Commission in Region 5, located in Bien Hoa, a few kilometers north of Saigon. Col. Walter F. Ulmer, the chief of the U.S. delegation, informed Captain Moushegian that he was to be "the delegation's expert on the Paris Peace Accords," and, in the formal meetings of the Region 5 Commission that followed, Moushegian advised and assisted both Ulmer and the deputy chief of the U.S. delegation.

Because Colonel Ulmer's Viet Cong counterpart "never came out of the jungle" to represent the Provisional Revolutionary Government, the commission's four deputy chiefs of delegation soon were meeting a few hours every other day around a square table covered with green felt. In discussing the intent and implementation of the Paris Peace Accords, the participants wrangled constantly over how the provisions should be interpreted. Little was achieved at the formal sessions. The topics of the meetings ranged from the significant (repatriation of American and South Vietnamese prisoners of war), to the ordinary (the ability of the North Vietnamese and Viet Cong delegations to travel freely throughout Region 5), to the absurd (whether the fans at the conference table adequately cooled the attendees). While the Provisional Revolutionary Government and the North Vietnamese generally supported each other, the Americans and the South Vietnamese were sometimes at odds, making it difficult to present a united front or to pursue a common strategy in the talks. Additionally, as the accords required any decision reached by the Joint Military Commission to be unanimous, one party's objection blocked any progress.

Captain Moushegian's role evolved over time. He assumed, in addition to his responsibilities as the legal adviser, the duties of principal liaison officer for the U.S. delegation. Thus, when the deputy chiefs of delegation stopped meeting formally due to a lack of measurable progress, the liaison officers were instructed to meet regularly to ensure that dialogue continued on the implementation of the accords. That said, "almost nothing was accomplished by the Joint Military Commission," in Moushegian's view, because "there were only eight weeks [and] the Viet Cong and North Vietnamese would agree to nothing, knowing that the United States was leaving Vietnam."

On 27 March 1973, the U.S. military headquarters stood down and the last American combat troops left Vietnam. The Four-Party Joint Military Commission also ceased to function, and Major Dommer, Captain Moushegian, and the other judge advocates working with it left for the United States. A new organization, the Four-Party Joint Military Team, now replaced the Joint Military Commission. From the perspective of the U.S. delegation, this new body had two main functions: locating and recovering the remains of Americans who had died in captivity and discovering the whereabouts of Americans still missing in action.

A lone Army lawyer now served in Vietnam, assigned as the legal adviser to the U.S. Delegation to the Joint Military Team. The first adviser was Maj. Charles R. Murray, who served with the team from the end of March until the middle of July. His replacement, Capt. Jerome W. Scanlon, Jr., arrived in July 1973. Met by Murray and a driver in a government sedan at Tan Son Nhut airport, they unexpectedly found themselves under fire en route to their quarters. Traveling in front of their vehicle had been a Vietnamese Army truck filled with

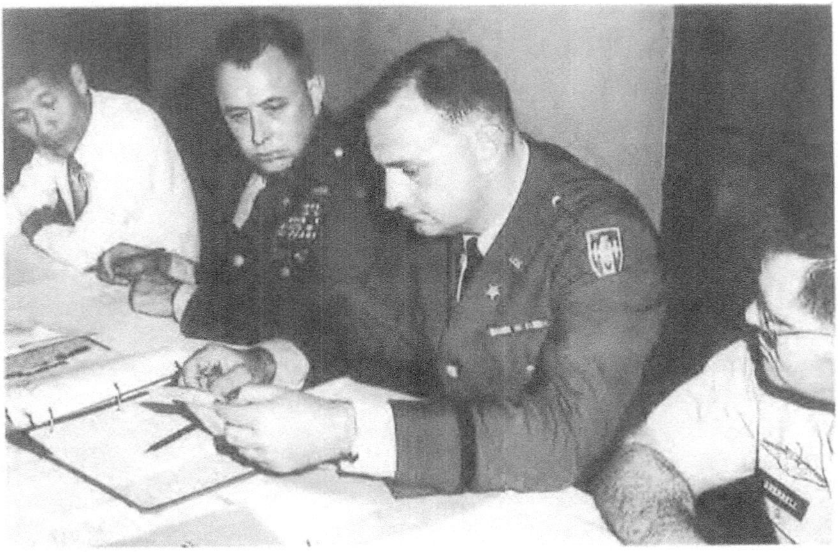

Capt. Jerome W. Scanlon, Jr., legal adviser to the Four-Party Joint Military Team, examines documents provided by the North Vietnamese at a 6 March 1974 meeting in Hanoi. Although a judge advocate, Scanlon wore general staff insignia at the direction of the chief of the U.S. delegation.

prisoners. When one of them jumped from the truck and ran past the sedan carrying Scanlon and Murray, a Vietnamese Army guard, with no hesitation, opened fire on the escapee. His bullets missed the prisoner, but struck the sedan. Fortunately, no one was hurt, and the prisoner was recaptured. Yet, as this was his first day in country, Captain Scanlon was sure "it would be a long year."[100]

Arriving without further incident at the Joint Military Team's offices, Scanlon assumed his duties at the Negotiation Division, advising Col. William W. Tombaugh, the chief of the U.S. delegation, on the rights and obligations of all parties under the Paris Accords. Gaining information concerning U.S. personnel who had died while prisoners or who remained missing in action required the compilation of files on missing Americans and the excavation of areas under North Vietnamese and Viet Cong control. The captain participated in all meetings of the Joint Military Team and reviewed files on missing persons prior to the release of these papers to the North Vietnamese or Viet Cong. The United States possessed information, obtained from prisoners of war already released, that particular individuals had been seen alive in North Vietnamese or Viet Cong custody. When the latter denied any knowledge of the missing person, the U.S. delegation would release the relevant information it held. Scanlon's task was to examine each file, ensuring not only that the information in it was accurate, but that any information disclosed was properly declassified. During his year in South Vietnam he journeyed by C–130 aircraft to Hanoi more than ten times. Captain Scanlon finally departed Vietnam in July 1974. His replacement, Maj. J. Lewis Rose, continued providing the same legal services. When Saigon fell on 30 April 1975, Rose was performing temporary duty in Hong Kong. He was the last judge advocate to serve in Vietnam.

Conclusion

How did the sixteen years of service by judge advocates in Vietnam affect the manner in which the Judge Advocate General's Corps would approach its delivery of legal services in the future? Institutionally, the Corps failed to view its years in Vietnam as a basis for engaging in any substantial modification of the way in which it had traditionally practiced military law. With the exception of an extensive effort to incorporate Vietnam lessons learned into both the Law of War training materials prepared and provided by the Judge Advocate General's School, little was done to capture the unique aspects of the Corps' Vietnam experience.

Looking back, however, it is clear that a metamorphosis in the role of the Army lawyer was under way. The efforts of individual judge advocates in Vietnam to do more than simply provide legal services identical to those offered in a peacetime garrison environment often contributed to mission accomplishment. Army lawyers proved to be principal players in dealing with such issues as the investigation and documentation of war crimes, the classification of detainees and treatment of prisoners of war, Law of War instruction, and the provision of advice to host nation authorities on a wide range of subjects.

Yet, while these initiatives were significant, it was the development of the 1974 DOD Law of War Program—a direct result of the Army's efforts to respond to the Peers Report on My Lai—that would most directly affect the future role of judge advocates in the Army. Whether these judge advocates realized it or not at the time—and most did not—successful implementation of the Law of War Program would require that they begin to communicate directly with commanders and their staff principals throughout the operational planning process, identifying and resolving issues of both a legal and a nonlegal nature.

Not until 1983, however, following the deployment of judge advocates to Grenada in Operation URGENT FURY, would the Judge Advocate

General's Corps fully appreciate the need to implement and institutionalize a process by which Army lawyers would be trained in and practice a newly conceived and formulated body of law directly applicable to the conduct of military operations. Although "operational law" was then a concept beginning to take shape in the minds of some judge advocates, it was not yet a reality. However, the Vietnam experience had laid the psychological and cultural foundation in the U.S. military for the transformation that would soon take place.

Notes

[1] Interv, author with Col Paul J. Durbin, 1 Jul 96, Historians files, Office of the Judge Advocate General (OTJAG), Department of the Army, Washington, D.C.

[2] Memorandum no. 10, Missions and Functions of MAAG Advisers, HQ, MAAG, 27 Jan 61, box 25, Records of the MAAG Adjutant General Division, Record Group (RG) 334, National Archives and Records Administration (NARA), College Park, Md.

[3] Vincent H. Demma, "The U.S. Army in Vietnam," in *American Military History*, rev. ed. (Washington, D.C.: Government Printing Office, 1989), p. 623; Ronald H. Spector, *Advice and Support: The Early Years, 1941–1960*, U.S. Army in Vietnam (Washington, D.C.: Government Printing Office, 1983), p. 291.

[4] Demma, "U.S. Army in Vietnam," p. 631.

[5] MAAG Vietnam Judge Advocate Responsibilities, Appendix I to Annex L (Legal), MAAG Vietnam OPLAN 61–61, 10 Feb 61, box 25, Records of the MAAG Adjutant General Division, RG 334, NARA.

[6] Interv, author with Durbin, 1 Jul 96.

[7] The Pentalateral Agreement also exempted all goods imported into Vietnam for use by the advisory group from Vietnamese customs and taxes. This special treatment reflected a belief that there would be only a small U.S. presence in Vietnam after 1950. As the American buildup began in the early 1960s, however, the United States and the Republic of Vietnam chose not to negotiate a Status of Forces Agreement like those used in Japan, Korea, and the Philippines. Consequently, all U.S. forces remained immune from Vietnamese criminal and civil law until the end of the war in 1975.

[8] Interv, author with Durbin, 1 Jul 96.

[9] Spector, *The Early Years*, p. 370.

[10] Interv, author with Durbin, 1 Jul 96.

[11] MAAG Vietnam OPLAN 61–61, Annex L (Legal).

[12] Interv, author with Durbin, 1 Jul 96. For an in-depth discussion of the Vietnamese legal system and its French structure, see George S. Prugh, *Law at War: Vietnam 1964–1973*, Vietnam Studies (Washington, D.C.: Government Printing Office, 1974), pp. 15–39.

[13] Interv, author with Col George C. Eblen, 23 Aug 96, Historians files, OTJAG.

[14] Interv, author with Col George C. Eblen, 9 Sep 96, Historians files, OTJAG.

[15] Ibid.

[16] Interv, author with Eblen, 23 Aug 96; Prugh, *Law at War*, p. 98.

[17] Interv, author with Lt Col William G. Myers, 15 Jul 96, Historians files, OTJAG.

[18] Ibid.; interv, author with Arthur H. Taylor, 2 Aug 96, Historians files, OTJAG.

[19] Intervs, author with Arthur H. Taylor, 2 Aug 96 and 21 Nov 96, Historians files, OTJAG.

[20] Interv, author with Alfred A. McNamee, 31 Jul 96, Historians files, OTJAG.

[21] Ltr, Prugh to Col Thomas H. Swan, 26 Apr 65, Historians files, OTJAG.

[22] Jeffrey J. Clarke, *Advice and Support: The Final Years, 1965–1973*, U.S. Army in Vietnam (Washington, D.C.: Government Printing Office, 1987), p. 119.

[23] Prugh, *Law at War*, p. 63.

[24] Clarke, *The Final Years*, p. 120.

[25] Prugh, *Law at War*, p. 63; interv, author with Maj Gen George S. Prugh, 31 Jul 96, Historians files, OTJAG.

[26] Prugh, *Law at War*, pp. 61–67; Clarke, *The Final Years*, p. 170.

[27] MACV Directive 20–4, *Inspections and Investigations of War Crimes*, 20 Apr 65, Historians files, OTJAG.

[28] Prugh, *Law at War*, pp. 72–73.

[29] Memo, HQ, MACV, 21 Jan 65, sub: Resources Control—Arrest, Search, and Seizure Laws in the Republic of Vietnam, Incl 2, p. 1, Historians files, OTJAG.

[30] Prugh, *Law at War*, p. 42; interv, author with Prugh, 31 Jul 96.

[31] Demma, "U.S. Army in Vietnam," pp. 638–39.

[32] George S. Eckhardt, *Command and Control: 1950–1969*, Vietnam Studies (Washington, D.C.: Government Printing Office, 1974), p. 54.

[33] Demma, "U.S. Army in Vietnam," p. 642.

[34] Prugh, *Law at War*, p. 88.

[35] Demma, "U.S. Army in Vietnam," p.649. For a perspective on small unit operations in Vietnam, see James McDonough, *Platoon Leader* (Novato, Calif.: Presidio Press, 1985), and David Donovan, *Once a Warrior King* (New York: McGraw-Hill, 1985). McDonough served as an infantry lieutenant with the 173d Airborne Brigade in 1970; Donovan served with a province advisory team in the Mekong Delta in 1969.

[36] Prugh, *Law at War*, p. 100.

[37] Ibid., pp. 10, 12.

[38] Ibid., p. 53.

[39] Interv, author with Col Edward W. Haughney, 27 Sep 96, Historians files, OTJAG.

[40] MACV Directive 20–5, *Prisoners of War—Determination of Eligibility*, 21 Sep 66, Historians files, OTJAG.

[41] Prugh, *Law at War*, p. 74; Guenter Lewy, *America in Vietnam* (New York: Oxford University Press, 1978), pp. 325, 348; Rpt, Clerk of Court, U.S. Army Judiciary, Convictions by General and Special (BCD) Courts-Martial of Offenses Against Vietnamese, 1965–1973, n.d., Historians files, OTJAG; *United States* v. *McGee*, CM 422412 (1969), (court admonished soldier after convicting him of involuntary manslaughter); *United States* v. *Hodges*, CM 420341 (1969), (sergeant cut off heads of dead enemy soldiers and posed for photographs with the corpses). See also *United States* v. *Lund*, *United States* v. *Francis*, CM 420181 (1969), (rape of Vietnamese female by two Americal Division soldiers); *United States* v. *Woods*, CM 416803 (1966), (soldier killed detainee allegedly on instructions of company commander); *United States* v. *Williams*, CM 419872 (1968), (soldier convicted of conduct prejudicial to good order and discipline for cutting off ears and index fingers of Vietnamese corpse); *United States* v. *Goldman*, CM 420332, 43 CMR 77, ACMR (1970), (company commander convicted for failing to report noncombat death of detainee and dereliction of duty in failing to protect Vietnamese female in custody of his unit); *United States* v. *Duffy*, CM 424795, 47 CMR 658 (ACMR 1973), (lieutenant convicted of involuntary manslaughter for causing Vietnamese to be taken to a woodline and shot with a rifle). For a more thorough discussion of American war crimes in Vietnam, see Lewy, *America in Vietnam*, pp. 323–441.

[42] Prugh, *Law at War*, p. 90.

[43] Ibid., pp. 108–09.

[44] Trip Report on Vietnam, Eugene T. Herbert, Department of State, 30 Dec 66, sub: Status of Civilian Contractors in Vietnam, box 1, HQ MACV Staff Judge Advocate, General Records, RG 472, NARA.

[45] Memo, Staff Judge Advocate, 1st Logistical Command, for Commanding General, USA Support Command, 13 Mar 67, sub: Court-martial jurisdiction over civilians, box 1, USARV Staff Judge Advocate Section, Administrative Office, RG 472, NARA.

[46] Interv, author with Haughney, 27 Sep 96; Gary D. Solis, *Marines and Military Law in Vietnam: Trial by Fire* (Washington, D.C.: Government Printing Office, 1989), pp. 99–100.

[47] Prugh, *Law at War*, p. 93.

[48] Staff Study, The Role of Civil Law in Counterinsurgency, HQ, MACV, May 65, Historians files, OTJAG.

[49] MACV Directive 25–1, *Claims*, 14 May 65, Historians files, OTJAG.

[50] Interv, author with Prugh, 25 Sep 96; Prugh, *Law at War*, p. 83; U.S. Public Law 90–521, sec. 3 (1968), amending Title 10, United States Code, sec. 2734.

[51] Prugh, *Law at War*, pp. 41, 49.

[52] Staff Study, Transfer of MACV Staff Elements to USARV, HQ MACV, 28 Nov 65, box 1, USARV Staff Judge Advocate Section, Administrative Office, RG 472, NARA.

[53] Prugh, *Law at War*, p. 50.

[54] Memo, USARV Staff Judge Advocate for USARV Assistant Chief of Staff, Comptroller, 12 Sep 68, sub: Mission of USARV SJA, box 1, USARV Staff Judge Advocate Section, Administrative Office, RG 472, NARA.

[55] Interv, author with Col John Jay Douglass, 19 Nov 96, Historians files, OTJAG.

[56] George S. Prugh, "Law Practice in the Vietnam War," *Federal Bar Journal* 27 (1967): 58.

[57] Prugh, *Law at War*, p. 83.

[58] Ibid., p. 62.

[59] Dennis R. Hunt, "Viet Nam Hustings," *Judge Advocate Journal*, Bull. no. 44 (July1972): 23.

[60] William M. Hammond, *Public Affairs: The Military and the Media, 1968–1973*, U.S. Army in Vietnam (Washington, D.C.: Government Printing Office, 1996), pp. 220–24.

[61] Ibid., p. 252.

[62] Ibid., pp. 258–59; Lewy, *America in Vietnam*, pp. 356–58; *United States v. Calley*, 46 CMR 1131 (ACMR 1973); interv, author with Col William G. Eckhardt, 8 Dec 98, Historians files, OTJAG. See also *Calley v. Callaway*, 382 Federal Supplement (F. Suppl.) 650 (1974).

[63] *Report of the Department of the Army Review of the Preliminary Investigations into the My Lai Incident* (Peers Inquiry), 14 Mar 70, I: 10–26; Hammond, *Military and the Media, 1968–1973*, pp. 244–45; Army Regulation (AR) 35–216, *Training: The Geneva Conventions of 1949 and Hague Convention No. IV of 1907*, 28 May 70; DOD Directive 5100.77, *DOD Program for Implementation of the Law of War*, 5 Nov 74.

[64] Interv, author with Lt Col Ronald M. Holdaway, 24 Jul 96.

[65] Judge Advocate General's Corps, U.S. Army, Personnel Directory, 1967, 1968, Historians files, OTJAG.

[66] Interv, author with Brig Gen Thomas R. Cuthbert, 2 Oct 96, Historians files, OTJAG.

[67] Ibid.

[68] Ibid.; USARV Suppl. 1 to AR 27–10, *Military Justice*, par. 20–15b, 15 Jun 70.

[69] Interv, author with Cuthbert, 2 Oct 96; interv, author with Col Burnett H. Radosh, 10 Jul 96, Historians files, OTJAG.

[70] Interv, author with Cuthbert, 2 Oct 96.

[71] Demma, "U.S. Army in Vietnam," pp. 672–83.

[72] Prugh, *Law at War*, p. 51.

[73] Ibid., p. 53.

[74] Ibid., p. 59.

[75] The first female judge advocate in Vietnam, Maj. Ann Wansley, served at U.S. Army, Vietnam, in 1966 and 1967. The second, Maj. Nancy A. Hunter, served at the 4th Transportation Command in 1970.

[76] For statistics on U.S. Army disciplinary actions in Vietnam from 1965 to 1972, see Prugh, *Law at War*, p. 154.

[77] Interv, author with Maj Gen Wilton Persons, 17 Sep 96.

[78] Prugh, *Law at War*, p. 107.

[79] MACV, *Command History, 1971*, II: x–9.

[80] Hammond, *Military and the Media, 1968–1973*, pp. 176, 382. See also Solis, *Trial by Fire*, p. 129.

[81] Interv, author with Col Barry P. Steinberg, 19 Aug 96, Historians files, OTJAG.

[82] Hammond, *Military and the Media, 1968–1973*, pp. 378–79.

[83] Ibid., p. 385.

[84] Given the Army's disciplinary problems in Vietnam, some concluded that the Uniform Code did not work well in combat situations. Generals Westmoreland and Prugh noted that the military criminal justice system "is too slow, too cumbersome, too uncertain, too indecisive, and lacking in the power to reinforce accomplishment of the military mission." William C. Westmoreland and George S. Prugh, "Judges in Command: The Judicialized Uniform Code of Military Justice in Combat," *Harvard Journal of Law and Public Policy* 3 (1980): 4.

[85] Prugh, *Law at War*, p. 85.

[86] DF, Chief, Foreign Claims Division, to USARV Staff Judge Advocate, 5 Nov 71, sub: Da Nang Ammunition Dump Cases, box 3, Records of the USARV Staff Judge Advocate, Foreign Claims, RG 472, NARA.

[87] Memo, USARV Staff Judge Advocate for Deputy Commanding General USARV, 22 Jan 71, sub: Foreign Claims Operations, box 3, Records of the USARV Staff Judge Advocate Section, RG 472, NARA.

[88] Ibid.

[89] DF, Chief, Foreign Claims Division, to USARV Staff Judge Advocate, 11 Oct 71, sub: Claims Against the United States Government, box 3, Records of the USARV Staff Judge Advocate Section, RG 472, NARA.

[90] Summary of After Action (SAA), Staff Judge Advocate, USARV/MACV Support Command, 1973, Historians files, OTJAG.

[91] Msg, Lt Gen Welborn G. Dolvin, CG, XXIV Corps, to Gen Creighton W. Abrams, COMUSMACV, 27 Aug 71, sub: US/VN Confrontations, box 3, Records of the USARV Staff Judge Advocate Section, RG 472, NARA.

[92] DF, Chief, Foreign Claims Division, to USARV Staff Judge Advocate, 12 Nov 71, sub: Confrontation at HQ XXIV Corps, box 3, Records of the USARV Staff Judge Advocate Section, RG 472, NARA.

[93] Interv, author with Col Donald A. Deline, 18 Sep 96, Historians files, OTJAG.

[94] Hammond, *Military and the Media, 1968–1973*, pp. 505–08. See also Keith W. Nolan, *Sappers in the Wire: The Life and Death of Firebase MARY ANN* (College Station, Tex.: Texas A&M University Press, 1995).

[95] Interv, author with Maj Gen William K. Suter, 25 Jul 96, Historians files, OTJAG.

[96] Interv, author with Honorable Royce C. Lamberth, 29 Jul 96, Historians files, OTJAG. The following paragraphs are based on this interview.

[97] William E. Le Gro, *Vietnam from Cease-Fire to Capitulation* (Washington, D.C.: Government Printing Office, 1981), p. 18.

[98] Prugh, *Law at War*, p. 49.

[99] Interv, author with Col Vahan Moushegian, Jr., 24 Jun 96, Historians files, OTJAG. The following account is based on this interview.

[100] Intervs, author with Lt Col Jerome W. Scanlon, 2 Aug, 28 Oct, and 30 Oct 96, Historians files, OTJAG. The following account is based on these interviews.

2

Grenada
Army Lawyers in Transition

Grenada was a real wake-up call for us.[1]

—*Capt. Gary L. Walsh*
Judge Advocate, 82d Airborne Division

Early on 25 October 1983, U.S. Army Rangers spearheaded Operation URGENT FURY, attacking Point Salines airport on the Caribbean island of Grenada in a daring parachute assault.[2] The Ranger mission was to seize this airstrip and then evacuate the several hundred American students attending St. George's University Medical School. A U.S. Marine Corps battalion landed at approximately the same time on the other side of Grenada, seizing the nearby Pearls airport. Units from the 82d Airborne Division arrived shortly thereafter to clear the island of enemy resistance and ultimately provide an occupation force. All told, URGENT FURY involved two battalions of some 650 Rangers from Fort Stewart, Georgia, and Fort Lewis, Washington; six battalions comprising roughly 6,100 paratroopers from the 82d Airborne Division, Fort Bragg, North Carolina; and a marine amphibious unit of approximately 2,000 marines. Additionally, Army, Navy, and Air Force special operations forces from various U.S. locations participated in this operation, as did a small number of personnel from the XVIII Airborne Corps.

Although the Rangers at Point Salines were opposed by elements of the 1,200-man Grenadian People's Revolutionary Army and some 700 armed Cuban construction workers at the airfield, enemy ground fire was suppressed fairly quickly, and Ranger casualties were relatively light.[3] The Marine heliborne assault against Pearls was unopposed, and elements of the 82d Airborne Division arriving later that same day also land-

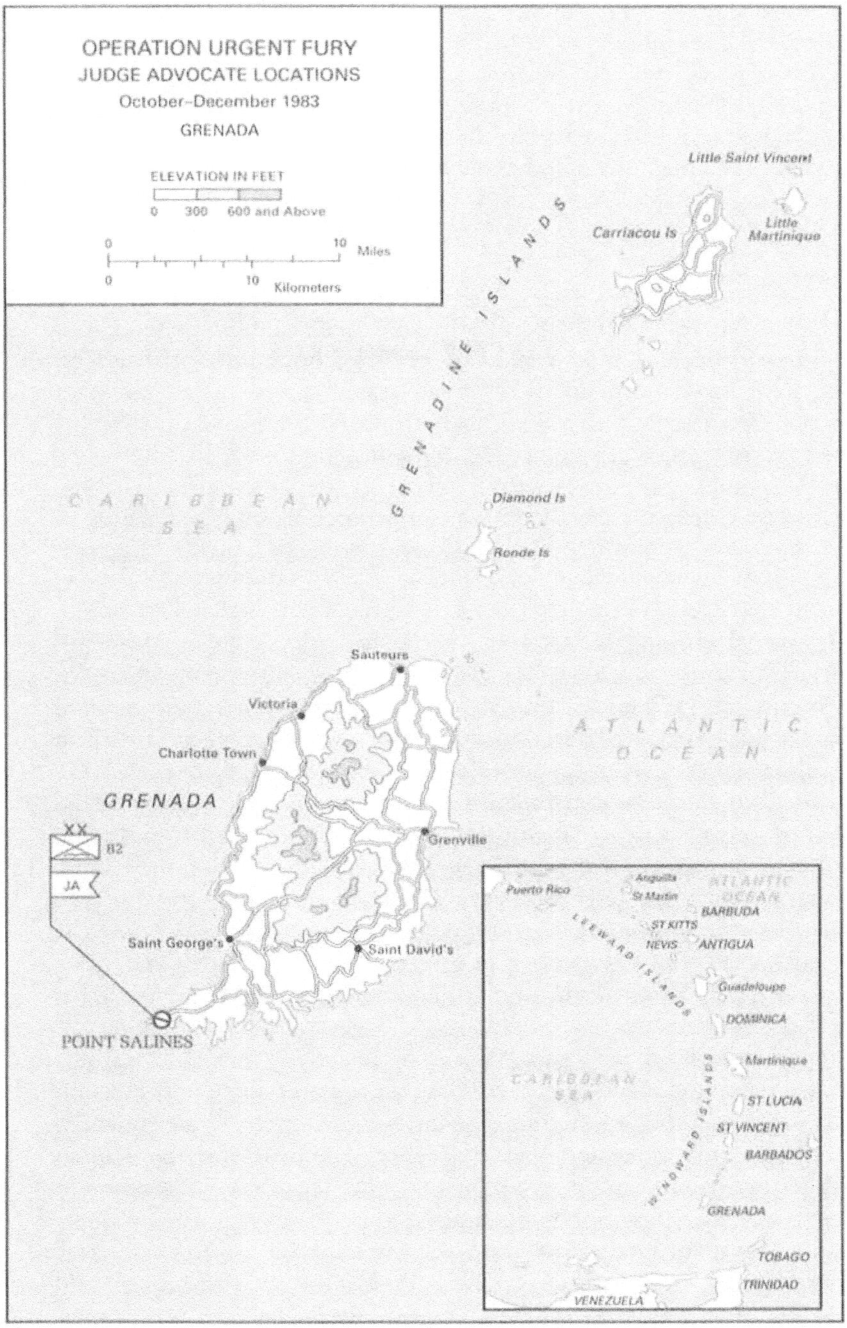
OPERATION URGENT FURY
JUDGE ADVOCATE LOCATIONS
October–December 1983
GRENADA
ELEVATION IN FEET
0 300 600 and Above
0 10 Miles
0 10 Kilometers
Little Saint Vincent
Carriacou Is
Little Martinique
GRENADINE ISLANDS
CARIBBEAN SEA
Diamond Is
Ronde Is
Sauteurs
Victoria
Charlotte Town
GRENADA
ATLANTIC OCEAN
XX
82
JA
Grenville
Saint George's
Saint David's
POINT SALINES
Puerto Rico
Anguilla
St Martin
ATLANTIC OCEAN
BARBUDA
ST KITTS
NEVIS
ANTIGUA
LEEWARD ISLANDS
Guadeloupe
DOMINICA
Martinique
CARIBBEAN SEA
ST LUCIA
ST VINCENT
BARBADOS
WINDWARD ISLANDS
GRENADA
TOBAGO
TRINIDAD
VENEZUELA

Map 4

ed with only minor opposition. Although some resistance from Cuban and People's Revolutionary Army personnel continued over the next few days, by 3 November 1983 hostilities had ended, and Vice Adm. Joseph Metcalf III, the commander of the joint task force with overall responsibility for URGENT FURY, passed military operations to Army Maj. Gen. Edward L. Trobaugh, the commander of the 82d Airborne Division, who became Commander, U.S. Forces, Grenada. *(Map 4)*

Background

What was the legal authority for the U.S. intervention in Grenada? Grenada had become independent in 1974. Although a British governor-general had continued in residence, he played only an advisory role, and Great Britain divorced itself from the affairs of state in this former colony. Since elections held in 1979, the Marxist-oriented New Jewel Movement had governed Grenada. Its leader, Maurice Bishop, had invited Cuba, the Soviet Union, and other Communist states to assist the young island nation. The Cubans provided considerable aid, including military advisers and troops. By 1983, however, a feud between two factions of the New Jewel Movement had led to bloodshed, and conditions on Grenada had deteriorated. Prime Minister Maurice Bishop was killed, the elected government collapsed, and a "Revolutionary Military Council" took power. When that council declared martial law, including "the closing of travel to or from Grenada, a news blackout, and a 24-hour shoot-on-sight curfew,"[4] the Reagan administration concluded that this turmoil had "converted the long-term problem of a Cuban presence in the Windward Islands into an immediate danger"[5] for the almost 1,000 U.S. citizens on the island. Consequently, the United States cited the protection of the lives of U.S. citizens, accomplished through a noncombat evacuation operation, as one legal basis for intervening in Grenada. In doing so, the United States justified URGENT FURY, in part, under international law, as a form of self-defense.[6]

Additionally, the Organization of Eastern Caribbean States (OECS)[7] made the decision at a 21 October meeting that conditions on Grenada posed a threat to regional security and stability. Consequently, the organization requested that Barbados, Jamaica, and the United States participate in a collective security force. This OECS invitation to join in a peacekeeping operation supported a 24 October 1983 request from the governor general of Grenada, Sir Paul Scoon, for a "peacekeeping force . . . in Grenada to facilitate a rapid return to peace and tranquility and also a return to democratic rule."[8] The OECS invitation to the United States to join a collective effort to ensure the peace and security of the Eastern Caribbean constituted a second legal basis for undertaking Urgent Fury. Implicit in the acceptance of such an invitation was a U.S. willingness to assist the OECS and Grenadian authorities in restoring a democratic government on the island.

Army lawyers, one of whom arrived with the first wave of the 82d Airborne on 25 October, would have a continuous presence in Grenada until 15 December 1983.[9] Initially, Army leaders expected these lawyers to focus only on specific issues related to the status and treatment of prisoners of war and civilian detainees. In addition, it was anticipated that judge advocates would deal with those types of administrative and criminal law matters routinely generated by the commands they supported at home station. In other words, a commander's expectation of the type of legal support to be provided by judge advocates was still being driven by a concept of the role traditionally played by Army lawyers.

Army judge advocates, however, had now been involved in the detailed review of OPLANs, pursuant to the My Lai–generated DOD Directive 5100.77 of 1974, for almost nine years and were far more aware of the potential for encountering legal issues impacting on the conduct of an operation. As a result of this increased awareness, lawyers on the ground in Grenada sought out and became involved in numerous issues, and the resolution of these matters proved critical to the success of Urgent Fury. Such issues included the preparation of rules of engagement and related guidance for both the combat and the peacekeeping phases of Urgent Fury; formulating a command policy on war trophies; advising on the treatment of captives; and advising the State Department on the preparation of a Status of Forces Agreement. Judge advocates also created a centralized procedure for paying claims for damaged and seized property; advised the Grenadian government on drafting domestic law; and provided liaison with various U.S. government agencies and other non-U.S. organizations such as the Caribbean Peacekeeping Force and the International Committee of the Red Cross. Finally, Army lawyers learned that they must be better prepared to provide timely and more comprehensive assistance to both deployed troops and their families back home. This required the creation of a Family Assistance Center at Fort Bragg designed to provide continued assistance with personal legal problems, including the drafting of wills and the execution of powers of attorney.

This account of the operations of the Judge Advocate General's Corps during Urgent Fury details the expanding "nontraditional" roles assumed by Army attorneys during the three phases of the operation: the "preparation and hostilities" phase, the "State of Emergency" phase, and the "stabilization" phase. Each section deals with what Army lawyers accomplished, both in Grenada and at Fort Bragg, in support of Operation URGENT FURY.

Phase I, Preparation and Hostilities (24–28 October 1983)

Judge advocates faced a significant handicap in planning for URGENT FURY: military lawyers had little knowledge of the projected operation until a few days prior to departure. The need for secrecy and security, in fact, hampered military planning at many levels, as the U.S. defense community's organization for rapid, joint contingency operations of this nature was still rudimentary. The first judge advocate on the ground in Grenada had little more than twelve hours' notice of his deployment.

The XVIII Airborne Corps' deputy staff judge advocate, Lt. Col. David McNeill, Jr., was the first military lawyer at Fort Bragg to learn of Operation URGENT FURY. He was briefed on the deployment by the corps deputy chief of staff on Sunday afternoon, 23 October, and told to research the "extent of 'martial law' powers on foreign soil."[10] Shortly thereafter, McNeill relayed this conversation to his chief, the corps' staff judge advocate, Col. Michael M. Downes. Although both Downes and McNeill were interested in the upcoming combat operation, they knew that no corps lawyers would be deploying to Grenada with the initial invasion force. General John W. Vessey, Jr., chairman of the Joint Chiefs of Staff, had decided against using the XVIII Airborne Corps as a follow-on unit, relieving the special operations troops after the initial assault. Too many "high-ranking officers on the ground," the chairman believed, would only "increase the perception that the United States was overreacting in Grenada."[11]

Consequently, Vessey "directed that only two battalions of the 82d Airborne be deployed to Point Salines airfield after the assault was finished."[12] Judge advocate tasks on the ground would thus be handled by elements of the 82d Airborne Division. Its staff judge advocate, Lt. Col. Quentin Richardson, had yet to be informed about the operation.

Downes contacted Richardson in the early evening hours of Monday, 24 October, relaying to a surprised Richardson that the 82d's movement into Grenada was expected the next day. Colonel Richardson had only that evening and the early morning hours of the next day to prepare for departure—and with travel on a C–141 scheduled for Tuesday, 25 October, at 10 A.M., he had little time to plan judge advocate operations. Due to the fact that judge advocates had rarely become involved in tactical operations in the past, their absence from the planning process in this instance was neither remarkable nor an oversight.

At 8 P.M., 24 October, the entire division was placed on alert, and all of its judge advocates reported for duty. At 10 P.M., as a member of the division's special staff, Colonel Richardson attended a division briefing on URGENT FURY at which members of the assault command post were identified. There, General Trobaugh, Commanding General, 82d Airborne Division, stressed that his Civil Affairs (G–5) staff section would be "particularly busy" in refugee matters, but he included no lawyer in his small assault command post.[13] Richardson, however, quickly convinced the chief of staff, Col. Peter J. Boylan, that the command group would require legal support from the beginning of the operation, and, as General Trobaugh was deploying, Richardson determined that the staff judge advocate must deploy as well. The G–3 (operations) officer was quickly told to include a lawyer in the initial command party. Although this decision, which resulted in another staff officer's staying behind, was unpopular in some quarters, Richardson later reflected that it "was the smartest thing I did."[14]

While Colonel Richardson was the lone lawyer in the command post, an officer who would later join the Judge Advocate General's Corps was working alongside him. This was Infantry 1st Lt. James E. Macklin, who, as an assistant G–3 operations officer, was primarily responsible for deploying the thirty-man assault command post to Grenada.[15]

While Richardson was working with the division staff, the 2d Brigade trial counsel, Capt. Glenn E. Murray, and Capts. Carlton L. Jackson and Clyde J. Tate, both division legal assistance officers, reported to the 2d Brigade headquarters. As this brigade was providing the battalions for the initial assault, these lawyers began the legal chores necessary to prepare soldiers for active operations, including preparing and executing wills and powers of attorney.[16] The possibility of combat made a will attractive to many young soldiers, most of whom had never cared to address the issue previously. Similarly, few had prepared those powers of attorney documents necessary to enable spouses or parents to negotiate paychecks and write checks.

At approximately 11:30 P.M., Richardson drew a weapon and protective mask and reported with his pack to division headquarters, later gathering C-rations, a poncho liner, a flak jacket, and a parachute. Along with these items, Richardson took Field Manual 27–10, *The Law of Land Warfare*, the *Manual for Courts-Martial*, paper, and several pens. This would have to serve as his legal library. At 10 A.M., 25

October 1983, he departed Pope Air Force Base on a C–141 aircraft with the command party.

The assault command post aircraft arrived at Point Salines airport around 2 P.M. Although the airport had been seized by the Rangers earlier that day, sniper fire could still be heard. Consequently, the plane circled, waited for this sniper fire to die down, and then landed about 3:15 P.M. For the next three days, Colonel Richardson, along with the rest of the command post, operated in the Point Salines terminal. Duty was two hours on, four hours off. No further contact was made with the enemy. Conditions were primitive, and the command post was without lights, water, or latrines. On 26 October, the day following his arrival, Richardson's real work began.

Prisoner of War Issues

Initially, Colonel Richardson focused his attention on the growing number of prisoners of war, detainees, and civilians in American hands. The U.S. military commanders had not expected to assume control over so many persons, so quickly. The Rangers had set up a prisoner of war/detainee camp on the previous morning, 25 October. The camp, located on a high hill to the rear of the airport terminal, consisted of recently abandoned sleeping quarters for Cuban construction workers. The Rangers had been relieved of their responsibility for the camp that afternoon by the Caribbean Peacekeeping Force. However, this force, consisting of some 300 peacekeepers from two OECS member states—Barbados and Jamaica—had no experience in administering such camps.[17] Food was in short supply; there was no electricity and very little water; and sanitary facilities were minimal. In addition, security was lax. No list of the camp's members had been compiled, nor had any attempt been made to classify them by status. Now, control and operation of the prisoner/detainee camp was to pass from the Caribbean Peacekeeping Force to the 82d Airborne Division.

The classification issue proved to be a thorny one. U.S. forces were opposed by Cuban military personnel and the Grenadian People's Revolutionary Army. Although there was no declaration of war accompanying Operation URGENT FURY—the United States and Cuba never stated that they were at war—Army lawyers determined that a de facto state of hostilities existed. This meant that common article 2 of the 1949 Geneva Conventions for the Protection of War Victims (GPW) applied. Under that article, neither a formal declaration of war nor a recognized state of war is required in order for the conventions to apply. Article 2 states that "the present Convention shall apply to all cases of declared war or *of any other armed conflict* which may arise between two or more of the High Contracting Parties, even if the state of war is not recognized by one of them [emphasis supplied]."

Consequently, Army lawyers at the Office of the Judge Advocate General in Washington, D.C., in conjunction with DOD and State Department authorities, determined that the 1949 GPW Convention applied and that all persons captured should be *treated* as prisoners of war. This was a significant decision. Individuals afforded prisoner of war status actually receive more rights and privileges than persons placed in a "retained" status or civilian internees. Thus, although the United States recognized the convention's differentiation between prisoners of war, retained personnel,[18] and civilian internees, all individuals taken into military custody were regarded as prisoners of war and treated as such until a more informed determination of their status could be accomplished.

As a result, the 440 Cuban civilian construction workers captured on Grenada were treated as prisoners of war, along with all other Cuban nationals and other persons captured by Caribbean Peacekeeping Force and U.S. personnel. Treating these workers as prisoners of war pending a decision on their actual status benefited commanders, as there was no time to classify these individuals during the brief hostility phase. Treating all Cubans and Grenadians taken into custody as prisoners of war also proved to be a wise public relations decision. As considerable criticism resulted from the U.S. intervention, this determination demonstrated that the United States was serious about meeting its responsibilities under international law.

When the 82d Airborne Division's acting provost marshal and a military police unit arrived on 28 October, Colonel Richardson advised them and the Caribbean Peacekeeping Force that the legal decision had been made to treat all those in the prisoner/detainee camp as prisoners of war, regardless of their actual status under the Geneva Conventions. Later, all detained persons were screened and classified as either POW, retained personnel, or civilian internees. In the meantime, the wounded and sick received care, and the Red Cross was invited to observe the detention conditions of all personnel. Wounded and sick Cuban personnel, as well as some Cuban medical personnel, were eventually repatriated to Cuba, and Cuban and Grenadian prisoners were permitted contact with their next of kin within seven days of capture.

Day-to-day operation of the prisoner of war/detainee camp passed from the Caribbean Force to the military police, with Richardson continuing to provide legal advice on the treatment of captured personnel. For example, he advised that the twenty to thirty Cuban medical personnel captured were entitled to "protected" status under the Law of War and must be permitted to minister to the sick and wounded in the camp. Richardson also advised that a list containing the name of each individual held in the camp was required. In both instances, his advice was followed. In fact, the utility of having judge advocates on the scene was demonstrated repeatedly in a variety of more minor matters. When, for example, one U.S. military intelligence officer told the prisoner of war camp commander that using Cuban evaporated milk to feed refugee babies "was in violation of the Geneva Convention," Richardson intervened, noting that such was not the case.[19]

Other Legal and Nonlegal Issues

Next to providing legal and practical advice concerning the administration of the prisoner/detainee camp, the most important task for Army lawyers in Grenada was the provision of legal support to their own troops. On 27 October Richardson's lone operation was bolstered by the arrival of Capt. Glenn Murray, trial counsel for the 2d Brigade, and Capt. Gary L. Walsh, trial counsel for the 3d Brigade. Murray arrived first, thus becoming the second judge advocate on the island. Walsh arrived a short time later that same morning.

While Colonel Richardson briefed Murray on issues that he might face as a judge advocate, Captain Walsh went directly on patrol with his unit in order to gain a feel for the operation. Prior to leaving Fort Bragg, Walsh had asked to draw an M16 from his headquarters company arms room. This request was refused. While other officers were issued rifles, "a .45-caliber pistol was considered sufficient protection for a JAG." But Captain Walsh put his handgun to good use. When his reconnaissance platoon took small-arms automatic fire from Cuban and People's Revolutionary Army elements, Walsh took cover in a ditch and returned fire, now regretting that he had not insisted upon receiving an M16 from the armorer. Walsh did not shy away from undertaking further military tasks as well, and when a battalion commander later asked him to lead a patrol escorting investigators to a helicopter crash site, Captain Walsh—still lightly armed with his .45—willingly accepted the mission.[20]

Colonel Richardson had expected that legal assistance for deployed troops would be limited. However, due to the rapidity of the deployment with little advance notification, he found that "long lines of soldiers were waiting to talk with JAG officers" about wills, debt payments, and other legal issues.[21] Additionally, as URGENT FURY was initiated close to an end-of-month payday, with few powers of attorney in effect, many spouses at Fort Bragg would soon be receiving servicemembers' paychecks without the ability to cash them.[22] Consequently, powers of attorney prepared by judge advocates in Grenada had to be returned to Fort Bragg as quickly as possible for use by military dependents.

Colonel Richardson was also called upon to solve problems in three other areas. About 5 P.M. on 26 October, Rangers rescued 240 American students from the St. George's medical school campus located near Point Salines airport. In the absence of the G–5, Richardson and the division command sergeant major organized the evacuation of these students to the United States. The last group of students was on a C–141 to Charleston Air Force Base, South Carolina, within four hours.

Richardson then learned of a substantial supply of food located at a Cuban camp about one mile from the Point Salines airstrip. The chief of staff ordered the logistics officer (G–4) to provide the needed vehicles and, protected by an escort provided by the operations officer (G–3), Richardson traveled to the camp and discovered enough food to feed 1,000 people for several weeks. Certain supplies were taken for the prisoner/detainee camp.

The Army's graves registration team did not arrive until Friday, 28 October. Two days earlier, however, the presence of bodies had become a health problem. When General Trobaugh queried his staff regarding the applicable health and legal requirements, Richardson advised that the dead should be buried, the graves marked with identifying information, and the names and locations recorded on a master written roster.

Lessons Learned by D+3

By the end of the day on 28 October, Army lawyers had already drawn specific conclusions from their three days in Grenada. First, judge advocates must be included in the planning of contingency operations from the beginning; lack of notice hinders preparation for potential legal problems. Additionally, giving correct and complete legal advice depends on understanding the nature and purpose of the deployment, as well as the legal authority for the U.S. military action in issue. Second, providing soldiers legal advice prior to deployment is critical. Legal assistance, however, will be a continuing need for both deployed troops and their families. Third, the operation had shown that Army lawyers must be prepared to solve or advise on nonlegal or quasi-legal operational issues, as well as legal matters associated with the conduct of the operation, and that many of these issues cannot be anticipated. Examples included the evacuation of U.S. nationals and the setting up of a graves registration system. Traditional issues of military justice were found to be of minimal concern initially; deployed soldiers were far too busy to get into serious trouble. Army lawyers did not render criminal law advice during their first few days on Grenada, except to provide guidance on potential nonjudicial proceedings for minor misconduct. Finally, the patrolling experiences of Captain Walsh demonstrated that judge advocates must expect to take on decidedly nonlegal responsibilities.

Phase II, State of Emergency, 29 October–15 November 1983

Although an official end to hostilities came with Governor General Scoon's proclamation of a State of Emergency on 1 November 1983, the fighting on Grenada was over by 29 October. The three days between that date and the official start of the State of Emergency was a twilight period during which U.S. forces maintained order and prepared for the official return of a lawful, democratic government in Grenada.

As additional legal expertise was required for the State of Emergency phase, Army lawyers from outside the 82d Airborne Division began arriving on 29 October.[23] Although the XVIII Airborne Corps had no follow-on mission in URGENT FURY, most of these new lawyers did come from that corps. As the corps was the 82d's higher headquarters and collocated with it at Fort Bragg, corps judge advocates had established a prior working relationship with the division.

Colonel Downes, the corps staff judge advocate, had quickly deployed corps lawyers Lt. Col. John P. Weber, Maj. Normand J. Hamelin, and Capt. Marc L. Warren to Grenada. Working together, corps and division judge advocates focused on the requisition and seizure of Cuban and Grenadian property, both personal and real; related claims by owners for compensation; criminal matters; and administrative law, to include the drafting of orders regulating the conduct of U.S. troops in Grenada. They also provided advice to the acting attorney general of Grenada on drafting a preventive detention ordinance and to U.S. State Department personnel on drafting a Status of Forces Agreement applicable to U.S. personnel in Grenada. Aside from the continuing need to provide legal assistance to troops and advice to those administering the prisoner of war camp, they also handled Law of War and other international law issues, including the alleged illegal bombing of a Grenadian hospital, the classification of detainees, the status of foreign diplomatic personnel, and issues surrounding the possession of war trophies. Additionally, judge advocates led the investi-

Left to right: *Capts. Doug Fletcher, Clyde J. Tate, and Jack T. Tomarchio, all 82d Airborne Division lawyers, Grenada, November 1983. The pro-American graffiti in the background was done by the local populace and appeared virtually overnight.*

gation of a friendly fire incident and drafted rules of engagement for the use of force by U.S. troops during the State of Emergency phase.

Law of War and Other International Law Issues

The 82d Airborne Division After Action Report for URGENT FURY records that "the law of war responsibilities [were] perhaps the most unpredictable and the most difficult with which to deal."[24] Given the lack of a declared state of war between the United States and Grenada or Cuba and the resulting complexity of the prisoner of war issue in Grenada, the division's comment is not surprising.

Responsibilities for Law of War issues in the State of Emergency phase passed to Major Hamelin, an international law specialist assigned to the XVIII Airborne Corps. Hamelin, who arrived on 31 October 1983, acted as a liaison to the U.S. embassy. By 4 November he had established a relationship with the principal outside organization observing U.S. conduct in Grenada, the fifteen-member team from the International Committee of the Red Cross. Working with the Red Cross, Hamelin ensured that Law of War principles were observed in the prisoner of war camp.

An important Law of War lesson for all judge advocates was their recognition that it was their singular responsibility to ensure that commanders complied with the Department of Defense Law of War Program, particularly its provisions requiring the investigation of any violations of the Law of War by or against U.S. military personnel.[25] This requirement resulted in two investigations. The first dealt with the bombing of a mental hospital near Fort Frederick in the southern section of the island. The second concerned the possible murder of a Marine Corps pilot who had landed his disabled helicopter near the small harbor of St. George's.[26] The investigating judge advocates quickly concluded that early involvement of the U.S. Army Criminal Investigation Command (CID) was the key to solving many of the more practical problems in both incidents.

On 4 November, during the evening command briefing, the involved judge advocates initially debated whether a preliminary investigation of the hospital bombing was required. At the time, the only sign of any violation of the Law of War was an ambiguous "news media report." But, as any credible report requires an inquiry, the decision was made to investigate, and Major Hamelin was tasked with the job.

The next day Hamelin visited the mental hospital with two CID agents. The men took photographs showing that, unlike the roofs of other hospitals in Grenada, there was no red cross marking on the hospital roof. In addition, the walls of the building were marked with the symbol of the People's Revolutionary Army (large red dots on a white background), and a common wall joined the mental hospital to the revolutionary army headquarters building at Fort Frederick. They also noted that the People's Revolutionary Army had positioned two antiaircraft batteries approximately fifty meters from the mental hospital, near the nurses' quarters, and that on 25 October, the day of the bombing, weapons were fired from inside the hospital and from the antiaircraft positions adjacent to its walls.

Based on this information, Hamelin concluded that there had been no violation of the Law of War. The findings of this initial investigation, later adopted by an investigative team sent by higher headquarters to examine reported war crimes in Grenada, demonstrated conclusively that no crime had occurred. The decision to conduct a prompt CID-supported investigation not only ensured command compliance with the law, but also helped correct the erroneous report of a "war crime." Later news media reports accurately reflected these findings.[27]

As for the Marine pilot, interrogation reports relating to a Grenadian national known as "Preacher Man" indicated that he had fired a full magazine from his AK-47 into the body of the *already* deceased pilot. While a violation of the Law of War, this did not constitute murder.[28]

Another unanticipated international law issue that arose was the treatment of foreign diplomats. A number of Cubans had fled to the Soviet embassy, where they sought asylum and safe passage out of Grenada. Additionally, the diplomats of embassies and consulates of governments unfriendly to the United States frequently claimed "immunity" from searches for weapons conducted by U.S. soldiers at checkpoints. These claims of immunity and the general status of diplomats required Army lawyers to examine the Vienna Convention on Diplomatic Relations.[29] They concluded that the convention did not preclude checkpoint searches of diplomatic personnel suspected of possessing guns or ammunition and that any such contraband discovered could be seized lawfully. Later, when many of these same diplomats—and their vehicles—were flown out of Grenada on C–141s, judge advocates again advised that the Vienna Convention did not prevent searches of diplomats or their property aboard these military aircraft; legitimate safety and security concerns took priority.[30]

Finally, Colonel Weber, the chief of administrative law for the XVIII Airborne Corps, advised the acting attorney general of Grenada on the contents of a preventive detention ordinance that would end the State of Emergency and set strict limits on the authority of foreign military personnel to apprehend, detain, search, and interrogate Grenadians. As U.S. and CPF personnel intended to continue manning roadblocks throughout the island in furtherance of public order, Weber drafted language for such an ordinance.[31]

Rules of Engagement

Rules of engagement (ROE) are the commander's rules for the use of force. They are shaped not only by the Law of War, but also by the politico-military situation on the ground. The rules used in the hostilities phase in Grenada were the responsibility of the operational commanders, and although the ROE link with the Law of War meant that Fort Bragg–based lawyers should have participated in their formulation, none did so, as judge advocates were not involved with the division's planning for URGENT FURY.

By 2 November fighting on Grenada was at an end, and U.S. forces had assumed a peacekeeping role. The hostilities-phase rules of engagement naturally had focused on using force against an opposing armed force. Peacekeeping during the State-of-Emergency phase, however, required new rules of engagement focused on restoring and then maintaining law and order. Guidelines on dealing with individuals carrying or using weapons were needed. Consequently, Captains Walsh, Murray, and Mark Winkler worked with corps judge advocates Major Hamelin and Captain Warren in writing new rules of engagement.[32] Lawyer involvement in the ROE process also flowed naturally from the requirements of the 1979 Defense Department directive establishing the Law of War Program.[33] As noted previously, initially promulgated in 1974 as a direct result of the Army's experiences in Vietnam, the directive not only required that U.S. armed forces comply with the Law of War, but it also required that they implement training programs to prevent Law of War violations. Also required was the "prompt reporting and investigation" of violations of the Law of War. The rules of engagement remained the responsibility of the commander, and their formulation resided in the domain of the operations officer (G–3). However, the legal expertise required for the interpretation of and compliance with the Department of Defense Law of War Directive logically resulted in military lawyers advising on—and actually writing—the rules of engagement for peacekeeping activities in Grenada.

These new rules of engagement required the application of "minimum force consistent with mission accomplishment." This mission was defined as "help[ing] restore and maintain law and order until such time as the host government can control the situation."[34] The rules covered the use of deadly and nondeadly force, targeting, self-defense, and warning shots, as well as procedures for controlling weapons, engaging snipers, and conducting searches. There was also an attempt to reduce the possibility of injury to "innocent bystanders" by severely restricting the use of force. For example, "great selectivity and precision" were required when using deadly force, and soldiers were to "aim, when possible, to wound, not to kill."[35] Similarly, weapons were not to be fired "except at clearly identified point targets," and "loudspeakers" were to be used "to persuade" a sniper to surrender before resorting to "well-aimed fire."[36]

These rules of engagement were used by 82d Airborne Division and XVIII Airborne Corps troopers after the cessation of hostilities, then by U.S. Forces, Grenada, and finally by the Military Support Element that remained in Grenada after combat troops had left.

Claims

Claims operations generated much goodwill among the populace in Grenada.[37] A major issue involved compensation for the requisition or seizure of private property by U.S. forces. In the initial stages of

URGENT FURY, the 82d Airborne Division had requisitioned or seized buildings, vehicles, and other property belonging to the Cuban and Grenadian governments, as well as to private individuals. Unfortunately, public and private property was used without a differentiation being made in ownership. In terms of private property, this meant that not only were requisitions or seizures made without the knowledge or approval of the command, but that no records were kept of what was seized or receipts given to the owners of the property. Consequently, when hostilities ended and Grenadians filed claims against the United States for loss of or damage to their private property, it was difficult to determine which claims merited compensation.

To protect the interests of both Grenadian claimants and the United States, Army lawyers moved quickly to inform the command on the law governing the seizure and use of public and private property in combat operations.[38] They advised that "enemy state" property may be used for any appropriate purpose. No compensation is payable for the use of such public property, and it need not be returned to the hostile government at the cessation of hostilities. For example, it is permissible to use government vehicles, fuel, and other movable property. Public buildings may also be occupied and used. The food Colonel Richardson seized on 26 October was but one small example of a seizure of enemy public property.

Military attorneys also informed the command that private property might be seized or requisitioned if necessary to accomplish a military mission. Thus, privately owned movable property, such as communications devices, vehicles, weapons, and ammunition, might be seized. Once hostilities ended, however, restoration of such seized property to its owners was required, and compensation must be paid for its use and any damage done to it. Thus, receipts should be given when property is seized and a record kept of all seizures. And while practically any type of private property might be requisitioned, judge advocates advised that requisitions be limited by the actual needs of the force, the extent of the country's resources, and humanitarian obligations—prerequisites not applicable to the seizure of public property. Military lawyers further advised that coercion was permitted, if necessary, to requisition articles, but noted that the requisition of private property should be made only under the order of the senior area commander and recommended the establishment of a centralized system and payment in cash. If there was no cash available, a receipt was to be given. Again, however, once hostilities ceased, seized and requisitioned private property had to be restored to its owners and compensation paid. Finally, compensation had to be made for any damage to such property resulting from its use by U.S. personnel.

On 28 October the Army was assigned sole responsibility for the adjudication and payment of all claims arising from URGENT FURY. Due to the number of claims, the commander of the U.S. Army Claims Service appointed Major Hamelin and Captain Warren as one-member claims commissions. As such, each could administratively settle a non-combat claim for personal injury, death, or property damage up to $2,500. The Claims Service also created a three-member commission, composed of Captain Warren, Colonel Weber, and a nonlawyer civil affairs officer, authorized to settle noncombat claims up to $25,000.

By 30 October claims inquiries were being made by individuals whose homes had been looted or vehicles damaged. Judge advocates instructed these claimants that the U.S. Claims Office would open toward the end of the week on the wharf near a popular restaurant. On 7 November, with the help of local Grenadian businessmen, a claims office with all three claims commissions in operation opened in a store-front location in downtown St. George's. By the end of the first week, claims judge advocates had received some seventy-five claims and had paid more than $5,000.

Captain Warren, assigned to the XVIII Airborne Corps, led the initial claims efforts in Grenada. Key to his success was the 1/4-ton Jeep. Wheeled transportation not only facilitated the investigation of claims, but also enabled judge advocates to gather valuable information to pass on to military intelligence officers. For example, while conducting an investigation, claims lawyers were informed of a hidden weapons cache. This information proved to be accurate, and it resulted in the first corps-level seizure of enemy AK–47 assault rifles, ammunition, and communications gear.

Similarly, Grenadians filing claims sometimes provided information concerning individuals termed to be "bad actors," or Marxist New Jewel Movement members still at large. A man entered the St. George's claims office one morning and advised Captain Warren and the claims legal clerks that a Grenadian named Chester Humphries, then wanted by the police, was outside on the street at that very moment. As there were no other U.S. personnel in the area and the claims office had no radio or telephone capabilities, Captain Warren quickly closed the office, and he and the legal clerks set off in pursuit of Mr. Humphries. During a short chase, Humphries drew his .38-caliber revolver, but seeing Warren's .45-caliber handgun and the legal clerks' M16 rifles leveled at him, he thought better of resisting. He was taken into custody and turned over to the local authorities. As a result of the Humphries episode and the constant high quality of information collected by claims lawyers, a counterintelligence soldier was subsequently assigned full-time to the claims office in St. George's to expedite the gathering of such data.[39]

Filing compensation claims quickly became popular, and the rapid payment of these claims appeared to garner goodwill for the United States. However, the Foreign Claims Act prohibited the payment of combat-related damage. This restriction caused political difficulties for Army claims lawyers, as there was a desire on the part of other U.S. government agencies to quickly rebuild Grenada. Eventually, the U.S. Agency for International Development implemented a program to fund and pay validated Grenadian claims for combat damage.

War Trophies and Soldier Conduct

The regulation of the general conduct of GIs, to include the taking of captured property as war trophies and the treatment of detainees, involved critical issues. On the subject of war trophies, Army regulations did provide some guidance for individual soldiers. For example, contrary to the perception of many soldiers and commanders, it was illegal to privately possess certain captured items, such as an AK–47.[40] To assist the command group in controlling the seizing of items as war trophies, judge advocates authored a directive entitled "Captured Enemy Property." Army Maj. Gen. Jack B. Farris, who succeeded General Trobaugh as commander of U.S. Forces, Grenada, formally published this as an order on 8 November.[41] The directive prohibited the confiscation of private property and reminded soldiers that wrongfully taking such property constituted a violation of the Uniform Code of Military Justice. Soldiers were advised as well that enemy property became the property of the United States and that wrongfully retaining such property also violated the Uniform Code. The directive set forth a "limited exception," however, providing that "captured enemy military clothing (i.e., hats, shirts, belts, trousers and insignia)" might "be retained . . . as souvenirs." Certain items of captured enemy individual military equipment (i.e., helmets, loadbearing equipment, canteens, mess kits, and ammunition pouches) could also be taken as souvenirs, while bayonets and firearms could not.[42]

Army lawyers drafted a second directive for General Farris' signature, dealing with the treatment of Grenadian citizens and other foreign nationals detained by U.S. forces. This directive did not address the detainees housed in the prisoner/detainee compound, as their treatment was dictated by the GPW Convention. Rather, this directive focused specifically on the treatment of detainees at U.S. and Caribbean Peacekeeping Force checkpoints. Signed on 15 November, it required that "detained citizens of Grenada [be treated] courteously and humanely at all times." U.S. forces were not to threaten Grenadians and were to "respect their persons, property, family rights, religious convictions, and customs." The directive emphasized that Grenadians must be treated "as citizens whom we wish to question, not as criminals," as most "are patriotic Grenadians who support the U.S. presence here." It went on to note that U.S. treatment "of each Grenadian must result in a reinforcement of his support of our mission." Additionally, soldiers were instructed to "employ only such minimum force as is absolutely necessary to detain a citizen," and were advised that "handcuffs, ropes, blindfolds, or other physical restraints will be used only when necessary to avoid flight or dangerous acts."[43]

General Farris' directives on war trophies and the treatment of detainees provided clear guidance to the soldiers on Grenada and furthered the goal of peacekeeping after combat operations ended. Moreover, these two directives were the forerunners of the General Orders no. 1 issued in subsequent contingency operations in Asia and Africa.

Criminal Law

With the exception of several nonjudicial punishment proceedings conducted under Article 15 of the Uniform Code, little criminal law was practiced during the initial stages of URGENT FURY. Captain Walsh, legal adviser to the 3d Brigade, one of the two 82d Airborne Division brigades on Grenada, did inform the division staff judge advocate, Colonel Richardson, however, that two incidents might possibly be referred for court-martial at a later date, one for looting and one involving a noncommissioned officer's striking a more senior sergeant. Consequently, Richardson requested that the division's chief of staff deploy a defense counsel to Grenada on 29 October. A U.S. Army Trial Defense Service attorney, Capt. Mark Winkler, arrived the next day.

Several other incidents were more serious in nature. One of these occurred on the afternoon of 27 October, when the 3d Battalion, 325th Infantry, an element of the 82d Airborne Division, requested air support against a suspected sniper position. The responding Navy aircraft inadvertently fired on the 2d Brigade command post. Seventeen soldiers were wounded, and one died. Captains Murray and Walsh immediately began a preliminary investigation, obtaining names, taking statements, and recording other key information concerning this friendly fire incident; the Navy would require this data in order to conduct an investigation.[44]

Phase III, Stabilization, 15 November–15 December

On 15 November 1983, Governor General Scoon, as head of the provisional government in Grenada, declared an end to the State of Emergency. The civilian government in Grenada was now able to function, although U.S. military and Caribbean Police Force elements were asked to remain to assist in maintaining public order. This stabilization phase ended on 15 December, when all U.S. combat troops departed Grenada. The two principal legal activities occurring during this phase were the negotiation of a Status of Forces Agreement and the continued adjudication of claims.

A Status of Forces Agreement normally affords a military force and its personnel certain forms of immunity from the criminal and civil jurisdiction of a host country, as well as exemption from its duties, taxes, and immigration and customs laws. Prior to the U.S. intervention in Grenada on 25 October, there existed, of course, no such agreement regarding the United States and its military personnel. It was thus essential that one be concluded as soon as possible. Although Department of Defense and Department of State representatives normally negotiate such treaties, their absence from Grenada resulted in Army lawyers' performing this function.

As early as 1 November, U.S. Ambassador to Grenada Charles A. Gillespie contacted Army lawyers to inquire as to whether a Status of Forces Agreement was actually needed. Under international law, a foreign combatant force engaged in hostilities does not require a Status of Forces Agreement. At the conclusion of hostilities, however, a status arrangement is generally required with any foreign force invited by a host country to remain on its territory. Consequently, Ambassador Gillespie tasked Major Hamelin with developing a draft status agreement that would be effective both during and after the State of Emergency proclaimed on 1 November 1983.

During the first week of November the U.S. embassy received a message from the State Department in Washington, recommending a prompt agreement on the status of American forces. The message included a draft of proposed diplomatic notes that were to be exchanged between the ambassador and the Grenadian governor-general, communicating a "waiver of jurisdiction" by the Grenadian executive. Major Hamelin modified the draft notes in order to provide coverage for civilian members accompanying the U.S. forces and their dependents, and then had the proposed notes retyped. These notes served as the basis for the Status of Forces Agreement when the State of Emergency ended on 15 November 1983. Under this agreement, U.S.

forces were specifically exempted from all forms of Grenadian duties, taxes, charges, and levies, as well as from immigration and customs requirements. Additionally, U.S. forces were accorded the same status provided the technical and administrative staff of diplomatic missions, a limited form of diplomatic immunity.[45]

During the stabilization phase, judge advocate efforts also concentrated increasingly on claims issues. By 12 December 1983, the claims operation had received 420 claims and had approved 240 of these, totaling more than $241,000.[46] Additional legal support during this period came from Lt. Col. Paul Seibold, chief of the Foreign Claims Division of the U.S. Army Claims Service. Seibold, who arrived on 15 November and departed on 23 November, reviewed the claims operation in Grenada and provided advice concerning the resolution of several high-dollar claims.

When Major Hamelin returned to Fort Bragg at the end of November, his work as a one-member commissioner was taken over by Capt. John Hinton, an XVIII Airborne Corps attorney who had arrived in Grenada on 28 November. Army claims personnel also worked with the Department of State and the Agency for International Development to obtain funds and to establish procedures to pay combat-related claims not otherwise compensable under the Foreign Claims Act. By late 1984 the United States had paid claims totaling nearly $2 million.[47]

In retrospect, Army lawyers recommended that a foreign claims commission be appointed prior to a deployment. However small the size of an operation, the claims function is important, manpower-intensive work, and, due to the need for damage surveys, it requires at least one mobile claims office. Another factor contributing to this recommendation was the recognized need to have claims filed as promptly as possible to ease the investigative process, which grows more difficult with the passage of time. The ability of U.S. Forces, Grenada, to rapidly establish a claims office and to broadcast its location and purpose through civil affairs offices and local media such as Spice Island Radio proved critical.[48]

Judge Advocate Activities at Fort Bragg, 25 October–15 December 1983

During URGENT FURY, Army lawyers at Fort Bragg focused on providing legal assistance to the family members of soldiers deployed to Grenada. While this was a traditional legal function, the rapid nature of the operation altered the tempo of the work dramatically. Much was required, especially in the area of wills and powers of attorney. Colonel

Richardson estimated that his office prepared some 1,500 powers of attorney and over 100 wills during the first seventy-two hours of URGENT FURY.[49] But, as URGENT FURY unfolded, the 82d's staff judge advocate's office was simply overwhelmed by the volume of requests, demonstrating the need for judge advocates to have units "Preparation for Overseas Readiness–qualified" in advance of future operations.

Within twenty-four hours of alert notification on 25 October, Army lawyers were participating in the newly opened Family Assistance Center, a facility on Ardennes Street at Fort Bragg, manned twenty-four hours a day. Judge advocates joined in family assistance briefings conducted there, informing family members of the available legal advice and how to obtain it. Army lawyers also contacted the branch manager of the bank on Fort Bragg and advised him that family members with powers of attorney would be negotiating soldiers' checks. As a result, the manager briefed his employees on powers of attorney and directed that any problems with transactions be brought to him immediately. He would then call Maj. Richard Gasperini, the division rear staff judge advocate, to resolve any difficulties.

Another area of judge advocate involvement was that of assisting in the operation of the site used to store the privately owned vehicles of deployed soldiers. Upon deployment, the 82d Division had implemented a preplanned storage arrangement administered by the division adjutant general that involved parking more than 1,000 vehicles in an open field. When these began to be vandalized, the division deputy staff judge advocate recommended additional security measures for the protection of the stored property in order to reduce the number of possible future claims. Later, when soldiers retrieved their vehicles, Army lawyers had drafted a claims form and developed a claims procedure that greatly accelerated the claims process.

Finally, the legal authority of rear detachment commanders, officers placed in charge of personnel and property left behind by deploying units, had surfaced as an issue. Their authority to take judicial, nonjudicial, and administrative action was greatly restricted,[50] as Army lawyers had determined that rear commanders were not legal "commanders" imbued with authority under the Uniform Code of Military Justice. Prior to this issue's becoming a serious problem, however, the redeployment of many units to Fort Bragg rendered it moot. Note was made, nevertheless, that had the deployment to Grenada lasted a longer period of time, Army lawyers would have been required to assist rear commanders in obtaining the authority under the Uniform Code of Military Justice necessary to maintain good order and discipline in their commands.[51]

Conclusion

Both the nature and the tempo of the deployment to Grenada presented novel legal challenges for the twelve or so judge advocates who served there from 25 October to 15 December. Most important, however, Grenada served as a watershed in the evolution of a formal recognition by the leadership of the Judge Advocate General's Corps that Army lawyers could no longer focus on performing traditional peacetime legal functions in what had become a contingency-oriented Army. As Colonel Richardson stated in his After Action Report: "You can only tell the CO [Commanding Officer] that he can't shoot the prisoners so many times. You reach a point at which, when the boss has run out of beans and bullets, has certain equipment requirements, and has the locals clamoring to be paid for property damage; you have to be prepared to provide the best possible legal advice concerning these issues as well."[52]

Captain Walsh, whose experience in Grenada included both legal work and combat, echoed Richardson's sentiments when he called URGENT FURY a "wake-up call" for the Judge Advocate General's Corps. That is, while Army lawyers in Vietnam certainly had handled a broad variety of legal and nonlegal issues, there had been no resulting institutional recognition that a deployed judge advocate must be prepared to advise on the many legal matters associated with the actual conduct of the operation itself. URGENT FURY had shown clearly that a continued failure to recognize and act on this fact was no longer acceptable. Additionally, the post-Vietnam establishment of a Defense Department Law of War Program, and the new responsibilities that it placed upon Army judge advocates, also served to underscore the obvious—that the role of the judge advocate must undergo a fundamental change if Army lawyers were to make meaningful contributions to future military operations.

In sum, the experiences of judge advocates in Grenada resulted in the Judge Advocate General's Corps' formal acknowledgment, as an institution, that judge advocates must be trained and resourced to provide timely advice on a broad range of legal issues associated with the conduct of military operations. Beginning in 1986, there was a concerted effort to reconfigure the corps' assets and training to meet these newly perceived challenges. Thus, URGENT FURY served as a catalyst for the development of a new military legal discipline referred to as "operational law," a compendium of domestic, foreign, and international law applicable to U.S. forces engaged in combat or operations other than war.[53] As Army lawyers were to learn over the next decade, operational law would become increasingly critical to deployed commanders, and judge advocates, a vital part of their operational staff, would enhance mission success as never before.

Notes

[1] Interv, author with Lt Col Gary L. Walsh, 5 Feb 96, Historians files, OTJAG.

[2] For additional detail on URGENT FURY, see Jonathan M. House, *The United States Army in Joint Operations, 1950–1983* (Washington, D.C.: U.S. Army Center of Military History, 1992). See also Bruce R. Pirnie, *Operation URGENT FURY: The United States Army in Joint Operations* (Washington, D.C.: U.S. Army Center of Military History, 1986); Ronald H. Spector, *U.S. Marines in Grenada, 1983* (Washington, D.C.: Headquarters, U.S. Marine Corps, 1987).

[3] Preintervention intelligence put People's Revolutionary Army strength at 1,200, backed by a militia of 2,000 to 5,000. Documents captured by U.S. forces during URGENT FURY, however, revealed actual People's Revolutionary Army strength as 500 to 600, plus a 2,000- to 2,500-man militia. Spector, *U.S. Marines in Grenada*, p. 3.

[4] Ronald Riggs, "The Grenada Intervention: A Legal Analysis," *Military Law Review* 109 (Summer 1985): 33.

[5] House, *United States Army in Joint Operations*, p. 168.

[6] For a detailed analysis of the legal bases for URGENT FURY, see Riggs, "The Grenada Intervention: A Legal Analysis," pp. 1–81. See also Thomas J. Romig, "The Legal Basis for United States Military Action in Grenada," *The Army Lawyer*, DA Pamphlet 27–50–148, April 1985, p. 1.

[7] Founded in 1981 to enhance regional cooperation, the OECS comprised the island states of Antigua-Barbuda, Dominica, St. Kitts-Nevis, St. Lucia, Montserrat, St. Vincent and the Grenadines, and Grenada.

[8] Statement by Assistant Secretary of State for Inter-American Affairs Langhorne A. Motley Before the House Armed Services Committee, 98th Cong., 1st sess. (1984), published by U.S. Department of State, Bureau of Public Affairs, as *The Decision To Assist Grenada*, Current Policy no. 541.

[9] Combat troops left Grenada on 15 December 1983 (although about 250 U.S. troops remained to perform police and peacekeeping functions). Judge advocates began redeploying to Fort Bragg as early as 4 November, with all Army lawyers returned by 15 December. The majority of Army lawyers participated in URGENT FURY for about two to three weeks.

[10] Memo, Staff Judge Advocate (SJA), XVIII Abn Corps, for Asst CofS, G–3/Director of Plans and Training, XVIII Abn Corps, 15 Dec 83, sub: Operation URGENT FURY—After Action Report and Lessons Learned, p. 1, Historians files, OTJAG.

[11] House, *United States Army in Joint Operations*, p. 174.

[12] Ibid.

[13] Memo, SJA, 82d Abn Div, for SJA, XVIII Abn Corps, 9 Nov 83, sub: After Action Report—URGENT FURY, p. 1, Historians files, OTJAG.

[14] Interv, author with Col Quentin Richardson, 4 Mar 96.

[15] Today, Department of the Army Field Manual 71–100–2, *Infantry Division Operations: Tactics, Techniques, Procedures*, calls for a judge advocate in the assault command post. In 1983, however, no Army operational document suggested that a lawyer had a place in it.

[16] Known as Preparation for Overseas Readiness (POR) or Soldier Readiness Checks (SRC).

[17] The Caribbean Peacekeeping Force, commanded by a Jamaican Defense Force colonel, consisted of an infantry company from Jamaica, "a strongly reinforced platoon"

from Barbados, and "a squad" from Antigua and Barbuda. Dominica, St. Lucia, and St. Kitts provided "detachments of armed police." Mark Adkin, *URGENT FURY: The Battle for Grenada* (Lexington, Mass.: D.C. Heath and Co., 1989), p. 219. The Caribbean Peacekeeping Force was not formally under U.S. command and was not included in U.S. planning for URGENT FURY. U.S. forces, however, later exercised de facto control over the peacekeeping force. House, *United States Army in Joint Operations*, p. 189.

[18] These are medical personnel and chaplains.

[19] 82d Abn Div After Action Report—URGENT FURY, 9 Nov 83, p. 7.

[20] Ltr, Lt Col Gary L. Walsh to author, 26 Feb 97; interv, author with Walsh, 15 Dec 98, both in Historians files, OTJAG.

[21] Thomas J. Feeney and Margaret L. Murphy, "The Judge Advocate General's Corps, 1982–1987," *Military Law Review* 122 (Fall 1988): 11.

[22] This was before the day of nearly universal "check-to-bank" or electronic fund depositing. Most soldiers received actual checks, which required their signatures to deposit or cash.

[23] A total of eight lawyers from the 82d Airborne Division deployed to Grenada; the remainder stayed at Fort Bragg.

[24] Memo, HQ, 82d Abn Div, for Cmdr, 82d Abn Div, 1 Dec 83, sub: After Action Report on Operation URGENT FURY, p. 1, Historians files, OTJAG.

[25] Department of Defense Directive 5100.77, *DOD Law of War Program*, 10 Jul 79, par. C–3, requires "violations of the law of war, whether committed by or against U.S. or enemy personnel" to be "promptly reported [and] thoroughly investigated."

[26] See Spector, *U.S. Marines in Grenada*, pp. 9–12.

[27] Ted Borek, "Legal Services During War," *Military Law Review* 120 (Spring 1988): 47–48; Feeney and Murphy, "The Judge Advocate General's Corps, 1982–1987," pp. 12–13.

[28] "Despoiling" the dead is a violation of Article 15, Geneva Convention for the Amelioration of the Condition of the Wounded and Sick in Armed Forces in the Field, 12 August 1949.

[29] *United States Treaties* 3:3227, reprinted in *International Legal Materials* 8 (1969): 679.

[30] Interv, author with Walsh, 15 Dec 98.

[31] Info paper, sub: Status of Forces Agreement, annex to memo, Chief, Admin Law Div, XVIII Abn Corps, for SJA, XVIII Abn Corps, sub: After Action Report/Lessons Learned—Operation URGENT FURY, n.d. [May 84], Historians files, OTJAG.

[32] Ltr, Lt Col Marc Warren to author, 4 Feb 96, Historians files, OTJAG.

[33] DOD Dir 5100.77, *DOD Law of War Program*, 10 Jul 79.

[34] HQ, U.S. Forces, Grenada, Rules of Engagement for Peacekeeping Operations, n.d., par. 2, Historians files, OTJAG.

[35] Ibid., par. 5.

[36] Ibid., par. 9.

[37] See Jeffrey L. Harris, "Grenada—A Claims Perspective," *The Army Lawyer*, DA Pam 27–50–157, January 1986, p. 7.

[38] Memo, author unknown, 82d Abn Div, 23 Nov 83, sub: After Action Report, Requisition and Seizure of Government and Private Property, Historians files, OTJAG.

[39] Interv, author with Lt Col Marc L. Warren, 26 Apr 96; ltr, Lt Col Marc L. Warren to author, 17 Dec 98, both in Historians files, OTJAG.

[40] Under Army Regulation (AR) 608–4, *Control and Registration of War Trophies and War Trophy Firearms*, 28 Aug 69, par. 6*f*, machine-gun-type weapons may not be retained as war trophies.

[41] General Order, unnumbered, 8 Nov 83, sub: Captured Enemy Property, Historians files, OTJAG. General Farris served as Commander, U.S. Forces, Grenada, from 8 November until 15 December, when the organization was inactivated.

[42] Ibid. In advising the command group on war trophies, Army lawyers looked for guidance to AR 608–4 and AR 700–99. On the issue of transporting captured property to the United States for use in the division museum at Fort Bragg, judge advocates followed AR 755–2, *Disposal of Excess, Surplus, Foreign Excess, Captured, and Unwanted Materiel*, 24 Jul 70. AR 755–2 has since been rescinded, but its subject matter is covered in AR 381–26, *Army Foreign Materiel Exploitation Program*, 27 May 91.

[43] General Order, unnumbered, 15 Nov 83, sub: Treatment of Detainees, Historians files, OTJAG.

[44] House, *United States Army in Joint Operations*, p. 197; interv, author with Walsh, 15 Dec 98.

[45] Info Paper, sub: Status of Forces Agreement (SOFA), annex to Memo, Chief, Admin Law Div, XVIII Abn Corps, for SJA, XVIII Abn Corps, sub: After Action Report/Lessons Learned—Operation URGENT FURY, n.d. [May 84], Historians files, OTJAG. The final exchange of diplomatic notes relating to the SOFA occurred on 12 March 1984.

[46] Claims work in Grenada was supported by the U.S. Army Claims Service, which continued claims operations long after judge advocates participating in URGENT FURY had returned to Fort Bragg. See Harris, "Grenada—A Claims Perspective," pp. 7–13.

[47] Ted Borek, "Legal Services During War," pp. 19, 50. Although Army lawyers returned to Grenada to administer a "combat claims program," this book ends its history of Army lawyers in URGENT FURY with the end of combat-troop involvement. For Army lawyer involvement after 15 December, see Harris, "Grenada—A Claims Perspective," p. 7.

[48] Memo for SJA, XVIII Abn Corps, 9 Mar 84, sub: Claims Operations in Grenada—After Action Report/Lessons Learned, Historians files, OTJAG.

[49] Memo, SJA, 82d Abn Div, for SJA, XVIII Abn Corps, sub: After-Action Report—"URGENT FURY," 9 Nov 83, p. 4, Historians files, OTJAG.

[50] Ibid., p. 6.

[51] Subsequently, under the provisions of AR 27–10, *Military Justice*, 22 Dec 89, it became possible to give a rear detachment commander (RDC) Uniform Code of Military Justice authority by designating that commander's detachment as a provisional unit with special court-martial convening authority. Such a provisional unit might be established upon the execution of a deployment contingency plan. AR 27–10 provided a solution to the RDC authority issue faced during URGENT FURY.

[52] Quoted in David E. Graham, "Operational Law—A Concept Comes of Age," *The Army Lawyer*, DA Pamphlet 27–50–175, July 1987, p. 10.

[53] Ibid. This article provides an analysis of the conceptual development and implementation of operational law within the Judge Advocate General's Corps.

3

Panama
Operation JUST CAUSE, 1989–1990

> After JUST CAUSE, I realized that you can't think of yourself in terms of a single role . . . [rather] you must think of yourself as a problem solver and be prepared to solve legal and non-legal problems quickly and with confidence.[1]
>
> —*Col. James J. Smith*
> *Staff Judge Advocate, 82d Airborne Division*
> *Operation JUST CAUSE*

In February 1988 federal grand juries in Miami and Tampa, Florida, indicted Panama's de facto leader, General Manuel A. Noriega, on numerous counts of drug trafficking. Thereafter, relations between Panama and the United States deteriorated steadily and, for the remainder of 1988 and into 1989, General Noriega and the National Assembly representatives whom he controlled became increasingly aggressive toward the United States and its military personnel in Panama. When, as a result, Congress cut off U.S. economic and military aid to Panama, Noriega turned to Cuba, Nicaragua, and Libya for money and weapons.

As Noriega's campaign of harassing U.S. citizens in Panama continued, President George H. W. Bush and others in his administration came to believe that the 30,000 Americans resident there were in danger, as was the operation of the Panama Canal. During Panamanian elections in May 1989, opposition candidates led by Guillermo Endara won by a three-to-one margin; Noriega, however, nullified the results and sanctioned violence against the winners.

Army General Maxwell R. Thurman, the new Commander in Chief, U.S. Southern Command (SOUTHCOM), accelerated the

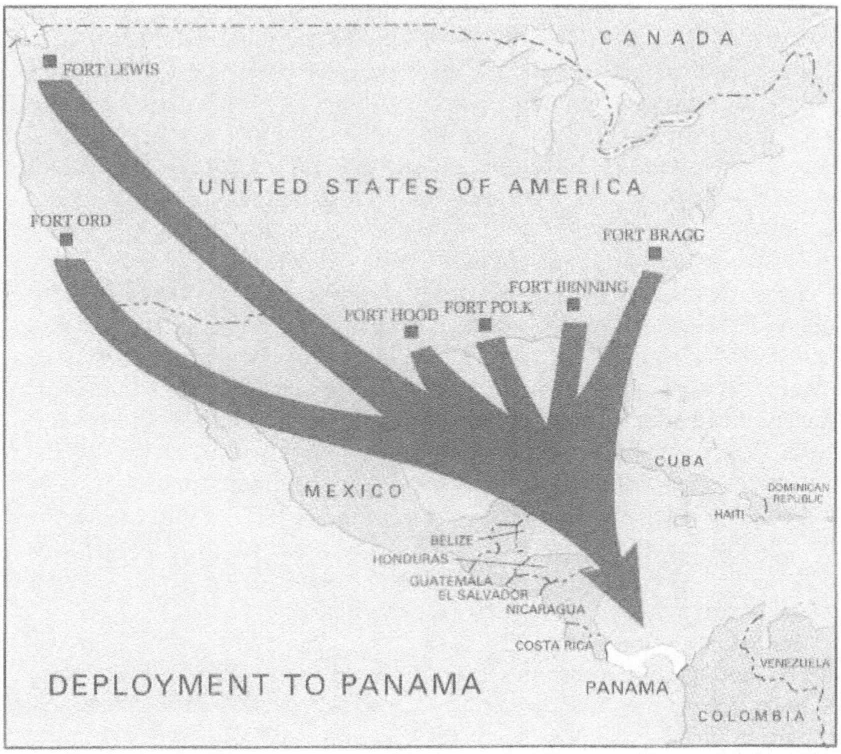

MAP 5

buildup of U.S. forces in Panama, bringing nearly 1,000 soldiers from the 7th Infantry Division (Light) and more than 700 troops from the 5th Infantry Division (Mechanized) to Panama in May. These soldiers, part of Operation NIMROD DANCER, initially provided a show of force and later played a significant role in the planning for Operation JUST CAUSE. In all, some 13,000 American troops, from all of the services, were in Panama before the start of hostilities.[2] They faced a Panama Defense Force of some 14,000 men, of whom at least 4,000 were well trained and well equipped for combat. Supplementing the Panamanian armed forces were civilian defense committees, called Dignity Battalions. These armed civilians gave Noriega and his key henchmen an additional tool for controlling Panamanian citizens.[3]

Exercising his authority under the Goldwater-Nichols Defense Reorganization Act of 1986, General Thurman selected Army Lt. Gen. Carl W. Stiner, the commander of the XVIII Airborne Corps, to take charge of Joint Task Force SOUTH, the multiservice task force that would ultimately execute Operation JUST CAUSE. Stiner took control of a conventional force that would grow to 22,000 soldiers, 3,400 airmen, 900 marines, and 700 sailors, and he accelerated planning for potential combat operations in Panama.[4] But while Stiner was in charge of all conventional forces, special operations units fell under Army Maj. Gen. Wayne A. Downing, the commanding general of the Joint Special Operations Task Force. These included the Rangers, Army Special Missions Units, and Navy Special Warfare Units. Both Generals Stiner and Downing reported directly to General Thurman.

The culmination of increasingly hostile relations between the United States and the Noriega regime came on 15 December 1989, when Panama's National Assembly declared that a state of war existed between Panama and the United States and Noriega proclaimed himself "Maximum Leader." The next evening a Panama Defense Force soldier shot at three American officers in an automobile; one, Marine 1st Lt. Robert Paz, was killed. His death precipitated Operation JUST CAUSE, authorized by President Bush on 17 December 1989. Its goals were "to create an environment safe for Americans there, ensure the integrity of the Panama Canal, provide a stable environment for the freely elected Endara government, and bring Noriega to justice."[5]

Just after midnight on 20 December 1989, some 700 U.S. Army Rangers parachuted onto the Rio Hato military base and Torrijos-Tocumen Airfield near Panama City. In these assaults on the Panama Defense Force, the Americans surprised and overwhelmed the defenders and, by first light, secured the airfields. Joint Special Operations Task Force units seized other key structures and terrain. At the same time, on the opposite side of the isthmus, a composite brigade from the 7th Infantry Division left its encampment. Its soldiers quickly secured key terrain around Fort Sherman Army Base, the city of Colon, Madden Dam, the Gatun Locks on the Panama Canal, and Coco Solo Naval Station.[6] *(Map 5)*

A few hours later, conventional force operations began near Panama City after the 82d Airborne Division followed the Rangers in an airdrop on Tocumen airfield. The remainder of the 7th Infantry Division landed at Howard Air Force Base and took position on the western side of Panama City. The 9th Infantry, part of the 7th Infantry Division, moved into Panama City to join with elements of the 82d. Finally, the 193d Infantry Brigade, joined by elements of the 5th Infantry Division (Mechanized) from Fort Polk, Louisiana, secured Fort Amador, La Comandancia (the headquarters of the Panama Defense Force), and the U.S. embassy. With La Comandancia in U.S. hands and Noriega in hiding, organized opposition ended quickly. By the end of the first day of JUST CAUSE, the Panama Canal had reopened.

By 22 December fighting was at an end in Panama City. American units in western Panama, however, continued operations to round up remaining enemy units. Except for isolated incidents, armed resistance to U.S. intervention ended by 24 December (D+4). The only remaining objective was the capture of Noriega, an effort led by the Joint Special Operations Task Force. After Noriega was discovered to have taken refuge in the residence of the papal nuncio, General Downing's special operations personnel ensured that he stayed there. After the Vatican signaled that his diplomatic immunity at the Nunciatura soon would be lifted, Noriega surrendered to U.S. forces on 3 January 1990. On the basis of outstanding federal indictments, Noriega was subsequently arrested by U.S. Drug Enforcement Administration agents and flown to Florida.[7]

With combat action diminishing rapidly, General Thurman initiated PROMOTE LIBERTY, his follow-on civil affairs operation. Consequently, a civil-military task force began concentrating on public safety, health, and law enforcement measures. Specific actions included distributing food and medical supplies; helping establish a new Panamanian police force, the Fuerza Publica; and working to develop public support for the Endara government.[8] By the time JUST CAUSE officially ended on 11 January 1990, Army units totaling some 27,000 soldiers had participated.[9]

Organization of Legal Services

The organization of legal services paralleled the command structure, with Army Col. John K. Wallace III serving as the staff judge advocate (SJA) to the SOUTHCOM commander in chief. *(Chart 5)* Wallace's legal operation, which consisted of himself and an Air Force and a Marine Corps judge advocate, was located at Quarry Heights in Panama City. The SOUTHCOM Army component, U.S. Army South (USARSO), headquartered to the northwest of Panama City at Fort Clayton, had its own staff judge advocate, Lt. Col. James S. Russell. Russell had thirteen Army lawyers in his operation, providing a full range of legal services to the command group and to Army soldiers and their families in Panama. The large number of judge advocates already in Panama allowed the joint task force legal adviser and judge advocates in Army units in the joint force to deploy to Panama with a minimum number of personnel and legal materials. *(Map 6)*

Although SOUTHCOM and USARSO judge advocates played important roles in JUST CAUSE, the authority given by General Thurman to Generals Stiner and Downing in planning and executing JUST CAUSE

CHART 5—LEGAL SERVICES IN OPERATION JUST CAUSE

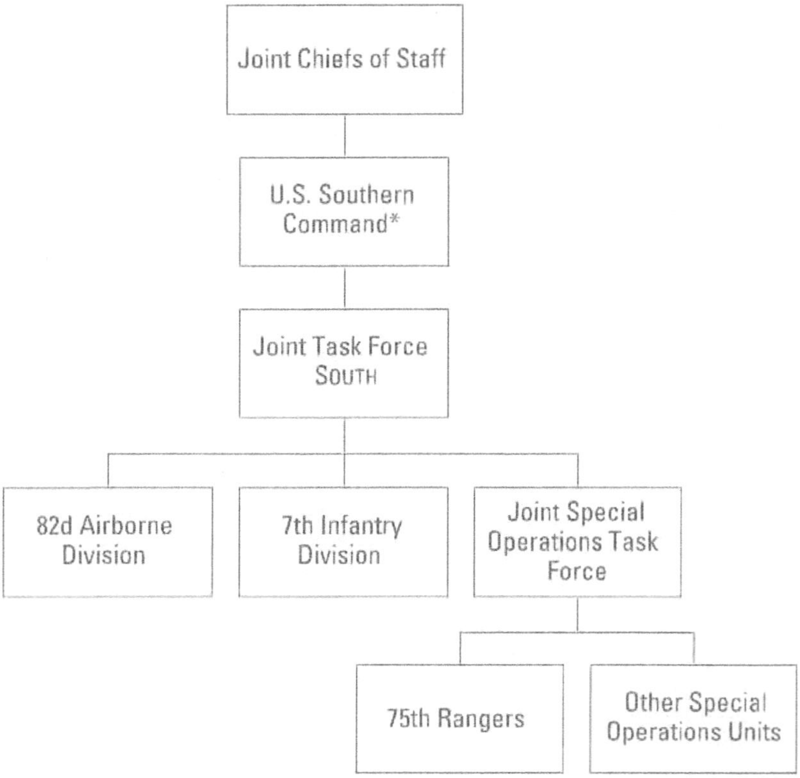

* Judge advocates at U.S. Army South, the service component of U.S. Southern Command, also supported legal operations.

resulted in their judge advocates becoming the key legal advisers during hostilities. Thus, as the military operation unfolded, Army lawyers deploying from Fort Bragg and Fort Ord provided direct legal support to units in Joint Task Force SOUTH and the Joint Special Operations Task Force. And, as almost all combat operations were conducted by units deploying to Panama, almost all operational legal advice came from judge advocates assigned to these units. The Fort Bragg lawyers, however, consulted frequently with those of SOUTHCOM, USARSO, and the 7th Infantry Division.

After assuming command of Joint Task Force SOUTH, General Stiner selected the XVIII Airborne Corps' staff judge advocate, Col.

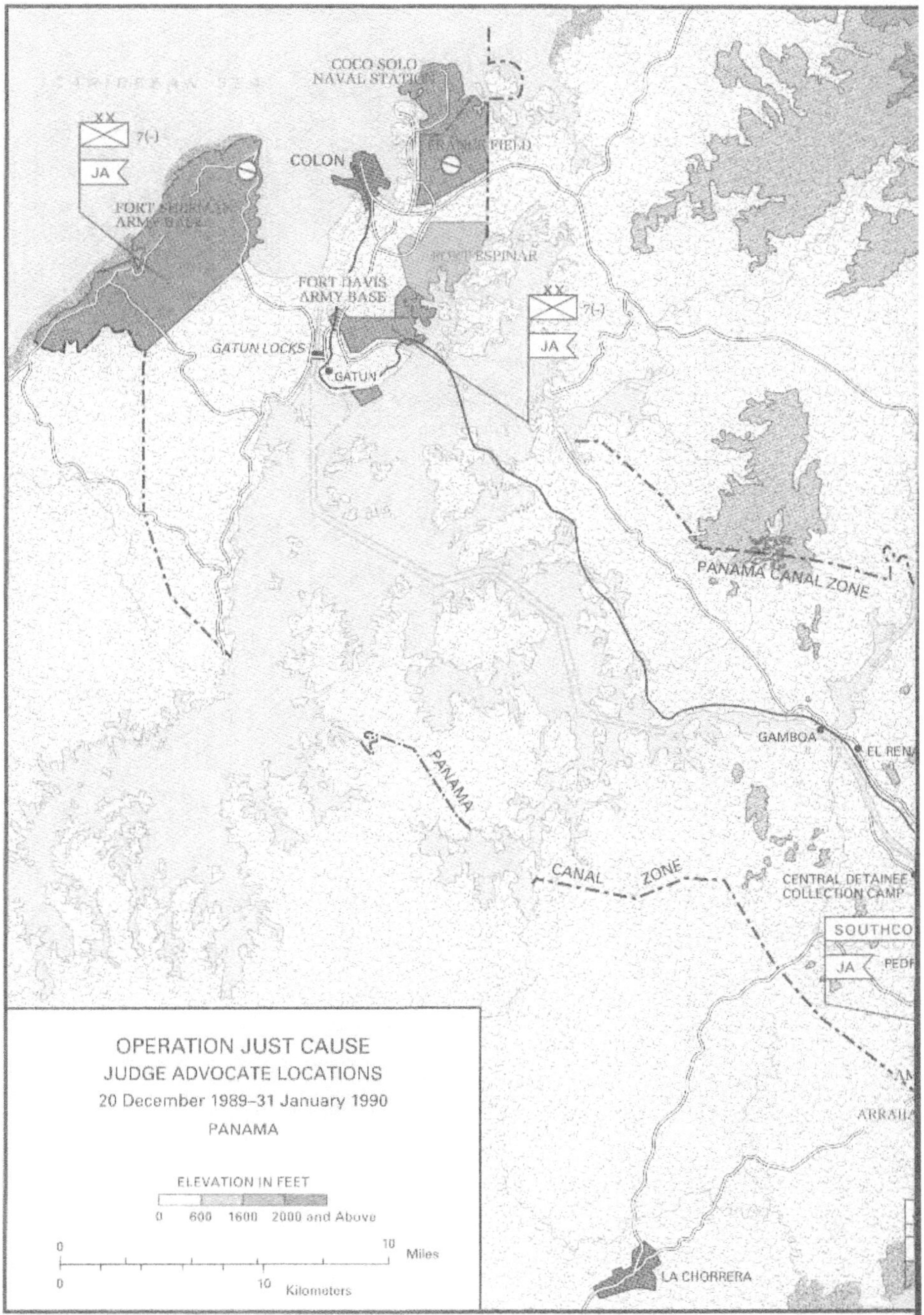
COCO SOLO
NAVAL STATION
XX
7(-)
JA
COLON
FORT DAVIS
ARMY BASE
GATUN LOCKS
GATUN
XX
7(-)
JA
PANAMA CANAL ZONE
GAMBOA
PANAMA
CANAL
ZONE
CENTRAL DETAINEE
COLLECTION CAMP
JA
LA CHORRERA
OPERATION JUST CAUSE
JUDGE ADVOCATE LOCATIONS
20 December 1989–31 January 1990
PANAMA
ELEVATION IN FEET
0 600 1600 2000 and Above
0 10 Miles
0 10 Kilometers

MAP 6

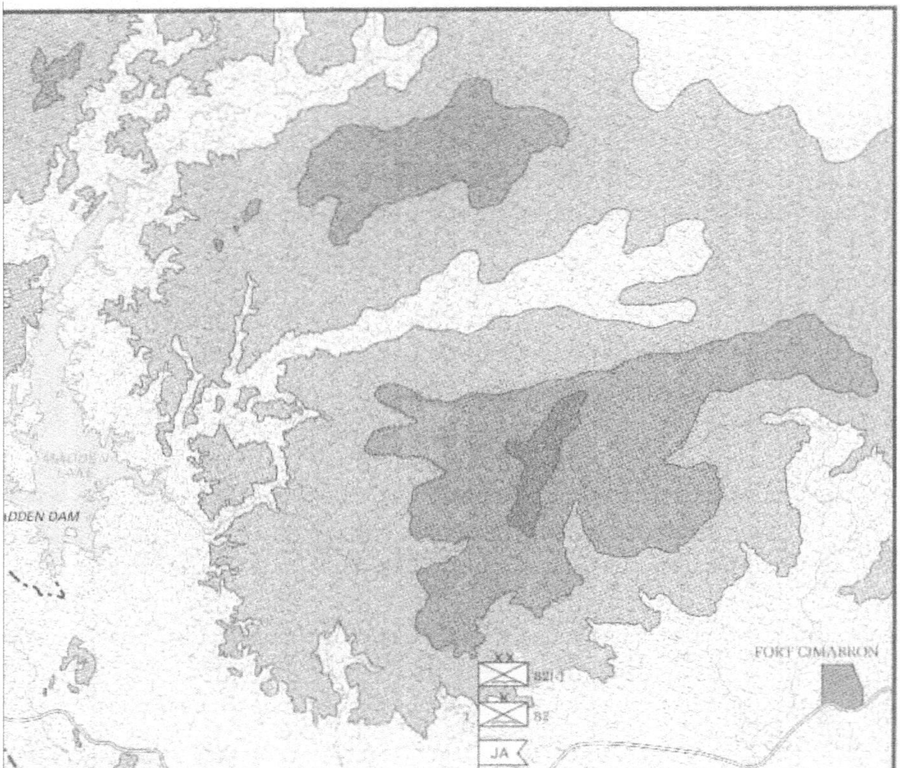

JTF SOUTH
USARSO
JA
TINAJITAS
ALBROOK AIR FORCE BASE
82(-)
1
82
JA
20 Dec
TORRIJOS/TOCUMEN
FORT CIMARRON
PANAMA VIEJO
PAITILLA
PANAMA
BALBOA
FORT AMADOR
FORT KOBBE
82(-)
1
7
JA
BAY OF PANAMA

John R. Bozeman, as the task force legal adviser. Given Stiner's existing relationship with Bozeman, this was a logical choice. However, Bozeman's background also made him ideal for the job. A combat veteran who had served in Vietnam with the 5th Special Forces Group (Airborne), Colonel Bozeman also had prior experience as an armored division staff judge advocate and had served in the Pentagon in the Office of the Judge Advocate General. Based on his knowledge of existing legal operations at U.S. Southern Command and U.S. Army South, Bozeman decided that a lean judge advocate operation would be sufficient. Consequently, he deployed with only his chief of operational law, Capt. Kevin H. Govern. Although Bozeman would later deploy Maj. John E. Baker and Capt. Michael T. Rudisill from Fort Bragg to serve as claims judge advocates, the number of corps lawyers in Panama remained small; only four judge advocates from the XVIII Airborne Corps participated in JUST CAUSE.[10]

The two major conventional Army units in Stiner's task force were the 82d Airborne and 7th Infantry Divisions. Like Colonel Bozeman, the staff judge advocate at the 82d, Lt. Col. James J. Smith, decided that a small legal operation was appropriate. Colonel Smith decided to deploy alone and was the only judge advocate with the division until 24 December, when Capt. William M. Mayes arrived. The arrival two days later of a third division judge advocate, Capt. Charles D. Luckey, gave the 82d Airborne Division the highest number of Army lawyers it would deploy during JUST CAUSE. At the 7th Infantry Division, Lt. Col. David A. Shull, the Fort Ord deputy staff judge advocate, deployed to Panama as the division staff judge advocate. Having already deployed to Panama some months earlier as the 9th Infantry's regimental judge advocate during Operation NIMROD DANCER, Colonel Shull was well prepared to return. He deployed with Capts. James M. Davis and Robin L. Johnson, affording each brigade a command legal adviser. Another 7th Infantry Division judge advocate, Capt. Michael E. Sainsbury, was already in Panama serving at Fort Sherman as the command judge advocate to the 3d Brigade, which had deployed from Fort Ord a few months earlier. Another Army lawyer, Capt. Norman F. J. Allen III, a member of the U.S. Army Trial Defense Service, was also in Panama serving as defense counsel for the brigade.[11]

Operation JUST CAUSE was characterized by a robust use of special operations. Consequently, judge advocates assigned to a number of Army special operations units under General Downing's command served in Panama.[12] General Downing's staff judge advocate, Lt. Col. Donald P. DeCort, and Maj. John M. Smith III provided legal advice to Army units in the Joint Special Operations Task Force, while Capt.

Philip W. Lindley, a judge advocate assigned to the 75th Ranger Regiment, served as the regiment's command judge advocate.

Judge Advocate Predeployment Activities

Based on the experience of Operation URGENT FURY in Grenada and on the extensive operational law training then being provided to Army judge advocates as an integral part of the rapidly evolving practice of operational law, Army lawyers first focused extensively on two areas in planning for JUST CAUSE: rules of engagement (ROE) and preparation for overseas movement. This predeployment activity reflected the radical change that had occurred in the role of judge advocates since the Vietnam era. There, Army attorneys played no part in drafting rules of engagement or in reviewing operations plans, and emphasis was not placed on providing legal assistance as an essential part of preparing soldiers for deployment. However, the emergence of operational law clearly reflected an institutional recognition by the Judge Advocate General's Corps that Army lawyers must be integrated into military operations at all levels and during all phases. It was no longer enough to be "just" a lawyer: a judge advocate now had to be an individual who could resolve the increasingly wide array of legal issues impacting on the conduct of an operation. In the predeployment phase of JUST CAUSE, this meant focusing on rules of engagement and the personal legal needs of the soldiers preparing to deploy.

Rules of Engagement

Unlike Operation URGENT FURY, for which little judge advocate planning was possible, Army lawyers began planning for a possible deployment to Panama almost a year and a half prior to the operation. Thus, in 1988 the SOUTHCOM staff judge advocate assisted in the development of rules of engagement for any future intervention. These were submitted for review and approval to Army Col. Fred K. Green, the legal adviser to the chairman of the Joint Chiefs of Staff. Then, following an aborted coup attempt by dissident elements of the Panama Defense Force in October 1989, Army judge advocates at the XVIII Airborne Corps, the 82d Airborne Division, the 7th Infantry Division, and at Army special operations forces began more serious preparations. First, they tailored the SOUTHCOM rules of engagement for use by their own units. When the SOUTHCOM commander directed the rewriting of his operations plan, however, all subordinate plans had to be revised. During this exercise, the 82d Airborne Division staff judge advocate, Colonel Smith, and the 7th Infantry Division deputy staff judge advocate, Colonel Shull, as well as the XVIII Airborne Corps staff judge advocate, Colonel Bozeman, and his deputy, Lt. Col. Patrick Finnegan, coordinated with the operational planners at the division, corps, and joint command levels and drafted a common rules of engagement annex for inclusion in each operations plan.

In tailoring the SOUTHCOM rules of engagement for Joint Task Force SOUTH, Bozeman and the other Army lawyers started with two basic propositions. The first was that a soldier might always exercise his right of self-defense, regardless of any restrictions on the use of force that might exist in the rules of engagement. Restrictions on the use of deadly force in Operation URGENT FURY had been criticized as infringing on the right of self-defense and, although no resulting harm had occurred to U.S. personnel during combat operations in Grenada, all concerned wished to avoid a similar problem in Panama. Thus, the rules had to be written so that a soldier would not hesitate to defend himself or others in his unit. The second basic ground rule was that the rules of engagement would adhere strictly to the Law of War and that particular emphasis would be placed on minimizing collateral damage and casualties.[13]

In addition to these two basic ground rules, General Stiner, the Joint Task Force South commander, raised five specific concerns that, operating together, resulted in more restrictive rules of enagement. These five concerns originated in a meeting involving Generals Stiner and Thurman and the SOUTHCOM staff, and Stiner subsequently tasked Colonel Bozeman to account for them in the proposed rules of engagement for the joint task force. This was a decision that was to have a significant impact on Bozeman and the other judge advocates (as well as on the Judge Advocate General's Corps as a whole), for the development of rules of engagement had been exclusively an operations (J–3/G–3) function. Army lawyers at that time ordinarily expected only to review operations work, not to actively participate in it.

Colonel Finnegan, Bozeman's deputy at the XVIII Airborne Corps, took the lead in revising the rules to incorporate the five concerns at issue. The first item related to armed civilians accompanying the Panama Defense Forces who, although they could be considered a hostile force and subject to attack, would perhaps surrender immediately if provided an opportunity. As Stiner wished these Dignity Battalion members to have such an opportunity, the rules of engagement included a warning requirement.

Stiner's second concern was the circumstances under which civilian aircraft could be targeted. He feared that commercial aircraft might transport enemy forces or supplies both into and around Panama. But if such a civilian plane was brought down erroneously, with a commensurate loss of life, the result would be a public relations disaster. Consequently, Colonel Bozeman suggested a heightened level of control for the approval of engaging civilian aircraft. No civilian aircraft, even if thought to be transporting General Noriega or other enemy personnel or supplies, would be engaged without the personal approval of General Stiner.[14]

A third issue concerned the use of indirect fire in populated areas. As such fire by its very nature has the potential to cause excessive collateral damage, Stiner wanted its use approved at an elevated level. As a result, only a ground maneuver commander in the grade of lieutenant colonel or above could approve the use of indirect fire in populated areas.[15]

General Stiner's fourth concern was regulation of air-to-ground attacks in populated areas, as such bombardments also ran the risk of causing extensive collateral damage. Thus, the rules of engagement provided that if civilians were present, close air support, white phosphorus bombs, and incendiary weapons could not be used without the approval of the task force commander.[16]

A fifth issue, brought to Stiner's attention by the Joint Special Operations Command (JSOC), concerned the treatment of individuals captured or detained during hostilities. The JSOC staff judge advocate, Colonel DeCort, suggested that every soldier be instructed to initially treat every Panamanian captured or detained as a prisoner of war (POW). This proposal was adopted, for it ensured compliance with the Law of War, eliminated the need for an on-the-spot decision concerning the type of treatment to accord a captive, and assured maximum protection for each detainee. Consequently, the rules of engagement required that all those detained or captured be treated as prisoners of war. Under these rules, captured Panamanians would receive treatment as prisoners of war until their actual status could be determined, even following their transfer to a detainee encampment at Empire Range.[17]

By the time Operation JUST CAUSE began on 20 December, rules of engagement for American combat units already in or deploying to Panama were firmly in place. Their basic thrust was to use maximum fire power, but to minimize collateral damage and suffering, so that when hostilities ceased, normal life could be resumed as quickly as possible. With this in mind, judge advocates at the 7th Infantry and 82d Airborne Divisions and at Army Special Operations Command provided instruction on the task force rules of engagement. Some Army lawyers went even further. Those at the 7th Infantry and 82d Airborne Divisions, for example, printed and distributed thousands of small cards containing the rules of engagement, enabling each soldier to have a written reference that could fit into his or her pocket.[18]

Preparation for Overseas Movement

As the result of operational law training, judge advocates serving with combat units in the late 1980s recognized that preparations for hostilities would be accompanied by a flurry of requests for wills and powers of attorney. They adopted various strategies to reduce such last-minute requests. A year prior to JUST CAUSE, for example, Colonel Smith had implemented a four-part program at the 82d Airborne Division. First, upon entering the division, soldiers were offered the chance to obtain wills and powers of attorney, and then each time their unit entered a new training cycle, they were again offered the opportunity to obtain these legal documents. Third, division legal assistance officers held a "One-Stop Will Program" every Friday and drafted and executed wills at the same time. Finally, soldiers were invited to arrange for their wills or powers of attorney by appointment or on a walk-in basis.

Although implementing this predeployment readiness program did reduce the demand for last-minute legal documents, soldiers still made requests. Thus, late on 18 December and early the next day, the chief of legal assistance, Capt. David V. B. Price; the deputy staff judge advocate, Maj. Joseph A. Russelburg; and the operational law attorney, Captain Mayes, executed wills and fill-in-the-blank general powers of attorney for paratroopers assembled in the personnel holding area at Pope Air Force Base. However, the fact that only twelve wills and approximately a hundred powers of attorney were prepared in the staging area demonstrated the success of preventive legal assistance.[19]

Similarly, at the 7th Infantry Division at Fort Ord, California, an aggressive preventive law program meant that as a battalion came into rotation as the most deployable unit, its soldiers were offered the opportunity to prepare wills and powers of attorney. Additionally, as brigades from the division had already deployed routinely to Panama in the months preceding JUST CAUSE, many soldiers in that division had up-to-date wills and powers of attorney. Consequently, as with the 82d, the 7th Infantry Division's aggressive approach paid dividends. During the final December deployment, only about 450 wills or powers of attorney were prepared for the 4,500 troops deploying to Panama. But, as these had to be completed in just two and one-half days, the division's lawyers worked round-the-clock at Fort Ord and Travis Air Force Base, where division troops gathered to board aircraft for their flight to Panama. While most documents were generated using special Legal

Automation Army-Wide System computer software, some were fill-in-the-blank documents.[20]

Overall, the Judge Advocate General's Corps' operational law emphasis on predeployment legal assistance produced positive results. In URGENT FURY, 40 to 60 percent of deployed forces required legal assistance. In JUST CAUSE, the percentage was much less—around 10 percent of those deployed.[21] Finally, both the 82d Airborne and 7th Infantry Divisions established family assistance centers similar to those in place during URGENT FURY. Judge advocates participated fully in these operations, providing family members with legal advice and assistance that eased the difficulties of separation from their deployed soldier spouses.

Judge Advocate Activities During Hostilities

The first Army lawyer to deploy to Panama from the United States with combat forces was Colonel Smith, the staff judge advocate of the 82d Airborne Division. An experienced attorney who had seen combat as an adviser to Vietnamese units some twenty years earlier, Smith was on the leading aircraft with the division commander, Maj. Gen. James H. Johnson, Jr. Smith carried a .45-caliber pistol, thirty rounds, his protective mask, and six meals, ready to eat. The side pocket of his battle dress uniform contained a microfiche *Manual for Courts-Martial*, a fiche handreader, and a notary seal, along with condensed versions of Army regulations on military justice, war trophies, and claims. He carried Field Manual 27–10, *The Law of Land Warfare*, in the top of his rucksack. Sleet and ice at Pope Air Force Base delayed the departure of his C–141, but once airborne, Colonel Smith donned his parachute, as did the rest of the assault command post. Then, at 3:36 A.M. on 20 December 1989, he jumped into the darkness. Smith and the other paratroopers came under fire; indeed, he had seen the flash of tracer rounds coming toward them before he had jumped from the plane. After landing in a marsh adjacent to Torrijos-Tocumen Airport, Colonel Smith made his way through the head-high vegetation to an assembly area near the runway. During this time, his biggest fear was not the enemy, but "getting shot by some private"; casualties from friendly fire were a very real possibility in the dark.[22]

By 4:30 A.M., Colonel Smith was in the assembly area, and within an hour he received his first legal questions. Some of the division's M551A1 Sheridan armored reconnaissance vehicles, dropped by parachute, had landed in the mud next to the airport. As they were too deeply mired to be driven, the aviation brigade commander asked if his

Lt. Col. James J. Smith, Staff Judge Advocate, 82d Airborne Division, the first Army lawyer to make a combat jump, in a photograph taken 12 January 1990 at Sicily Drop Zone, Fort Bragg, North Carolina, during "jump back" from Panama

pilots could take control of Panama Defense Force helicopters and use them to extract these vehicles. Additionally, the division logistics officer (G–4) wanted to know if civilian cars and trucks located at the airfield could be taken and used by division troops. Most of the 7,000 paratroopers now on the ground were without vehicles, but transport was needed if the division was to move rapidly in the next hours and days. Knowing that the Law of War permitted the confiscation of enemy military equipment and that certain civilian private movable property necessary for mission success could be seized, Colonel Smith advised that the helicopters and vehicles could be taken immediately.[23]

By 10 A.M. Colonel Smith was busy visiting the airport's main passenger terminal, which housed the civilian detainee and prisoner of war collection point. Military intelligence personnel were questioning the detainees and prisoners. Anticipating legal issues, Smith realized that most of the 400 detainees were passengers trapped in the airport because of the ongoing hostilities. Consequently, Colonel Smith counseled accelerated interrogation and early release of these men and women. Otherwise, the humanitarian protections afforded detained persons under international law would require that they be provided food and perhaps other care—something the division lacked the ability to do. By 3 P.M. the division was releasing civilian detainees, and its civil affairs (G–5) staff was transporting them to downtown Panama City.

At a briefing for General Johnson later that evening, Colonel Smith learned that American soldiers operating near Tinajitas, two miles north of Panama City, had been fired upon by enemy soldiers. The Panamanian troops had been positioned in a temple, which the U.S. paratroopers had damaged in an ensuing firefight. Now, after the fact, Smith was asked if a cultural site could be targeted if used by the enemy. Explaining that enemy misuse of the temple had resulted in the loss of its protected status under the Law of War, Colonel Smith advised that firing on the structure had been lawful.[24]

Early on the morning of 21 December Smith got a few hours' sleep—his first in two days. Later that morning he advised the division logistics officer concerning the disposition of Panamanian military and civilian bodies being stored in a Marriott Corporation freezer at the airport and coordinated the transfer of enemy prisoners of war from the division to the XVIII Airborne Corps. Lt. Col. Daniel K. McNeill, the acting chief of staff and head of division operations (G–3), made no decisions related to civilians without conferring with Colonel Smith. At the 82d Airborne, the staff judge advocate was expected to solve legal, quasi-legal, and other related problems quickly and with confidence.[25]

In contrast to the 82d Airborne Division, the 7th Infantry Division had not been in combat since the Korean War; its deployment of some 4,500 soldiers to Panama was its first large-scale operation in more than thirty years. Since the division's top lawyer remained at Fort Ord, it was the division's deputy staff judge advocate, Colonel Shull, who deployed with the division from Travis Air Force Base on 20 December. Captain Johnson, the 2d Brigade judge advocate, climbed aboard a C–5 aircraft that same morning. Both Shull and Johnson landed in Panama after Tocumen airfield had been secured. Colonel Shull then went to Fort Amador, while Captain Johnson stayed with her brigade headquarters at Tocumen. Captains Sainsburg and Allen, the two 7th Division attorneys already in Panama, remained at Fort Sherman.

Captain Johnson was not the only lawyer to deploy initially with the brigade—a defense counsel, Capt. Robbie W. Bare, also deployed. Johnson, however, provided all legal advice to the brigade's commanders and their staffs. From the beginning, she faced a number of issues requiring both sound legal analysis and common sense. Her first legal question concerned the treatment of enemy dead. A number of uniformed Panamanian bodies had been discovered by brigade personnel. What should be done with them? After coordinating with other Army lawyers, Captain Johnson advised that the bodies should be placed in the refrigerators in the Eastern Airlines facility at Tocumen airport. This safeguarded them from animals, prevented their deterioration in the heat, and aided in future identification—all actions ensuring compliance with the applicable Geneva Conventions. Shortly thereafter, Captain Johnson also responded to an inquiry from the Rangers. They were concerned about damage done to a Panama Defense Force medical treatment clinic at Rio Hato. The Rangers had been fired upon by persons in the clinic and had damaged it when returning fire. Recognizing that the Panamanian actions had caused the clinic to lose its protected status, Captain Johnson informed the Rangers that their return of fire—and the resulting damage—was permissible.[26]

Other judge advocates serving with the 7th Infantry Division, like the Trial Defense Service's Captain Allen, also combined sound legal advice with common sense. Thus, when the executive officer at 9th Infantry proposed setting up his regiment's headquarters in a building near a church, Allen advised against it. While it was true that the Law of War did not prohibit using this building, it appeared unwise to position a potential military target so close to a protected building. As a result, the command set up its headquarters elsewhere. Over the next few days Captain Allen remained at the brigade tactical operations center, answering legal questions when they arose, but also "doing anything that needed doing."[27]

While judge advocates at the 82d Airborne and 7th Infantry Divisions handled both legal and nonlegal issues, legal operations at the XVIII Airborne Corps also were under way. Colonel Bozeman had arrived at Howard Air Force Base about 8:30 A.M. on 20 December, although his departure from Fort Bragg—like that of hundreds of soldiers—had been delayed by ice on the wings of his C–141 aircraft. Bozeman went directly to Fort Clayton, the location of the emergency operations center for the joint task force. Bozeman was the lone judge advocate at the center until later that day, when Captain Govern, a member of Bozeman's office at Fort Bragg, arrived in Panama and joined him.[28]

Originally, Colonel Bozeman had intended that Captain Govern assist him at the joint task force, but the volume and intensity of issues facing Col. Lawrence Brede at the 16th Military Police Brigade

(Airborne) resulted in a change of plans. Deploying from Fort Bragg as part of the task force, this brigade was responsible for prisoner operations, traffic management, and force protection. It also was to maintain law and order throughout the urban areas of Panama during the ensuing weeks or months that Panama would lack a functioning police force. The 16th needed its own judge advocate, and Captain Govern's prior association with the brigade during Operation HAWKEYE a few months earlier made him the logical choice to serve as its legal adviser. Colonel Bozeman first believed that Govern could both work for him at Joint Task Force South and serve as the military police legal adviser. Once on the ground, however, Captain Govern's duties at the 16th Military Police Brigade occupied so much of his time that Bozeman detailed him to the brigade as its full-time judge advocate. A USARSO judge advocate, Capt. Louis E. Peraertz, replaced Govern as Colonel Bozeman's assistant in the joint task force emergency operations center.[29]

Post-Hostilities Judge Advocate Activities

Following the fighting between American and Panamanian forces, Army lawyers faced a variety of challenges, including issues involving detainees and prisoners of war, law and order, the status and treatment of foreign diplomatic personnel, foreign claims, military justice, and war trophies.

Prisoners of War

During the first days of JUST CAUSE, Americans captured or otherwise took into custody some 4,100 individuals. Some of these were members of the Panama Defense Force and Dignity Battalions. An assortment of criminals was also detained, many of whom had been caught looting homes and businesses in Panama City. Since U.S. troops had also seized two mental hospitals during the combat phase of JUST CAUSE, a number of mentally ill persons were being held for their own protection and that of the local populace. Finally, some prominent individuals, including ministers in the Noriega government, were also detained.

Under the supervision of USARSO engineers, the Central Detainee Collection Camp was quickly built at Empire Range. The camp had water and electricity, good road access, and ample open areas for tents. At the same time, the jungle and the Panama Canal acted as natural barriers to any escape. When an American unit captured an enemy soldier or took a civilian into custody, it brought that person to the camp. A combat captive or a looter generally was "tagged" with documents or other written information reflecting the circumstances under which he had been seized. However, some detainees had no documentation, and their status was unclear. As the 1949 Geneva Convention Relative to the Treatment of Prisoners of War (GPW) affords prisoners of war certain rights and privileges, determining the status of each person held at Empire Range was critical. Until such a determination occurred, the decision was made to treat each detainee as a prisoner of war. This approach, first adopted in Grenada during Operation URGENT FURY, benefited commanders and soldiers in the field, in that they were not required to make decisions of this nature during hostilities. As a result, no detainee was misclassified during the fog of war, and, as every detainee was afforded the best possible treatment, such an approach demonstrated the U.S. resolve to meet its responsibilities under international law.

Some four to five days into JUST CAUSE, Colonel Russell asked Maj. Richard B. Jackson, the senior defense counsel at U.S. Army South, if he would serve as the legal adviser to the captain in charge of the Detainee Collection Camp. Jackson agreed. When he arrived at Empire Range to take up his new duties, he discovered that detainees whose status had been determined were divided into five categories: mental patients, criminals, civilian internees, officers, and enlisted personnel. Recognizing, however, that it was critical to properly classify those individuals whose legal status was still not clear, Major Jackson consulted Field Manual 27–10, *The Law of Land Warfare*.

Reaffirming that Article 5 of the GPW Convention required that a three-person tribunal determine the status of detainees, Jackson, along with a representative of the camp commander and a military intelligence officer, began the process of sorting out the individuals in question. In determining their status, Major Jackson and the two other officers acted as a de facto Article 5 tribunal, examining any paperwork accompanying the detainees and questioning the individuals concerned when necessary. They also used a "black-white-gray list" provided by military intelligence. Those Panamanians and other foreign nationals known to have engaged in torture or drug trafficking were on a black list; those believed innocent of wrongdoing were on a white list; those suspected of misconduct, or who were the subject of an investigation, were on a gray list. Within a week, all those detained at Empire Range had been classified by Major Jackson and his two colleagues. Although the tribunal operated without being officially designated by a higher headquarters—it was not, for example, formally established by regulation, nor were Major Jackson or his colleagues appointed by orders—this did not affect the value or effectiveness of the tribunal.

With all detainees classified, the camp commander and his staff began readying for the release and repatriation of certain detainees. Since the GPW Convention requires that prisoners of war be repatriated without delay at the end of active hostilities, and as resistance to U.S. troops in Panama had effectively ended by 24 December, it was necessary to make decisions regarding whom to release first and at what pace. Prior to any decisions, however, President Endara's government requested custody of all detainees. Consequently, Major Jackson, working with a SOUTHCOM plans (J–5) officer, spent the first two days of January 1990 reviewing the entire list of 4,100 detainees. At this time, Jackson and his colleague, drawing upon information made available to them from various sources, identified only about 100 individuals who merited continued detention. Based on their recommendation, these persons were retained; the other 4,000 detainees were handed over to the Panamanian government. Most of these individuals returned to their homes. Others, after taking an oath of allegiance to the new Endara government, formed the nucleus of the Fuerza Publica, the new Panamanian police force.[30]

Meanwhile, for the roughly 100 detainees who remained at Empire Range, it was necessary to devise a system for determining who had a continuing interest in detaining a specific individual and the basis for such detention. The result was a unique solution—the creation of the Judicial Liaison Group to work with the new Panamanian government's Attorney General's Office and the Fuerza Publica. Army judge advocates, including the XVIII Airborne Corps' Captain Govern and USARSO's Capts. Rachelle M. Hayes and Michael P. Nido, Jr., prepared files on each detainee, and these were submitted to the liaison group. The files were then used in determining the disposition of the remaining detainees. Some were sent to the United States for prosecution, but most were turned over to the Panamanian government.[31]

Major Jackson and his fellow defense counsel, Capts. Joe T. Cravens and Matthew B. Devlin, also advised the camp commander on the treatment of detainees. Article 13 of the GPW Convention required the humane treatment of prisoners of war, to include the provision of adequate housing, food, and medical care. When a delegation of the International Committee of the Red Cross (ICRC) visited Empire Range on D+9, 29 December, Jackson met with its members. The Red Cross was pleased with the operation, noting U.S. compliance with both the spirit and the letter of the law. The Americans had, for example, taken the local diet into account by providing rice and beans as a supplement to the meals, ready-to-eat (MRE), provided each detainee. Additionally, medical care was provided to detainees on a nondiscriminatory basis. Thus, wounded Panamanian soldiers were evacuated on the same aircraft as U.S. forces and were provided the same degree of medical care as that of American sick and wounded. Finally, Army lawyers had advised that the Geneva Prisoners of War Convention protected the privacy of those held at Empire Range. Consequently, any public display of prisoners was prohibited. The media were allowed to tour the detention facility, but no roster of those in the camp was provided and no photographs were permitted. However, if they so desired, detainees could speak with the press.

Law and Order

No U.S. law enforcement officials had been present in Panama between May and December 1989. On the second day of JUST CAUSE, however, the Department of Justice decided to reestablish federal law enforcement operations and, on 24 December, with the approval of the U.S. embassy in Panama City, thirty-five Justice Department attorneys, as well as Federal Bureau of Investigation and Drug Enforcement Administration agents and members of the U.S. Marshals Service, arrived in Panama. These civilian law enforcement personnel received direct military support from Colonel Wallace, the SOUTHCOM staff judge advocate. Wallace worked closely with these civilian officials, functioning principally as a liaison between them and SOUTHCOM in the apprehension and transfer to the United States of Panamanians who had been indicted for drug trafficking. He also participated in the activities of the Judicial Liaison Group.[32]

Wallace's involvement with the Department of Justice and the activities of judge advocates at Empire Range meant that American military personnel were assisting civilian law enforcement authorities in apprehending Panamanians for prosecution, separating indicted persons from other detainees, interviewing detainees who might be witnesses in the United States, and collecting and protecting documents pertaining to potential defendants. While recognizing that the Posse Comitatus laws restricting military assistance to civilian police authorities had no extraterritorial application, Army judge advocates nevertheless carefully considered the scope of Army involvement with Department of Justice operations in Panama. Thus, although a military intelligence unit became the custodian for criminal evidence, judge advocates decided that when military authorities took custody of any Panamanian wanted for extradition by the Federal Bureau of Investigation or Drug Enforcement Administration, no military interrogation of that individual would occur. This would prevent any later complaint by the accused that he had been improperly questioned by the military prior to being turned over to civilian law enforcement authorities.[33]

While the end of hostilities had made possible substantial American military assistance to civilian police officials, the transition from combat to maintaining law and order also brought a need for modified rules on the use of force by American forces. As a result, judge advocates wrote new rules of engagement for four scenarios in which U.S. troops might find themselves: manning defensive positions, setting up and working roadblocks, stopping looting, and clearing buildings.

With regard to defensive positions, the new SOUTHCOM rules on the use of force required that perimeter limits be clearly marked and that troops be authorized to fire warning shots to deter violators. If warnings failed, minimum force then could be used to detain civilian infiltrators. Similarly, U.S. troops setting up and manning roadblocks were authorized to use varying levels of force. Any vehicle attempting to force its way past a roadblock could be halted and, if necessary, disabled. Proportional force, up to and including deadly force, could be used to repulse an outright attack or to respond to the threat of such an attack.

Of particular interest to commanders—and their lawyers—was the recurring scenario at a roadblock involving a car driving toward a checkpoint, then making a U-turn and driving in another direction. Those manning the roadblock tended to conclude that vehicles avoiding the checkpoint were demonstrating "hostile intent." But the joint task force staff judge advocate, Colonel Bozeman, realized that some Panamanians might simply be attempting to avoid a long line of traffic and a long wait. As initially drafted, the rules of engagement called for a warning shot to be fired and, if the car or truck did not stop, a disabling shot. If the vehicle failed to stop, it then could be attacked using deadly force. Colonel Bozeman believed, however, that it was likely that those in the automobile would not hear a warning shot and that any subsequent disabling shot might miss its mark or otherwise be ineffective. Recognizing that these two possibilities meant that deadly force might be used against individuals innocent of wrongdoing, Bozeman advised a modification to the proposed rules of engagement.

His practical suggestion was that a "chase vehicle" be present at a roadblock whenever possible. For a checkpoint dealing with traffic from one direction only, only one chase vehicle would be needed; traffic from both directions would require two chase vehicles. The idea was for the vehicles to be located at a distance from the checkpoint. If an approaching car or truck made a U-turn, the chase vehicle could then move into the road and block its retreat. If the car or truck continued on a path toward the blocking chase vehicle, this would constitute much clearer evidence of hostile intent. Colonel Bozeman's chase vehicle proposal was incorporated into the final version of the rules of engagement.

Stopping looters and looting also required supplementary rules of engagement. Although some looting was anticipated, the high level of stealing from homes and shops, particularly in Panama City during the first few days of JUST CAUSE, was unexpected. Commanders and judge advocates agreed that using deadly force against looters was ill-advised, but warning shots could be fired under the rules and looters could be aggressively pursued. Those caught by U.S. troops would be detained until they could be transported to Empire Range or the Fuerza Publica.[34]

Finally, as the Americans moved through both Panama City and the countryside in search of remnants of the Panama Defense Force and Dignity Battalions, supplementary rules of engagement were needed for clearing buildings. Again, troops were required to warn the occupants to exit the building and, if necessary, to fire warning shots to hasten their exit. Damage to medical, religious, and historical sites was to be avoided unless these sites were used to mount attacks against U.S. troops or civilians. Additionally, the rules required that damage to nonmilitary government buildings and dwellings be minimized and that private property be respected.[35]

While policy decisions were being made at Colonel Bozeman's level, Captain Govern wrestled with the practical aspects of law and order. The 16th Military Police Brigade, to which he was attached for duty, was primarily responsible for controlling riots and disorder in Panama. Thus, in addition to providing legal assistance to all brigade soldiers, Govern advised its command group on administrative and criminal law matters. Much of his work focused on law and order issues, and he remained in Panama and continued providing judge advocate support to Operation PROMOTE LIBERTY until his return to Fort Bragg on 13 February 1990.[36]

Another law and order issue requiring judge advocate advice was the "Muskets for Money" program, in which an invitation was extended to the local populace to exchange their weapons for cash. Television programming in the United States, notably that of the Cable News Network, announced that a "money for guns" program was under way in Panama long before it was in effect. This complicated matters considerably, as commanders who saw or heard these television news reports naturally queried their judge advocates. At Joint Task Force SOUTH, for example, Colonel Bozeman responded to questions about buying guns for cash by explaining that establishing such a program and using appropriated monies to fund it would require legal authorization. While Bozeman anticipated that such authority would be forthcoming, it was not. This caused difficulties, since Panamanians who had heard of the new money-for-guns program were already bringing their rifles and pistols to American soldiers. "It wasn't smart to say, 'Come back later [as] we don't have the money now,' so soldiers were taking the weapons and giving the people receipts written on candy bar wrappers and MRE cartons. It was too risky to turn someone with a weapon away, with instructions to return later."[37]

Although Colonel Bozeman thought that this issue might be an XVIII Airborne Corps logistics (G–4), provost marshal, or even a joint task force intelligence (J–2) matter, he stepped into the vacuum and wrote a policy for the corps and task force. Pursuant to the policy, the person turning in the weapon was to be paid on the spot as quickly as possible, so the weapon could not be used against friendly forces. Class A agents, individuals authorized to disburse government funds, saturated the city, each with $60,000 or more, ready to buy weapons. The money-for-guns program was a major success, at least in terms of the quantities of weapons acquired. Prices ranged from $25 for a hand grenade and $100 for a pistol to $125 for a rifle and $150 for an automatic weapon or mine. By the end of Operation JUST CAUSE, 8,848 weapons had been purchased at a cost of $811,078.[38]

Status and Treatment of Foreign Diplomats

Judge advocates participating in JUST CAUSE knew that the status and treatment of diplomats might become an issue. First, operations in Panama City, the capital and seat of government, made contact with foreign diplomats likely. Second, after Noriega fled to the Vatican Nuncio's residence, intelligence reports indicated that other Panama Defense Force personnel under criminal indictment might also attempt to avoid capture by fleeing to foreign embassies. The likely safe havens would be six diplomatic locations: the Vatican's Nunciatura; the Cuban, Nicaraguan, and Libyan embassies; the Cuban ambassador's residence; and the Peruvian chargé d'affaire's residence. Consequently, Joint Task Force South established cordoned areas around these locations and implemented rules of engagement for all diplomatic sites and personnel.

In implementing these rules, American troops, acting in the interest of public safety, stopped vehicles and searched their interiors and trunks. The practice was not enthusiastically received by the diplomatic community, but, for the most part, it was accepted. Missteps occurred, however. On 28 December, for example, members of the 82d Airborne Division detained the Cuban ambassador and his escort and, despite vociferous protests by the two men, took them to Joint Task Force SOUTH headquarters. General Stiner ordered their immediate release.

The next day U.S. troops in the El Dorado section of Panama City forced their way into a house, later identified as the residence of the Nicaraguan ambassador, after a U.S. citizen who had provided reliable tips in the past informed the military authorities that a cache of weapons was in the building. At 5:30 P.M. troops from a battalion of the 9th Infantry surrounded the house, made three verbal requests that its occupants exit the building, and then fired two bursts of rifle fire as a warning. No one came out. Then, a man drove up to the building. Though he could produce no diplomatic credentials, the man identified himself as the Nicaraguan ambassador and, claiming that the building was his residence and that he had diplomatic status, demanded that the American soldiers leave.

Believing that there was a reasonable basis to think that there were weapons in the home, the battalion commander decided that a search of the residence should be made. Following an approval of this action by Colonel Smith, the 82d Airborne Division's staff judge advocate, notification of the impending search was made to the joint task force emergency operations center. Colonel Bozeman, on duty at the center, immediately recognized that if the building was a diplomatic residence, no search should occur. Consequently, he telephoned the U.S. embassy, provided the address and description of the house, and asked if the building was in fact a diplomatic residence. He was informed by the U.S. embassy that it was not, and based on this information Colonel Bozeman concluded that the man claiming to be an ambassador was simply attempting to provide a cover for a safe house. The search went forward, with soldiers entering the building about 7 P.M.

Shortly thereafter, it was learned that the individual in question was the Nicaraguan ambassador and that the building was his residence. After learning of the status of the building, Southern Command directed that the joint task force personnel cease their search and leave the premises immediately. The Americans did so, leaving behind some 38 weapons, including 6 rocket-propelled grenades, 3 hand grenades, and assorted ammunition. Leaving the weapons and ammunition in place was an overreaction to the instructions from higher headquarters to cease the search of the building. This aspect of the incident was overlooked, however, in the harsh news media criticism of the search that followed. Entering and searching the diplomatic residence was clearly a violation of the 1961 Vienna Convention on Diplomatic Relations.[39]

Claims

While military personnel claims from soldiers in the 82d Airborne and 7th Infantry Divisions continued to be handled by their judge advocates, overall responsibility for foreign claims fell to the Joint Task Force South judge advocates. The presence of the USARSO Claims Office, with its extensive claims experience in Panama, afforded these deployed judge advocates valuable support. Recognizing the need for additional support for the joint task force, however, Colonel Bozeman deployed Major Baker and Captain Rudisill from the XVIII Airborne Corps. They arrived during the first week in January and served as a claims team until their return to Fort Bragg at the end of the month.[40]

Although Bozeman had anticipated that there would be a significant number of claims filed by Panamanians against the United States and had coordinated with the U.S. Army Claims Service for the appointment of a foreign claims commission, a major issue had not been anticipated: paying claims for the battlefield taking and use of private property. Under the Law of War, U.S. forces could lawfully seize private movable property when so required by military necessity. As noted earlier, acting under this authority, the 82d Airborne Division had seized scores of rental cars and privately owned vehicles in order to quickly move the ten miles from Tocumen airport to Panama City. These seized vehicles were immediately "modified"—windshields kicked out, doors torn off—to make them combat ready. After the troops no longer needed the cars and trucks, they simply parked them and walked away.

Although some 82d Airborne Division officers had used 3-by-5 cards to record what had been seized and by whom, some did not. Those who did sent one card to their battalion logistics officer (S–4) and left a duplicate with the seized car or truck. Major Baker and Captain Rudisill thus started their claims work by gathering information on seized vehicles. Using the laptop computers they had brought with them, the two claims judge advocates checked with battalion logistics officers and recorded information from the 3-by-5 cards on file. Baker and Rudisill also established an impound lot for all seized vehicles, so that following the order to turn in all battlefield-seized cars and trucks, these vehicles could be moved to one central location. Once undocumented vehicles began arriving, they also were included in the claims database. The final tally listed more than 300 seized cars, trucks, and buses, as well as three Marriott in-flight kitchen trucks. To aid in servicing commercial aircraft, a scissors-like assembly on these trucks permitted their cabs to be raised some fifteen feet off the ground. This feature had made them particularly attractive to troops, for the added height could be used as an observation point.

Complicating the issue of battlefield seizures was the question of how such claims would be paid. International law required payment for certain private property seized or requisitioned, including battlefield seizures, and the Army intended to meet this obligation. But what monies would be used? One suggested approach was to make use of the contract process to pay for the use of and damage to seized cars and trucks. Thus, for example, rental cars seized by troops at the airport and then used for the military mission would be considered an irregular procurement—a contract that could be ratified later by appropriate contracting authorities. The problem inherent in this approach was the fact that no agreement had ever existed between the property owner and the person taking the property. Thus, the contracting process could not be used. Similarly, an early proposal to use claims monies to pay for battlefield seizures was also abandoned. Paying monies under the Foreign Claims Act requires a finding of negligence or some other tortuous behavior on the part of agents of the United States. A requirement also exists that the damage to or destruction of the property in question be non-combat related. As the seizures in issue were intentional in nature and combat related, payment under the Foreign Claims Act did not appear to be a viable option.

After examining the U.S. statutes governing DOD Operation and Maintenance (O&M) monies, Major Baker came to the conclusion that these funds could be used to pay for battlefield seizures undertaken as an operational necessity. Colonel Bozeman and U.S. Army South concurred, and the joint task force assistant chief of staff for logistics approved the use of O&M funds for all claims related to private property seized by U.S. forces during hostilities.

Claims for battlefield seizures involved more than those for cars and trucks, however. Barrier material (bricks, wire, wood), some services (international telephone calls), and even some apartments and storefronts had also been seized. Unit headquarters were set up in stores, and, in one instance, an apartment overlooking the Vatican compound in which Noriega had taken refuge was seized. The Army would later pay rent to the owner. In another instance, a Panama Defense Force soldier who had lost part of his arm in the fighting requested compensation for the seizure of his car by U.S. troops. Deciding that the Law of War afforded no exception for the seizure of an enemy soldier's private property, the judge advocates honored his claim.

An issue related to claims made for battlefield seizures was that of claims lodged for war damages. As noted, under the Foreign Claims Act, no monies may be paid for damages incurred incident to combat. As a matter of policy, however, the United States desired to compensate Panamanian citizens and others living in Panama for such damage. In the aftermath of Operation URGENT FURY, the Department of State had made money available to an Army claims operation in Grenada for the payment of claims for war damage. Some now suggested that a similar procedure be created in the aftermath of JUST CAUSE, with the U.S. Army Claims Service administering a claims program of this nature. The Department of State, however, determined that giving the new Panamanian government grant funds with which to compensate its citizens for war damages was preferable to the establishment of an Army-run program. In order to prevent any misunderstanding on the part of the Panamanian public, USARSO lawyers ensured that this approach was widely publicized.[41]

Military Justice

Despite its short duration, Operation JUST CAUSE produced a number of courts-martial. *United States* v. *Bryan* was the highest profile case, no doubt because of its combat scenario. On 23 December 1989, soldiers of the 3d Battalion, 504th Parachute Infantry, 82d Airborne Division, were manning a checkpoint at a roadblock near Madden Dam when they were approached by a carload of five Panamanians. They ordered the occupants out of the vehicle and, in searching it, they discovered a tear-gas grenade, a machete, and an unexpended rifle cartridge.

The Americans had just ordered the Panamanians into prone positions when a hand grenade exploded. After the blast, the paratroopers opened fire on the Panamanians. Then 1st Sgt. Roberto E. Bryan, who was about seventy yards away, ran toward the roadblock when he heard the grenade explode and called for a halt to the firing. When one of the soldiers advised the sergeant that a Panamanian was continuing to move, Bryan opened fire, joined by many of the other soldiers. After the firing ceased, the sergeant saw that several of the Panamanians were dead. What followed next was a matter of dispute. An American soldier began dragging a wounded Panamanian toward the medics. When this soldier advised the sergeant that the Panamanian was still alive, Sergeant Bryan fired the remaining rounds in his M16 rifle into the wounded man. There was testimony to the fact that the Panamanian offered no resistance; however, Sergeant Bryan stated that the man had made a movement with his arm, indicating that he might be reaching for a weapon. The fact that Bryan had shot and killed the Panamanian was not in dispute, however. He was charged with unpremeditated murder.

The court-martial proceedings, held at Fort Bragg some six months after the 82d Airborne returned from Panama, were difficult from the outset. Sergeant Bryan's former battalion commander recommended that all charges against him be dismissed. However, the officer conducting the pretrial investigation decided that, despite the conflicting testimony and a lack of physical evidence, there was sufficient evidence to support a prosecution for voluntary manslaughter. Consequently, he recommended trial by court-martial. However, he also included the following in his investigating officer's report: "No court would ever find the accused guilty and if I were a court member, knowing the evidence as I do now, I would find him not guilty."[42]

When Bryan's brigade commander indicated that he intended to dismiss the charges, the division commander withdrew the former's authority to take any action and directed that the case be forwarded to him. He then decided to send the case to the XVIII Airborne Corps commander, who referred the unpremeditated murder charge to trial by general court-martial.

The prosecution, spearheaded by Capts. Charles D. Luckey and Mark J. Gingras, both judge advocates in the 82d Airborne Division, argued that the man who was killed posed no immediate threat and that Sergeant Bryan had acted in retaliation for the grenade attack that preceded the man's death. Their case rested principally on the testimony of a few soldiers and an infantry lieutenant who supported the government theory that the killing was nothing more than a summary execution. The defense team of Capt. Steven A. Lamb and a civilian attorney countered that, in the chaos following the grenade explosion, Sergeant Bryan had acted in self-defense and in the defense of his fellow soldiers. On 31 August 1990, after a five-day trial, a panel of officers and enlisted soldiers deliberated two hours before acquitting the Panamanian-born sergeant of all charges.

In retrospect, the conflicting testimony and strong community support of Sergeant Bryan made a successful prosecution highly unlikely. Additionally, as the trial occurred in the midst of the 82d's rapid deployment to Saudi Arabia, and with the very real possibility that the division would go into combat against Iraq in the near future, no one was anxious to second-guess a decision made by a senior noncommissioned officer in the heat of battle. Yet those at the highest levels of command, including their judge advocates, believed that possible violations of the Law of War could not be ignored.

In addition to *United States* v. *Bryan*, there were other military justice incidents of note. Capt. Robin Johnson, for example, advised on the appropriate criminal action to be taken against soldiers from the 7th Infantry Division involved in a sham firefight in which a Panamanian woman was killed. On 25 January 1990, Pfc. Mark A. McMonagle and two other soldiers went to a bar and brothel in Panama City. One soldier carried a pistol, which he placed on a table in the bar. While

McMonagle and the two soldiers were in the club, they were alerted that the military police were nearby, and they hid in a back room. The pistol left on the table was missing when they returned to the bar.[43]

McMonagle and one of the soldiers then agreed to stage a sham firefight so that they could claim that the pistol was lost during the course of the fight. All three soldiers fired their M16 rifles into the air and told members of their unit responding to the gunfire that they had been fired upon by men with AK–47 rifles riding in a black Toyota. They also claimed to have received fire from a rooftop. As the sham firefight escalated, a Panamanian woman was shot and fatally wounded.

Captain Johnson drafted charges against McMonagle and his compatriots and arranged with the command for the appointment of an officer to conduct an investigation under Article 32, Uniform Code of Military Justice. At a general court-martial held at Fort Ord, California, a few months later, McMonagle was found guilty of murder and other offenses. His sentence, approved by the division commander, was a dishonorable discharge and seven years' confinement. The two other soldiers involved in the sham firefight were also court-martialed. Although convicted of offenses less serious than McMonagle's, they too were punished with confinement, forfeitures, and bad conduct discharges.[44]

From the other side of the bar, the 7th Infantry Division's Captain Allen defended a military police lieutenant charged with maltreating prisoners. When four captured Panamanians failed to answer questions to his satisfaction, the lieutenant separated one man from the group, took him away, and then fired a round from his weapon, thereby suggesting to the three remaining prisoners that he would shoot those who did not cooperate. Although court-martial charges were preferred, Captain Allen persuaded the command to handle the matter through the nonjudicial punishment process.

War Trophies

An ongoing issue was the retention of enemy property as souvenirs, and judge advocates were often called upon to advise on this matter. All enemy property captured by particular units was sent to Letterkenny Army Depot, Pennsylvania, and later returned to the appropriate units. As to individual war trophies, Colonel Bozeman discovered that this was "an enormously emotional area for commanders." Some wanted no trophies taken; others said take it all.

Initially, Colonel Bozeman had looked to the war trophies directive used during Operation URGENT FURY for guidance. His plan was to copy this directive, making minor modifications appropriate for

Panama. Bozeman quickly discovered, however, attempting to write a war trophy policy at the Joint Task Force level was more complicated. Individual war trophies were not the problem—the issue was enemy property to be taken back for museums, unit trophy rooms, and unit training. For example, Special Operations units might be able to demonstrate a legitimate need for possessing AK–47s—and for placing these automatic rifles on their unit property books. Other units, however, had no such requirements. Bozeman also was unprepared for the complexity of writing rules on war trophies applicable to all of the services.

Colonel Bozeman's problem was solved, however, when the Department of the Army determined that all unit war trophies would be shipped to Letterkenny for demilitarization and later distribution. This made the unit war trophy issue a matter for resolution another—and later—day.

Bozeman also encountered problems in the area of individual war trophies. He had assumed that commanders would subscribe to the view that organizational equipment—flags and bayonets, for example—would be acceptable individual war trophies. He learned, however, that a substantial number of commanders believed that soldiers either should not be allowed to bring back any item, or, at most, only a single item.

In the end, Joint Task Force SOUTH adopted a war trophy policy similar to that used in Grenada. This meant that soldiers could retain captured enemy clothing (hats, shirts, belts, trousers, insignia) and certain items of enemy personal military equipment (helmets, canteens, mess kits, ammunition pouches) as souvenirs. Initially, this list also included Panamanian flags. But, as the legitimate government continued to use the same flag carried by Noriega's forces, the taking of Panamanian flags from captured Panama Defense Force facilities was quickly prohibited.

Joint Task Force South policy on war trophies was not uniformly applied. For example, Maj. Gen. Carmen J. Cavezza, the 7th Infantry Division commander, decided that no individual war trophies would be permitted. As the unit was known as the Bayonet Division, however, Cavezza declared that each soldier would receive a bayonet. As enough bayonets already had been seized, each soldier received one upon his return to the United States.[45]

In sum, before soldiers redeployed from Panama to the United States, clear guidance on war trophies had been published, and thorough inspections by the chain of command, military police dogs, and customs officials prevented the taking of most unauthorized items. No one doubted, however, that a few automatic weapons and other similar contraband made its way to the United States with some American soldiers.

Conclusion

When President Bush announced on 3 January 1990 that, with the capture of Noriega, General Stiner's task force had attained all of the objectives established for JUST CAUSE, judge advocate involvement in this operation began to wind down. When JUST CAUSE ended on 11 January, the participation of Army lawyers in the operation ended as well.

But what had been learned? How did judge advocate involvement in JUST CAUSE differ from that of Army lawyers in Vietnam or in Operation URGENT FURY? The fundamental difference was the emphasis that the Judge Advocate General's Corps had placed on the newly developed practice of operational law. Judge advocates in Operation JUST CAUSE were far better prepared than their colleagues who had participated in URGENT FURY. Army lawyers in Vietnam and Grenada had engaged in a number of operational law activities, but had done so in an unstructured manner, and as individuals. Now the corps as an institution had developed and implemented a comprehensive operational law program. Particularly successful were predeployment legal assistance programs and judge advocate involvement in operational planning and the drafting of rules of engagement. Judge advocates like Colonel Shull, who had deployed initially to Panama with a regimental task force and had then returned to Panama as a division staff judge advocate, and Colonel Smith, who parachuted onto Torrijos-Tocumen airport as part of the first assault, demonstrated that lawyers had become an integral part of a commander's combat team.

But there was little time to reflect upon what had been learned in JUST CAUSE, as the largest combat deployment since Vietnam—DESERT SHIELD—began within months of the termination of hostilities in Panama. Once again, the operational law skills of Army lawyers were about to be tested. And once again, judge advocates at Fort Bragg would be the first to deploy.

Notes

[1] Interv, author with Col James J. Smith, 30 May 96, Historians files, OTJAG.

[2] Ronald H. Cole, *Operation JUST CAUSE: The Planning and Execution of Joint Operations in Panama* (Washington, D.C.: Joint History Office, Office of the Chairman of the Joint Chiefs of Staff, 1995), p. 6.

[3] Ibid., p. 37.

[4] Ibid., pp. 2, 6, 11.

[5] Thomas Donnelly, Margaret Roth, and Caleb Baker, *Operation JUST CAUSE* (New York: Lexington Books, Inc., 1991), p. xi.

[6] Ibid.

[7] Ibid., pp. 62–63.

[8] Ibid., p. 67.

[9] Cole, *Operation JUST CAUSE*, pp. 2–3.

[10] Intervs, Dr. Robert K. Wright, Jr., with Col John R. Bozeman, 26 Jun 90; author with Col John R. Bozeman, 19 Aug 97; author with Col John E. Baker, 7 Jan 97, all in Historians files, OTJAG.

[11] Memo, SJA, 82d Abn Div, sub: After Action Report—Operation "JUST CAUSE," n.d., p. 1; intervs, author with Smith, 30 May 96; with Maj. Robin L. Johnson, 6 May 97; with Capt. Norman F. J. Allen III, 17 Mar 97, all in Historians files, OTJAG.

[12] Classified missions and Army legal involvement therein are not discussed in this work.

[13] Interv, Wright with Bozeman, 26 Jun 90. The rules of engagement for the hostilities phase of JUST CAUSE are reprinted in the Judge Advocate General's School, U.S. Army, *Operational Law Handbook (JA 422)* (Charlottesville, Va.: The Judge Advocate General's School, 1996), p. 8–9.

[14] The rules of engagement stated that civilian aircraft would "not be engaged without approval from *above Division level* unless it is in self-defense" (emphasis supplied). See also Working Paper, The Judge Advocate General's School, 26 Feb 90, sub: Operation JUST CAUSE After Action Seminar, pp. 10–11, Historians files, OTJAG.

[15] Interv, Wright with Bozeman, 26 Jun 90.

[16] Ibid.

[17] Ibid.

[18] Working Paper, JUST CAUSE After Action Seminar, p. 47.

[19] Executive Summary, JUST CAUSE After Action Seminar, Ft. Bragg Special Operations Forces Conference, 4–6 Feb 90, n.d., p. 1, Historians files, OTJAG; Gerald P. Kohns, [82d Abn Div] Legal Assistance During Operation JUST CAUSE—Lessons Learned, n.d., p. 2, Historians files, OTJAG.

[20] Working Paper, Operation JUST CAUSE After Action Seminar, pp. 40–41.

[21] Ibid., p. 35.

[22] Interv, author with Smith, 30 May 96.

[23] Ibid.

[24] Ibid.

[25] Ibid.

[26] Interv, author with Johnson, 6 May 97.

[27] Interv, author with Allen, 17 Mar 97.

[28] Interv, Wright with Bozeman, 26 Jun 90.

[29] Center for Military Law and Operations, JUST CAUSE After Action Seminar, 26 Feb 90, p. 78, Historians files, OTJAG. Intervs, Wright with Bozeman, 26 Jun 90; author

with Bozeman, 19 Aug 97; author with Maj. Kevin H. Govern, 6 May 97, Historians files, OTJAG.

[30] Interv, author with Lt Col Richard B. Jackson, 17 Mar 97, Historians files, OTJAG.

[31] Working Paper, JUST CAUSE After Action Seminar, p. 78; interv, Wright with Bozeman, 26 Jun 90.

[32] Working Paper, JUST CAUSE After Action Seminar, pp. 73, 164–78.

[33] After Action Report—Operation "JUST CAUSE," p. 167; interv, Wright with Bozeman, 26 Jun 90.

[34] Interv, Wright with Bozeman, 26 Jun 90.

[35] Cole, *Operation JUST CAUSE*, pp. 66–67.

[36] Interv, author with Govern, 6 May 97.

[37] Interv, Wright with Bozeman, 26 Jun 90.

[38] Cole, *Operation JUST CAUSE*, pp. 65–66.

[39] Interv, author with Bozeman, 19 Aug 97; Cole, *Operation JUST CAUSE*, pp. 60–61; After Action Report—Operation "JUST CAUSE", pp. 138–45.

[40] Interv, author with Col John E. Baker, 7 Mar 97, Historians files, OTJAG. All material in this section on claims, except the last sentence, is from this interview.

[41] After Action Report—Operation "JUST CAUSE", p. 92.

[42] J. Paul Scicchitano, "What's So Special About the Bryan Case," *Army Times*, 13 Aug 90, p. 13.

[43] Interv, author with Johnson, 6 May 97.

[44] *United States* v. *McMonagle*, 34 M.J. 852 (A.C.M.R. 1992); 38 M.J. 53 (C.M.A. 1993); *United States* v. *Gussen*, 33 M.J. 746 (A.C.M.R. 1991); *United States* v. *Finsel*, 33 M.J. 739 (A.C.M.R. 1991).

[45] After Action Report—Operation "JUST CAUSE", pp. 169–70; intervs, Wright with Bozeman, 26 Jun 90; author with Bozeman, 19 Aug 97.

4

Army Lawyers in DESERT SHIELD, 1990–1991

What did I learn in Saudi Arabia? That the Army is an amazing organization when given a mission to do. It can do anything. And I learned that I was both a soldier and a lawyer.[1]

—*Capt. Patricia A. Martindale*
Judge Advocate, XVIII Airborne Corps
Southwest Asia, 1990–1991

On 2 August 1990, Iraqi tanks rolled into Kuwait; Saddam Hussein, the Iraqi ruler, proclaimed Kuwait a province of Iraq. That same day the United Nations (UN) Security Council condemned the invasion; U.S. President George Bush, joined by a large majority of the nations of the world, announced that Iraqi aggression must be opposed. Foreseeing the need for direct military action, the United States took the lead in developing and coordinating a multinational coalition of armed forces to liberate Kuwait and deter any possible Iraqi moves against Saudi Arabia and other Persian Gulf states. On 6 August the UN General Assembly imposed an embargo on Iraq, and Operation DESERT SHIELD began the following day. By the time this operation had come to a close and Operation DESERT STORM had commenced in January 1991, the U.S. Army had moved 295,000 soldiers, 2,000 tanks, 4,000 heavy trucks, and thousands of other vehicles into Saudi Arabia, making DESERT SHIELD the largest deployment of American troops and materiel since the Vietnam War.[2]

Overall responsibility for American operations against Iraq fell to the U.S. Central Command (CENTCOM), one of five geographically oriented unified combatant commands. Responsible for U.S. forces in

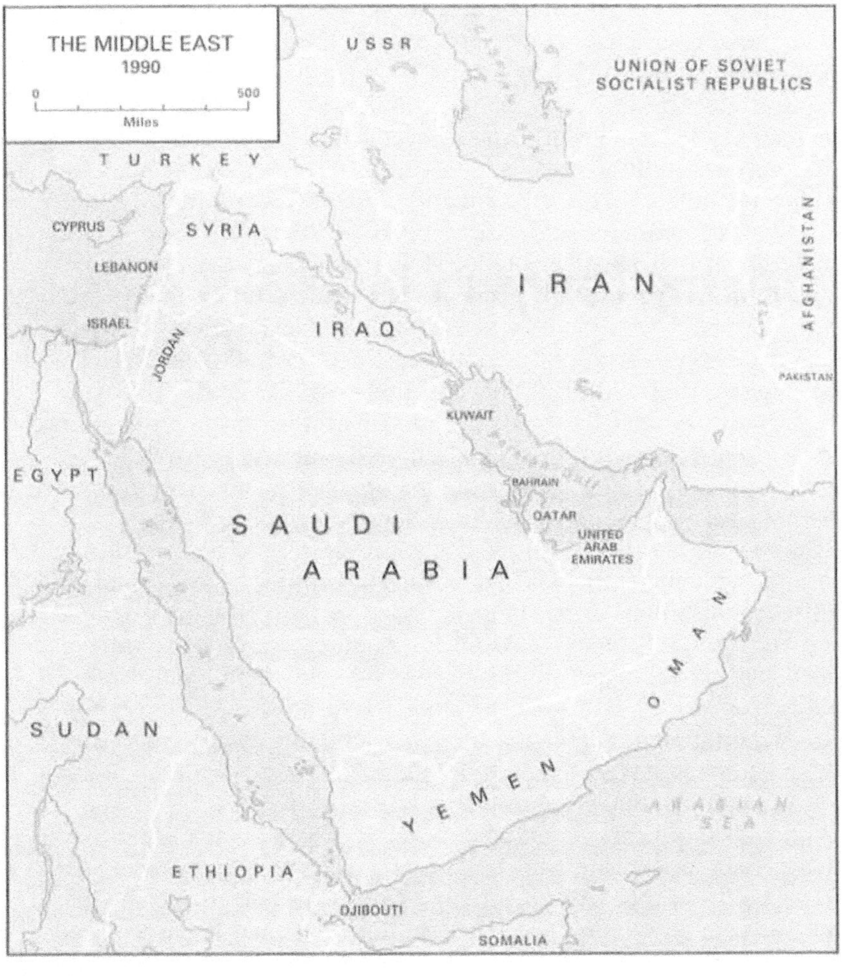

MAP 7

Northeast Africa, Southwest Asia, and the Persian Gulf, CENTCOM's commander in chief, Army General H. Norman Schwarzkopf, exercised authority over all U.S. Army, Navy, Air Force, and Marine elements in its geographic area of operations. Schwarzkopf's command was an operational headquarters, responsible for the buildup of all American forces in the Persian Gulf and the planning for ground, sea, and air combat operations against Iraqi forces. *(Map 7)*

The U.S. Central Command also was an allied headquarters, exercising operational control not only over American forces but also over British and French units in the theater. However, Schwarzkopf did not exercise authority over all armed forces in the area. Arab members of the coalition, ranging in size from a 40,000-man Egyptian force to a 300-man Afghan police force, were led by Saudi Lt. Gen. Prince Khalid ibn Sultan. In practice, the Arab coalition members followed CENTCOM's lead in operational matters, but never formally ceded authority to Schwarzkopf.

A combined military organization with one overall commander and clear lines of authority over the roughly 700,000 troops from twenty-eight coalition countries would have best satisfied the unity-of-command principle. It also would have reduced the confusion that sometimes resulted from having two independent commanders in the theater. For political reasons, however, this was not possible. Nevertheless, although it might have been better organized, the coalition proved to be effective.[3]

In contrast to the informal lines of authority that existed between Central Command and the other coalition partners, the internal command and control structure was much more defined. U.S. Army Forces Central Command (ARCENT), as the major subordinate Army component of the unified command, functioned as the higher headquarters for all Army ground troops. While it had no forces assigned to it during peacetime, with the start of DESERT SHIELD Army Central Command began receiving troops from units stationed in the continental United States. Thus, on 1 September ARCENT commander Lt. Gen. John J. Yeosock had overall responsibility for 31,000 soldiers in Saudi Arabia, primarily from the XVIII Airborne Corps. By 15 January, the day before U.S. and coalition forces launched their attack on Iraq and Kuwait, Army Central Command had grown to more than 245,000 soldiers, including two corps, a major support command, and seven divisions.[4]

Although the theater command structure was similar to that used some thirty years earlier in Southeast Asia, with Army Central Command akin to U.S. Army, Vietnam, the army involved was significantly different in nature. The conscription army had been replaced with a highly trained all-volunteer force. The Iraqis also faced an army with a new fighting doctrine, AirLand Battle, and with new weapons and equipment, including the Abrams main battle tank, Bradley infantry fighting vehicle, Apache attack helicopter, and Patriot air defense system.[5] Although untested, this force would soon prove itself to be one of the most formidable military expeditions ever fielded.

Organization of Legal Services

The organization for legal services paralleled the command structure. The CENTCOM staff judge advocate, Army Col. Raymond C. Ruppert, was the chief legal adviser to General Schwarzkopf. Ruppert's staff,

Chart 6—Legal Organization of U.S. Army Units (with Staff Judge Advocates), Persian Gulf, February 1991

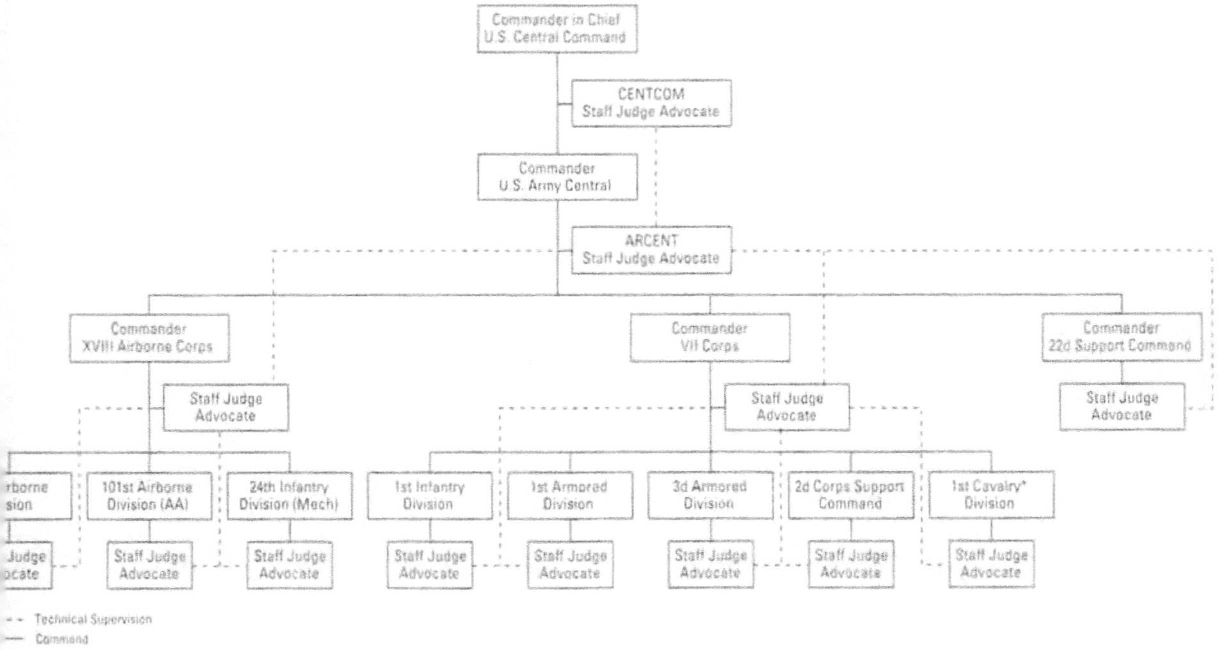

valry Division was transferred from VII Corps to the theater reserve at the beginning of the ground war on 24 February 1991

me Army judge advocates were in theater with U.S. Special Operations Command and, although supporting U.S. Central d, were not under its combatant command authority.

composed of attorneys and clerical personnel from the Army, Navy, Air Force, and Marine Corps, provided legal advice to the CENTCOM commander and his staff. Because their principal mission was formulating the theater's legal policy rather than delivering legal services, the number of military attorneys at Central Command was relatively small. Colonel Ruppert and the five other CENTCOM lawyers were not involved in trying courts-martial or providing legal assistance; these tasks were the responsibility of CENTCOM's component legal staffs and the judge advocates and lawyers at subordinate units. *(Chart 6)*

The ARCENT staff judge advocate, Col. Tonu Tommepuu, was the legal adviser to ARCENT's commander, General Yeosock. Tommepuu's staff, composed of Army lawyers and support personnel, provided legal services to Yeosock and his staff. ARCENT judge advocates also were responsible for implementing CENTCOM's legal policy and for providing legal direction and guidance to ARCENT's three major subordinate units: XVIII Airborne Corps, VII Corps, and ARCENT Support Command (SUPCOM). In theory, Central Command made legal policy for the entire theater, with the Army component overseeing the implementation of that policy for Army personnel at the corps and division levels. As a practical matter, however, legal policy was not formulated at Central Command without informal discussions with all senior service component judge advocates. Thus, in addition to coordinating with Air Force, Navy, and Marine attorneys, Colonel Ruppert routinely spoke with the XVIII Airborne Corps staff judge advocate, Colonel Bozeman; the VII Corps staff judge advocate, Col. Walter B. Huffman; and the ARCENT Support Command staff judge advocate, Lt. Col. William R. Hagan. Moreover, Bozeman and Huffman routinely consulted with the staff judge advocates of the divisions assigned to their respective corps. Such consultations guaranteed that legal policy was not created in a vacuum. *(Map 8)*

Although Colonels Bozeman and Huffman had overall responsibility for the delivery of legal services in the two Army corps, the seven subordinate divisions and VII Corps' 2d Corps Support Command each had its own legal staff. At XVIII Airborne Corps, Colonel Bozeman worked closely with Lt. Cols. James J. Smith and Theodore P. Littlewood and Col. Malcolm H. Squires, Jr., the staff judge advocates for the 82d Airborne Division, 24th Infantry Division (Mechanized), and 101st Airborne Division (Air Assault), respectively. At VII Corps, Colonel Huffman oversaw the activities of Lt. Cols. John D. Altenburg, Jr., John T. Burton, Warren D. Hall III, and Philip E. Lower, the staff judge advocates for 1st Armored Division, 3d Armored Division, 1st Infantry Division, and 1st Cavalry Division, respectively. Huffman also

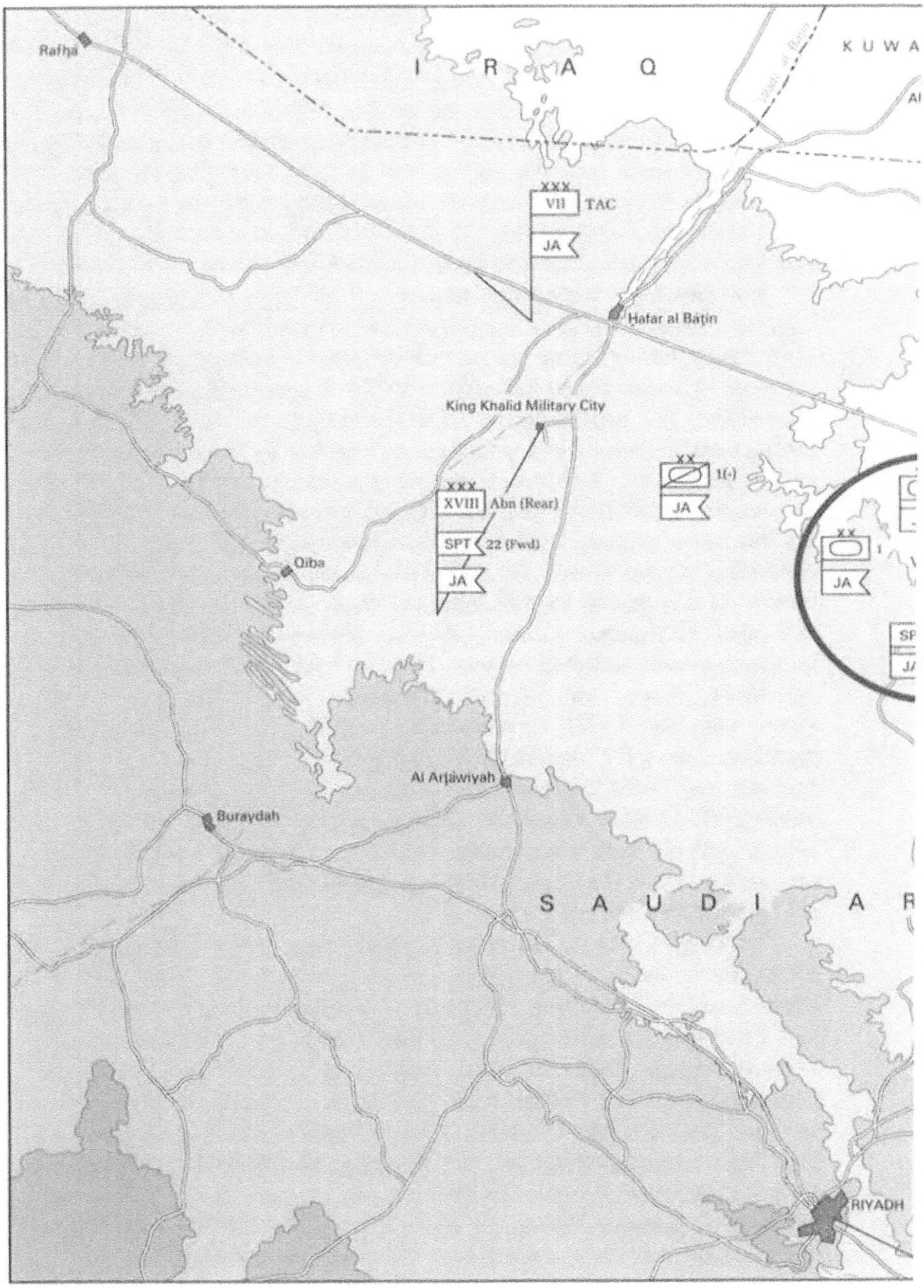
Rafha
I R A Q
K U W A
VII TAC
JA
Hafar al Batin
King Khalid Military City
XVIII Abn (Rear)
SPT 22 (Fwd)
JA
1(-)
JA
JA
Qiba
Al Artawiyah
Buraydah
S A U D I A R
RIYADH

MAP 8

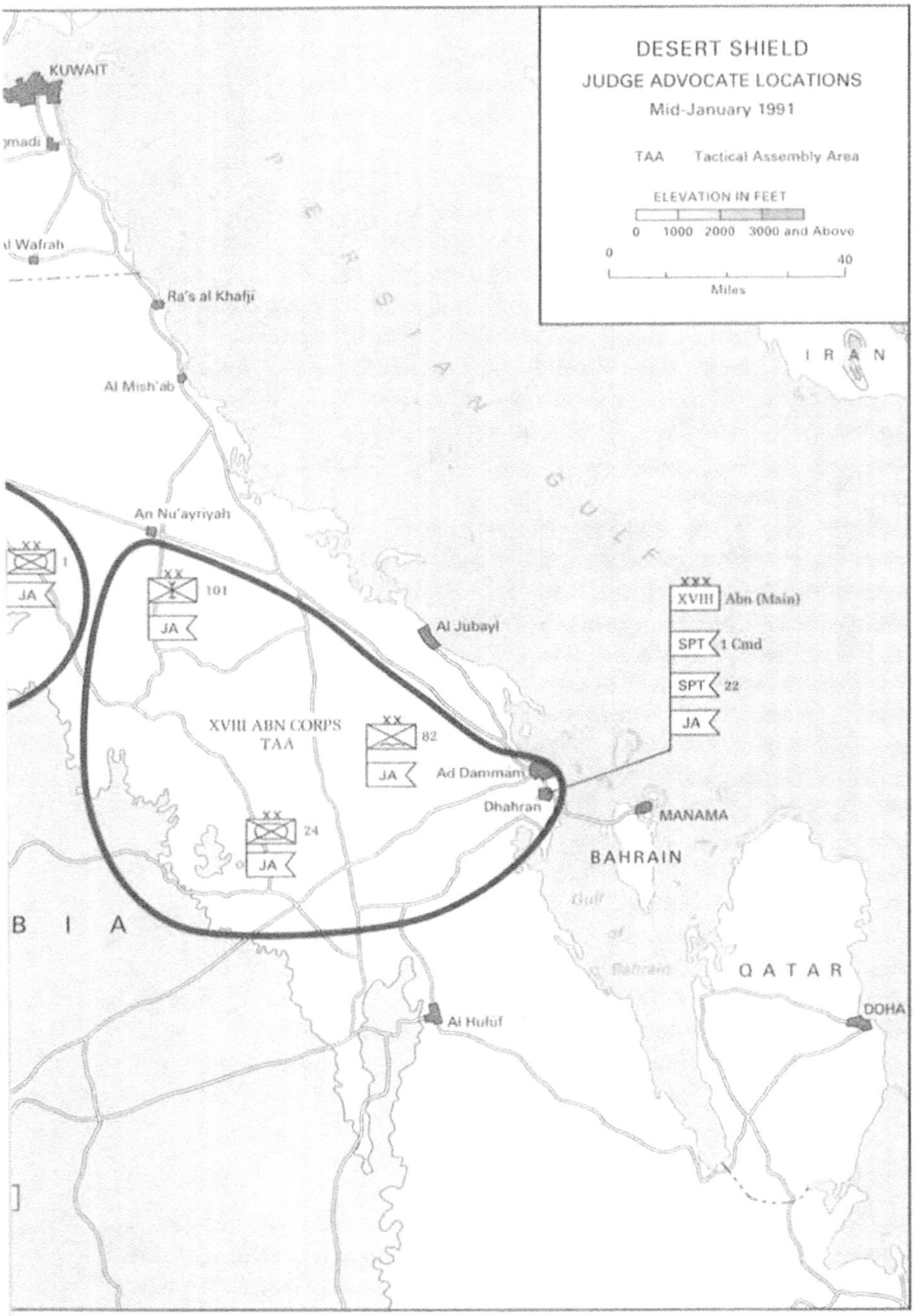
DESERT SHIELD
JUDGE ADVOCATE LOCATIONS
Mid-January 1991
TAA Tactical Assembly Area
ELEVATION IN FEET
0 1000 2000 3000 and Above
0 40
Miles
KUWAIT
Ra's al Khafji
Al Mish'ab
An Nu'ayriyah
PERSIAN GULF
IRAN
XXX
XVIII Abn (Main)
SPT 1 Cmd
SPT 22
JA
XX
1
JA
XX
101
JA
Al Jubayl
XVIII ABN CORPS
TAA
XX
82
JA
Ad Dammam
Dhahran
MANAMA
BAHRAIN
XX
24
JA
B I A
Gulf of Bahrain
QATAR
DOHA
Al Hufuf

provided guidance and support to Lt. Col. Edward W. France III, staff judge advocate for the 2d Corps Support Command. The technical chain of command thus ran from these division-level staff judge advocates through the corps and Army Central Command to U.S. Central Command.

Finally, judge advocates belonging to the U.S. Army Trial Defense Service, although collocated with the corps and divisions, were not part of the same technical chain of command. Thus, Trial Defense Service personnel at Fort Bragg continued to supervise trial defense counsel such as Capt. John I. Winn, who deployed with the first lawyers from the XVIII Airborne Corps and the 82d Airborne Division. With the arrival of VII Corps units from Europe in Saudi Arabia, however, Lt. Col. Michael C. Chapman became the regional defense counsel, assuming overall responsibility for all trial defense attorneys in Saudi Arabia. Chapman reported directly to the chief of the Trial Defense Service, in Falls Church, Virginia.

How did Army lawyers structure their legal organizations? Generally, the structure adopted depended on individual preference and the type and volume of legal work. But in certain cases the expertise of a particular lawyer also impacted on the organization of legal services. The 1st Armored Division, for example, created a "military law" branch, with responsibility for both military justice and operational law, as an experienced trial attorney with extensive operational law experience was best qualified to oversee both these functions. There was, however, no doctrinally "correct" organization for legal services. For example, the CENTCOM staff judge advocate, who had no deputy, had four distinct areas of practice in his office: Operational Law, Administrative and Fiscal Law, Military Justice and Claims, and International Law. *(Chart 7)* While Army Central Command also was organized in four divisions, they were quite different: Operational and International Law, Administrative Law, Military Justice, and Contracts and Fiscal Law. The ARCENT Administrative Law Division had three branches—legal assistance, claims, and administrative law. At VII Corps, Colonel Huffman organized his office differently, and at 1st Armored Division, Colonel Altenburg's office was different still.

U.S. Central Command

The multiservice nature of Central Command naturally was reflected in its legal staff, all of whom were seasoned judge advocates. Colonel Ruppert, the top lawyer at the command, had prior experience as a rifle platoon leader and military intelligence officer in Vietnam and as an

CHART 7—LEGAL ORGANIZATION AT U.S. CENTRAL COMMAND, FEBRUARY 1991

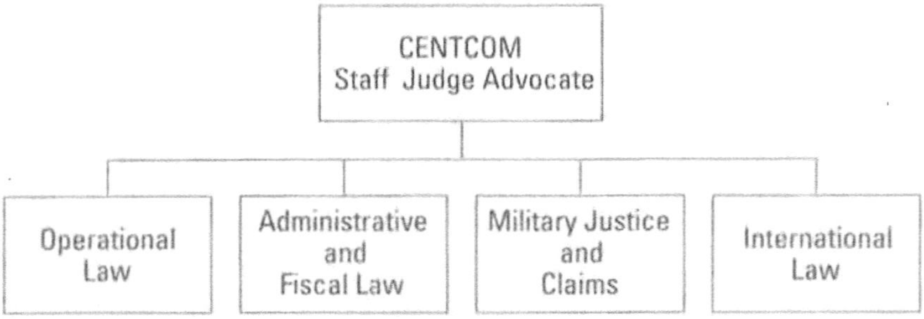

infantry division staff judge advocate. Maj. Douglas K. Fletcher, the only other Army lawyer at Central Command, was a fiscal law expert and adviser to CENTCOM's special operations component. Rounding out the six-person office were two Air Force lieutenant colonels, a Navy commander, and a Marine major.

When General Schwarzkopf decided that it was time to deploy a forward headquarters to Saudi Arabia, Colonel Ruppert had to decide whom to send. Given the lack of an existing infrastructure for deploying forces, CENTCOM forces would be purchasing many supplies and services. Because Colonel Ruppert believed that this would pose complex fiscal law issues, he first chose Major Fletcher to deploy. Accompanying Fletcher would be Lt. Col. Bernard E. Donahue, an Air Force lawyer and the office's international law expert. These two judge advocates flew to Riyadh, Saudi Arabia, on or about 10 August.

Two weeks later, after the decision was made to deploy the rest of the headquarters, Colonel Ruppert left for Southwest Asia. He was followed shortly thereafter by two other CENTCOM judge advocates, Maj. John C. Harris, a Marine lawyer, and Comdr. Walter L. Jacobsen, a Navy judge advocate. Ruppert decided that his one remaining lawyer, Air Force judge advocate Lt. Col. William J. Camp, would remain at MacDill Air Force Base, Florida. Camp would serve as the link between the deployed judge advocates and rear operations at MacDill, and would also deal with those those legal issues that required legal research to be conducted in the United States. Camp was not alone, however. Reserve judge advocates were mobilized to replace those attorneys who had deployed from CENTCOM's legal office at MacDill Air Force Base.[6] Additionally, CENTCOM legal operations in Riyadh were supplemented with the mid-October 1990 arrival of Army Capt. Robert A. Burrell, who came directly from Hawaii and the 25th Infantry Division (Light).[7]

At CENTCOM headquarters, Colonel Ruppert and his legal staff did not deal with legal assistance, military justice, or claims matters. After deploying to Saudi Arabia in August 1990, however, Ruppert reorganized his legal operation to address these areas of the law, as CENTCOM would be formulating a uniform theater legal policy for these subjects. Thus, the headquarters lawyers, working in Riyadh at either the Saudi-CENTCOM underground complex or a nearby office building, focused on legal policy relating to administrative, fiscal, and operational law, as well as military justice, legal assistance, and claims.

Standards of Conduct and Discipline

A key issue for CENTCOM was the establishment of theaterwide standards of conduct for CENTCOM personnel, both military and civilian. Although punitive orders and regulations had been used since Vietnam to supplement the major offenses punishable under the Uniform Code of Military Justice, General Schwarzkopf now proposed a general order that would restrict conduct "generally permissible in western societies." Based on his knowledge of Saudi culture and Islamic law, Schwarzkopf believed that prohibiting certain activities was "essential to preserving U.S.-host nation relations and the combined operations of U.S. and friendly forces," even though such activities were lawful in the United States.

Consequently, General Schwarzkopf directed Colonel Ruppert to draft a "General Order no. 1" regulating the behavior of all CENTCOM forces in the Persian Gulf. As intoxicating beverages were illegal under Saudi law, Schwarzkopf instructed Ruppert to include a ban on alcohol consumption in the order. Recognizing Saudi cultural sensibilities in the area of sexually suggestive literature, he also directed that General Order no. 1 contain a prohibition on the possession of such material. General Schwarzkopf also directed his staff judge advocate to address the taking of war trophies.[8]

While still in Florida, and using as a guide a MACV directive published in the 1960s as well as a similar directive drafted for use in Panama, Colonels Ruppert and Camp drafted General Order no. 1. The final product, entitled *Prohibited Activities for U.S. Personnel Serving in the USCENTCOM A[rea] O[f] R[esponsibility]*, was published in Riyadh on 30 August 1990. It applied to all U.S. personnel in Southwest Asia. Violations of its provisions by military personnel were criminal offenses under the Uniform Code of Military Justice; civilians disobeying General Order no. 1 faced administrative sanctions. The order listed nine general categories of prohibited activities. These included possessing a privately owned firearm, entering a mosque (by a non-Moslem),

General H. Norman Schwarzkopf, Commander in Chief, U.S. Central Command, and Col. Raymond C. Ruppert, Staff Judge Advocate, U.S. Central Command, Safwan, Iraq, March 1991

gambling of any kind, exchanging currency other than at the official exchange rate, and possessing or defacing archeological artifacts.

Of principal interest, however, was the fact that for the first time U.S. personnel were forbidden to possess, make, sell, or consume "any alcoholic beverage."[9] Given the popularity of beer, wine, and other alcoholic products among American troops and the fact that no other American theater-level commander had ever previously attempted to ban alcohol, this was a radical step. While subsequent events demonstrated that the prohibition promoted good order and discipline among U.S. troops, this result was unintended. Rather, General Order no. 1 banned alcohol consumption due to General Schwarzkopf's belief that mission success would require a demonstrable respect for Saudi culture and Islamic law.[10]

General Order no. 1 also made unlawful the possession of any "pornographic" or "sexually explicit" photograph, videotape, book, or magazine. Schwarzkopf was aware of the Saudi views on sex and women and, as sexually suggestive films and literature were not permitted in Saudi Arabia, U.S. personnel also would not be allowed to possess them. Agreeing that in a combat zone military necessity outweighed any constitutional right to freedom of expression, Colonels

Ruppert and Camp defined the terms "pornographic" and "sexually explicit" to include not only images of "human genitalia" or "uncovered women's breasts" but also any display of "human anatomy" implying such images. Thus, General Order no. 1 specifically banned bodybuilding magazines, swimsuit editions of periodicals, and lingerie or underwear advertisements and catalogs.[11]

Finally, the order gave clear guidance on the taking of war trophies. Reminding U.S. military and civilian personnel that enemy private property could never be confiscated, General Order no. 1 next noted that under the law all captured enemy public property would become U.S. property. As a limited exception, however, U.S. personnel would be allowed to retain items such as enemy hats, shirts, belts, insignia, canteens, mess kits, helmets, and ammunition pouches as trophies. The retention of Iraqi bayonets also was prohibited, although this ban would later be rescinded after subordinate commanders persuaded General Schwarzkopf that U.S. troops should be permitted to keep these particular items.[12] Apart from this modification, however, General Order no. 1 remained unchanged for the duration of Operations DESERT SHIELD and DESERT STORM.

A related issue was whether Central Command would establish an in-theater confinement facility. As General Order no. 1 promulgated a disciplinary policy for CENTCOM's area of operations, should the command also provide for a military prison in Saudi Arabia for those individuals sentenced to jail for violating the order or for other serious misconduct? When the XVIII Airborne Corps requested the establishment of an in-country facility for a maximum of twenty-five prisoners, Colonel Ruppert advised that no legal obstacle existed to creating such a facility. Because he was familiar with the Long Binh jail in Vietnam, Schwarzkopf understood that a confinement facility required a sizable number of military police. Anticipating that large numbers of Iraqis might be captured, General Schwarzkopf wanted military police to be available to guard prisoners of war rather than soldiers. Consequently, he decided that any Americans ordered confined by sentence of a court-martial would be returned to the United States.[13]

Operational Law

Even before they departed by plane for Southwest Asia, the command's lawyers had wrestled with rules of engagement for the ground, sea, and air forces in the theater of operations. Although DESERT STORM's large-scale ground combat operations might suggest otherwise, the first questions on the use of force concerned American naval forces. In support of UN Security Council Resolution 661, imposing an embargo on the import and export of all commodities originating in Iraq and occupied Kuwait, the United States and Great Britain deployed warships to the Persian Gulf and the Red Sea. While cautioning Colonel Ruppert "not to start World War III," General Schwarzkopf instructed him to draft rules of engagement governing the interception of inbound and outbound commerce from Iraq and Kuwait. As Ruppert recollected, this was "very much contrary to doctrine . . . as lawyers aren't supposed to write Rules of Engagement." However, he and CENTCOM's Navy judge advocates drafted guidelines for stopping and searching ships on the high seas. The lawyers coordinated closely with the joint command's operations section in arriving at a finished product.

Army Col. Fred K. Green, the legal adviser to the chairman of the Joint Chiefs of Staff, also consulted with the CENTCOM lawyers concerning how to refer to the naval operation. The decision was made to avoid the use of the word "blockade." Since a state of war did not exist between the United States and Iraq, and since a blockade is an act of war under international law, using this term "might give the Iraqis an excuse for expanding the scope of the war." Similarly, the phrases "maritime interdiction" or "quarantine operation" were also avoided, lest this operation be associated with the crisis atmosphere that marked naval operations undertaken during the Cuban missile crisis of 1961. Ultimately, the name selected was "maritime intercept operation."[14]

After President Bush approved the deployment of combat forces to Saudi Arabia on 7 August 1990, legal work focused on disseminating Staff Memo 846–88, *Peacetime Rules of Engagement for U.S. Forces*, to deploying forces. Promulgated by the Joint Chiefs of Staff, these rules were the starting point for all rules of engagement issued by U.S. land, sea, and air forces.[15] With the 82d Airborne Division and XVIII Airborne Corps now identified as the first Army units to deploy to Southwest Asia, Colonel Ruppert conferred with the corps staff judge advocate, Colonel Bozeman, to ensure that sufficient copies of these peacetime rules were available at Fort Bragg. As more and more U.S. forces arrived in Southwest Asia, CENTCOM judge advocates monitored rules of engagement published by subordinate units, ensuring uniformity throughout the area of operations. The active role played by these judge advocates demonstrated the ever-increasing participation of lawyers in the development of rules of engagement. Line officers planning combat operations were still responsible for formulating rules governing the use of force; however, they sought the involvement of lawyers. At U.S. Central Command, this meant that judge advocates did more than simply advise on and review rules of engagement—they also wrote them. This fact served as proof certain of the rapidly evolving role of the deployed judge advocate as an operational lawyer and, as such, a key member of the commander's staff.

Fiscal Law Issues

As American and coalition forces arrived in the Persian Gulf, individuals, corporations, and governments supportive of the allied cause began donating money and property for the war effort. Japan, for example, donated almost 2,000 four-wheel-drive vehicles, water trucks, refrigerator vans, and fuel trucks.[16] But what legal authority did General Schwarzkopf and other commanders have to accept these "gifts"? Major Fletcher, the principal action officer for fiscal law issues, learned that U.S. law permitted only the General Services Administration to accept gifts of property for the Defense Department's use. Similarly, only the secretary of the treasury could accept donations of money. Monies given by individuals or governments could not be used by CENTCOM forces until the Treasury Department formally accepted these funds.

Gifts of property, however, were another matter. While explaining that there could be no ownership of such gifts until acceptance by the General Services Administration, Central Command's judge advocates advised that donated property could be used pending acceptance. Thus, brightly painted Japanese sport utility vehicles helped theater logistical organizations make up for shortages of assigned U.S. military vehicles.[17] Fortunately, the lack of Defense Department authority to accept gifts was resolved in a matter of weeks when, on 1 October 1990, Congress enacted legislation allowing the secretary of defense to accept money and property given in connection with defense operations from any person, foreign government, or international organization. Although the requirement that the Defense comptroller must approve any gifts meant that General Schwarzkopf and his subordinate commanders could not accept donations of money and property directly, the decision-making authority now resided in the Department of Defense.[18]

In addition to clarifying the issue of gifts, Central Command's judge advocates also advised on the conclusion of a formal agreement for the use of resources in Saudi Arabia. Such an arrangement with Saudi Arabia for host nation support would replace the unstructured and decentralized contracting activities then taking place. Initially, U.S. forces spent thousands of dollars on goods ranging from brooms, washbasins, and showers to tents, food, and fuel. On 10 September, however, King Fahd made the commitment to provide comprehensive support to all allied forces and organizations located in Saudi Arabia.

By the end of October, a Department of Defense negotiating team, headed by Army Maj. Gen. James W. Ray and including an attorney from the department's General Counsel's Office, assisted by CENTCOM's lawyers, reached an "understanding" with the Saudis. The latter agreed to pay the costs of all contracts entered into by the United States prior to 30 October 1990 and pledged to pay for all freshly prepared meals, water, and fuel. The Saudis agreed as well to construct needed facilities and to provide transportation within Saudi Arabia. As the Saudi government wished to avoid a formal arrangement, no written agreement was ever signed; the negotiated "understanding" took the form of a de facto agreement. The result was a uniform support arrangement for all of Central Command that substantially reduced the legal obstacles facing ARCENT and XVIII Airborne and VII Corps contract attorneys.[19]

War Crimes Issues

Iraqi forces committed a large number of war crimes against Kuwaiti citizens. Acting in accordance with Department of Defense Directive 5100.77, 10 July 1979, which designated the Army as the DoD executive agent for war crimes matters, and Army Chief of Staff Regulation 11–2, May 1975, which tasked the Judge Advocate General with investigating war crimes, CENTCOM's lawyers in Riyadh collected evidence of Iraqi misconduct. War crimes included the taking of Kuwaiti hostages and their forcible deportation to Iraq, and the murder, rape, and inhumane treatment of Kuwaitis (particularly suspected members of the resistance) and third country civilians. Evidence existed that the Iraqis tortured prisoners of war, used prisoners and civilians as human shields in combat operations, pillaged Kuwaiti civilian hospitals, and looted Kuwaiti cultural property.[20]

Central Command judge advocates reported their findings on Iraqi war crimes to the International Affairs Division of the Office of the Judge Advocate General. By December 1990, however, Iraqi abuses in Kuwait had continued at such a pace that additional legal support was required. As a result, two Army Reserve judge advocate international law detachments were called to active duty. The 199th Judge Advocate General (JAG) Detachment, headed by Lt. Col. Lee E. Haworth, was mobilized on 15 January 1991 and deployed to Saudi Arabia a short time later. Accompanying Colonel Haworth were Maj. David R. Tyrrell, Maj. Alan Overton, Capt. Mark P. Brewer, and Capt. Lauren L. Hafner. Twelve Kuwaiti civilians—nine men and three women—also deployed with the detachment. All the Kuwaitis had been studying at American colleges and universities when the Iraqis attacked and occupied their country. Now, eager to help in the fight to free Kuwait, they were voluntarily returning to the Persian Gulf to serve as interpreters. Over the

Maj. Mark P. Brewer, center, *and two Kuwaiti interpreters, Kuwait City, March 1991. Brewer and his colleagues collected riveting proof of Iraqi war crimes, including evidence that the Iraqis tortured Kuwaitis by burning them with red hot metal coat hangers, shaving their skin with carpenters' planes, and drilling into their elbows and knees with electric drills.*

next few months the dedication and hard work of these volunteers made it possible for the judge advocates of the 199th JAG Detachment, under the supervision of the CENTCOM staff judge advocate, to collect extensive information on war crimes committed in Kuwait.[21]

A second Army Reserve unit, the 208th JAG Detachment, consisting of six reserve judge advocates and commanded by Lt. Col. James G. Hergen, was mobilized in January and manned the War Crimes Documentation Center in the Office of the Judge Advocate General. In addition to evidence obtained from their colleagues at Central Command, the 208th's lawyers gathered information on war crimes from other sources, including other U.S. government agencies, the media, and private parties.[22]

Prisoner of War Issues

In late fall 1990, General Schwarzkopf concluded that Central Command lacked the in-country assets to accept and care for prisoners of war (POWs). Consequently, he determined that all prisoners captured by U.S. and allied forces would be transferred to Saudi control.

Although an international agreement to this effect was not required by international law, U.S. Army policy did call for such an agreement prior to the transfer of prisoners.[23] As a result, a U.S. negotiating team met with the Saudi government to reach agreement on the terms that would govern the transfer, custody, and administration of such prisoners. Colonel Ruppert, participating as a member of the team, advised that the United States must retain the right to inspect any Saudi-run camps containing transferred prisoners. Placing Iraqi prisoners in Saudi custody did not relieve the United States of its responsibilities under the Law of War; the United States had a continuing obligation to ensure that Saudi treatment of enemy prisoners complied with the 1949 Geneva Convention Relative to the Treatment of Prisoners of War (GPW). When the United States and Saudi Arabia reached an understanding on prisoner transfer issues, at the request of the Saudi military's senior legal adviser, Colonel Ruppert personally briefed the Saudi Red Crescent, the Saudi Foreign Office, and additional members of the Saudi military establishment on their obligations under the GPW Convention. Later, American inspectors did visit Saudi-operated prisoner of war camps.[24]

Central Command lawyers also drafted documents dealing with the transfer of prisoners between the allied contingents. On 31 January 1991, for example, the CENTCOM provost marshal and his British counterpart at British Forces Middle East signed "An Arrangement for the Transfer of Enemy Prisoners of War and Civilian Internees from the Custody of the British Forces to the Custody of the American Forces." Under the terms of this two-page protocol, "American Forces" agreed to accept all individuals captured or interned by the British, while the British agreed that the United States would later transfer these prisoners to Saudi control.[25]

Legal Assistance and Claims

As the duration of DESERT SHIELD grew from weeks to months, and as the end of the calendar year approached, Colonel Ruppert realized that a theaterwide tax program for U.S. forces would be needed. Ruppert tapped Maj. Robert A. Burrell, who had recently joined CENTCOM's legal operations in Riyadh, to establish this program. Over the next few months Burrell worked with legal assistance officers throughout the theater to ensure that every U.S. citizen would have the necessary federal and state tax forms available for filing by the 15 April deadline.[26]

At the same time, CENTCOM's judge advocates, working through the Office of the General Counsel at the Defense Department, sought relief from the 15 April filing requirement. As a result, President Bush signed an executive order designating Southwest Asia as a "combat zone." The tax relief benefits conferred by this presidential decree were later enhanced by congressional action. While these events ended the need for a large-scale tax program, CENTCOM lawyers had no way of anticipating if—or when—Congress or the president would take action in this matter. Thus, although hundreds of boxes of federal and state tax forms were ordered, shipped, and arrived in Dhahran, few were ever used.[27]

The command's lawyers also established a theater claims policy. Initially, the Air Force exercised single-service claims responsibility for Saudi Arabia. This meant that all military personnel claims were forwarded to the Air Force for approval, regardless of the service involved, as well as all foreign claims. Working from the premise that the service with the predominant presence should exercise single-service responsibility for claims, however, the CENTCOM lawyers, after coordinating with the service component judge advocates general, recommended that the Army assume single-service responsibility for Saudi Arabia, the Navy for Bahrain, and the Air Force for Oman. The Defense Department agreed. ARCENT judge advocates subsequently implemented the transition from the Air Force to the Army, with the exception of those large-dollar claims requiring action in the United States. Thus the senior Army headquarters became the approval authority for all military personnel and foreign claims filed in Saudi Arabia.[28]

U.S. Army Forces Central Command

Some twenty Army lawyers served in the ARCENT staff judge advocate's office, providing guidance to legal operations conducted at the two corps and ARCENT Support Command (later designated as the 22d Support Command). They also directly supported ARCENT headquarters and units that were without lawyers at echelons above corps. Since Colonel Tommepuu and two captains were the only active duty military lawyers with Army Central Command, the foundation of legal operations was built on a large number of Army Reserve judge advocates. *(Chart 8)*

A significant number of reservists was assigned to Third U.S. Army's headquarters at Fort McPherson, Georgia, and when Third Army deployed as Army Forces Central Command, its reserve personnel were some of the first mobilized and deployed. Thus, the ARCENT deputy staff judge advocate and chiefs of the Operational and

CHART 8—LEGAL ORGANIZATION AT U.S. ARMY FORCES CENTRAL COMMAND, FEBRUARY 1991

International Law, Administrative Law, and Military Justice Divisions were all reserve judge advocates. Later, an active duty contract law expert, Lt. Col. Frankie D. Hoskey, joined the ARCENT judge advocates as the head of the Contracts and Fiscal Law Division. Hoskey's presence, however, was the exception: Reserve lawyers were the rule when it came to those judge advocates performing legal work at the senior Army headquarters in theater. Initially, ARCENT judge advocates lived and worked in the Royal Saudi Land Forces Building in Riyadh. Later, however, they moved to the modern duplexes of Eskan Village, located some fifteen kilometers away.

Administrative Law and Claims Issues

The Administrative Law Division, headed by Lt. Col. Leslie K. Mason, had three branches. Maj. Bernard A. Pfeiffer was the chief of the Administrative Law Branch; Capt. Daniel L. Winand headed the Claims Branch; and Maj. Larry Daniels was the chief of the Legal Assistance Branch. All four were reservists: Mason was a civilian Army attorney at Redstone Arsenal, Alabama; Pfeiffer and Winand were civilian Army attorneys who worked at Fort Benning, Georgia; and Daniels was a state court judge from Maryland.

After his arrival in September, Major Pfeiffer advised on routine matters such as administrative eliminations, reports of investigation, line of duty investigations, and reports of survey. Perhaps his most significant legal problem, however, concerned the deadline for filing Standard Form 1555, Confidential Financial Disclosure Reports, and Standard Form 278, Public Financial Disclosure Reports. Some seven hundred personnel in the theater were required to file one or both forms. Although the filing deadline for Form 278 was 15 April, all Forms 1555 had to be filed by 31 October. Completing this form requires detailed financial information, including stockholdings, sources of income, and other data usually obtainable only from tax records. As most of those required to file the Form 1555 had not brought financial records with them, many requested an extension of the filing time for the duration of Desert Shield. Under the applicable statute, however, no authority existed for granting such an extension. Consequently, while concluding that there should be an exception for service in a combat zone, Major Pfeiffer advised those concerned that the filing deadline must be met. Those required to file were located, provided with forms, and advised how to complete them. The entire process "was particularly difficult given the fluid nature of the theater."[29]

Captain Winand, who, like Pfeiffer, had arrived in September, was designated as the Theater Army claims judge advocate. Recognizing that he first had to establish a theater claims policy, Winand conferred with the U.S. Army Claims Service in Fort Meade, Maryland. The result was the creation of framework within which all claims filed by soldiers would be investigated at the unit level, with adjudication done by claims judge advocates at the divisions. Thus, except for those units directly supported by ARCENT judge advocates, these military personnel claims were not handled by Winand. Rather, he focused on claims filed by Saudi and other non-U.S. citizens against the United States. In adjudicating these claims, Captain Winand worked closely with the Air Force, as that service initially exercised single-service responsibility for foreign claims in Saudi Arabia. Before the Department of Defense gave the Army single-service responsibility for Saudi Arabia, all foreign claims filed against the Army were completed on Air Force paperwork and forwarded by Winand to the Air Force's foreign claims operation in Riyadh. The Air Force would either approve or deny the foreign claim, based on Captain Winand's recommendation. Claims for less than $5,000 were approved without comment; claims for larger amounts required a telephone call or a face-to-face meeting between Winand and his Air Force counterpart.[30]

The claims filed by Saudi citizens were varied. The largest claim involved damage to a cement factory near Hafar al Batin. As it was located along a likely Iraqi invasion route, the factory was transformed into a defensive perimeter by troopers of the 101st Airborne Division. Unfortunately, damage to heavy equipment and buildings caused by the soldiers resulted in a claim for some two million riyals—over $500,000. An investigation established U.S. liability, and the United States was prepared to pay the claim. Prior to payment, however, the Saudi government agreed to assume all claims submitted by its citizens against the United States and, as the agreement covered the period of time during which the cement factory had been damaged, the Saudi government paid this claim as well.[31]

Just as in Vietnam, there were claims for losses and damage to livestock. In Southwest Asia, however, these claims were for camels rather than water buffalos. Late in 1990, for example, a Saudi citizen filed a claim for six camels that were killed when they wandered onto a 105-mm. howitzer range operated by the 82d Airborne Division. There was some suspicion that the animals had been intentionally targeted, but the claim was paid based on a finding of negligence.

Even when the cause of the loss was intentional rather than negligence, however, the claim was paid. In February 1991 a soldier on guard duty fired his M203 grenade launcher at a camel. This intentional killing earned him an Article 15 punishment, and the United States reimbursed the aggrieved owner for the camel's fair market value. Captain Winand had determined that value earlier by visiting the local market, or "suk," to study the pricing structure for the different kinds of camels. Some were bred for meat, some for milk, and some for racing. Furthermore, while a healthy male camel usually was worth more than a female, the reverse was true if the female was pregnant. By the end of DESERT SHIELD, judge advocates dealing with foreign claims fully understood the sophisticated pricing structure for camels.[32]

Military Justice

The first challenge for ARCENT's criminal lawyers was sorting out the jurisdiction exercised over Army Reserve units. A number of reservists arrived in theater with little more than orders calling them to active duty and directing them to report to their mobilization stations. As military discipline requires that every soldier be attached to a general court-martial convening authority, the issue was whether jurisdiction should be exercised along command lines or by geographic area. At first a command-line framework was favored; however, as theater-level reserve component support units were increasingly dispersed in Saudi Arabia, the distance between these units and the support command headquarters in Dhahran made an area jurisdiction structure more appropriate.

Another early issue was whether Operation DESERT SHIELD should be considered a "war" when determining the punishments for certain military offenses. Under Article 85, for example, desertion in time of war is punishable by death, whereas the maximum peacetime punishment for this offense is five years. The Office of the Judge Advocate General advised that, despite the purpose and scope of DESERT SHIELD, in the absence of a congressional declaration of war or a factual determination by the president (usually done by executive order) that a state of war existed, the punishment for this particular offense could not be increased.

Because of the general ban on alcoholic beverages and the typical soldier's lack of contact with the civilian population, disciplinary problems were relatively minor. But some misconduct did occur, usually related to such military offenses as fraternization, the wrongful discharge of a weapon, dereliction of duty, and the violation of a lawful order. Given the command's status as the senior Army headquarters, however, the attorneys focused more on ARCENT's supervisory responsibility for Army criminal law activities rather than on prosecuting and defending cases. Thus, Maj. David H. Brunjes, a Georgia reservist with extensive criminal law expertise, spent much of his time gathering data on the numbers and types of courts-martial and nonjudicial punishment activity occurring at subordinate general and special court-martial jurisdictions. He also taught military justice classes to reserve units and provided instruction on the wide range of administrative options to nonjudicial punishment and courts-martial.[33]

Army Forces Central Command prosecuted only one court-martial during DESERT SHIELD and DESERT STORM. Paradoxically, this court-martial probably generated the highest interest of any criminal case occurring in Southwest Asia. Capt. David S. Wiggins, a 1984 graduate of the U.S. Military Academy who also had received his medical degree and flight surgeon training at Army expense, was stationed at Fort Hood, Texas, when Operation DESERT SHIELD began. Ordered to deploy to Saudi Arabia, Wiggins applied for conscientious objector status and, when this was denied, filed suit in U.S. District Court for the Western District of Texas, requesting a preliminary injunction to bar the Army from sending him to Southwest Asia. The district court denied the injunction and, unsuccessful in avoiding deployment, Captain Wiggins flew to Saudi Arabia, where he was assigned duties as a physician in a troop clinic at Log Base BRAVO near King Khalid Military City.

Shortly thereafter, Wiggins shaved his head, went on a hunger strike, and unilaterally "resigned" from the Army. On 14 January 1991, he removed his uniform and, wearing only his underwear, obstructed traffic in a busy intersection by sitting down in the middle of the road and refusing to move. His act of protest disrupted traffic on one of the busiest supply routes in Saudi Arabia at a critical time during the preparations for the ground war. As a result of his behavior, Captain Wiggins was charged with failing to repair to his place of duty, disobeying the order of a superior officer, and conduct unbecoming an officer and gentleman.[34]

Capt. Edward B. Cottingham, a reservist from South Carolina assigned to ARCENT headquarters, faced a number of problems in prosecuting *United States* v. *Wiggins*. The lack of a confinement facility in theater meant that the accused could not be placed in confinement pending his trial. Thus, while preparing for war, his unit also had to watch over Captain Wiggins. From Cottingham's perspective, the biggest challenge was arranging for the appearance of witnesses. The pretrial investigation under Article 32 of the Uniform Code and the subsequent general court-martial proceedings were held in Riyadh. Most of the witnesses, however, were located in Dhahran or in King Khalid Military City, more than 200 dusty truck miles away. Despite these travel difficulties, however, all in-theater witnesses did appear. Arranging for the accused's wife to travel to Saudi Arabia to appear as a witness proved to be so difficult, however, that Captain Cottingham requested that the convening authority dismiss the affected charge. At his general court-martial, Captain Wiggins was convicted. Although Cottingham argued that Wiggins should be sentenced to two years' imprisonment, the military judge, Col. Frank B. Ecker, Jr., disagreed. He sentenced Captain Wiggins to forfeit all pay, to pay a fine of $25,000, and to be dismissed from the service. On appeal, the Army Court of Military Review upheld the findings and sentence.

22d Support Command

ARCENT Support Command, as it was called before being renamed the 22d Support Command (SUPCOM) in December, did not even exist until Desert Shield was under way. Thus, as at Army Forces Central Command, legal operations at the support command were handled primarily by reserve component judge advocates. Only the staff judge advocate, Lt. Col. William R. Hagan, and one captain in the support command's legal office had been on active duty before the command's deployment. Although experienced as a lawyer in airborne, armored,

Staff judge advocates in Southwest Asia, 11 February 1991. Kneeling, left to right: *Col. Malcolm H. Squires, Jr., 101st Airborne Division; Col. John R. Bozeman, XVIII Airborne Corps; Lt. Col. John T. Burton, 3d Armored Division.* Standing: *Col. Walter B. Huffman, VII Corps; Col. Tonu Tommepuu, ARCENT; Col. Raymond C. Ruppert, CENTCOM; Lt. Col. Warren D. Hall III, 1st Infantry Division; Lt. Col. James J. Smith, 82d Airborne Division; Lt. Col. William R. Hagan, 22d SUPCOM; Lt. Col. Philip E. Lower, 1st Cavalry Division; Lt. Col. John D. Altenburg, Jr., 1st Armored Division.*

mechanized, and light infantry divisions, Hagan thought of logistics in terms of a division support command. He soon discovered that lawyering for logisticians at echelons above corps, combined with the task to "form a nonexistent general court-martial convening authority from scratch," was to be his biggest challenge ever.[35]

After packing a footlocker with copies of current regulations, as well as an 1896 edition of William Winthrop's two-volume *Military Law and Precedents* and a 1943 edition of Field Manual 27–5, *Military Government and Civil Affairs*, Colonel Hagan departed for Dhahran. Arriving on 21 September 1990, the forty-fifth day of DESERT SHIELD, he joined members of the recently mobilized 46th JAG Detachment (International Law) who had deployed several days earlier from Boston, Massachusetts. With the subsequent arrival of reserve lawyers from the 207th JAG Detachment (Contract Law) and the 21st Theater Area Command (Continental U.S. Augmentation), Hagan's office numbered thirteen attorneys. It was aided by about an equal number of lawyers in subordinate or "downtrace" units. Organized into five divisions—Military Justice, Civil and Operational Law, Legal Assistance and Claims, Contracts, and Forward Support–King Khalid Military City—Colonel Hagan's office provided guidance and support to some 75,000 soldiers in fourteen special court-martial convening authorities.[36]

Contract and Fiscal Law Issues

Contract and fiscal law was the number-one priority for SUPCOM judge advocates for several months. Because U.S. forces needed far more than they were able to bring with them, they negotiated a large number of contracts for supplies and services. Hagan's Contracts Division, headed by Lt. Col. Robert P. Lowell, a reserve judge advocate from Massachusetts, provided advice and support on contract matters. The first major problem confronted by Lowell and his colleagues, however, was one of fiscal law. The operational requirement for a heliport was critical, since the Saudis did not have adequate airfield space to accommodate the large number of attack, scout, transport, and utility helicopters arriving in Southwest Asia. The estimated cost for such a heliport was $1 million or more, Legally, however, the spending of Army Operation and Maintenance (O&M) funds for "unspecified minor construction" was limited to $200,000. As building projects exceeding this limit required congressional authorization, did this mean that the heliport could not be built without specific legislative authority? The SUPCOM judge advocates determined that the secretary of defense could legally exceed the existing $200,000 limit by authorizing the additional expenditure of funds as "emergency construction." This was not a good solution to the problem, however: the law required a 21-day waiting period after a secretarial authorization of this nature, and the near-combat tempo of DESERT SHIELD required immediate construction of the heliport.

While the 22d SUPCOM commander, Maj. Gen. William G. Pagonis, conferred daily by telephone with the chief of engineers, Lt. Gen. Henry Hatch, the Contracts Division researched the issue and wrote a fiscal law opinion setting forth an approach for resolving the matter. Based on his discussions with the Corps of Engineers and the legal opinions of the SUPCOM lawyers, General Pagonis determined that the heliport did not fall under the statutory provisions governing minor military construction. Accordingly, it was not subject to the O&M expenditure cap applicable to such construction. DESERT SHIELD was an operation, not a training exercise. Paving the desert was a project more akin to building bunkers or constructing antitank revetments.

As limits on spending O&M funds did not apply to real-world operations or to combat-related military construction, no bar existed to building the helipad. When Lester Edelman, the chief counsel of the Corps of Engineers, concurred in the SUPCOM legal opinion, the authority for the heliport—and other similar projects—was established. Also firmly established was General Pagonis' confidence in the judge advocates in his command.[37]

Legal Assistance

With some 70 percent of the support command's soldiers from the Army Reserve or National Guard, the provision of legal assistance to these individuals was a major concern. Most units had undergone extensive preparation for movement, with wills and powers of attorney executed for all who desired them, but rapid mobilization resulted in more legal assistance problems than those incurred in active component units. Frequently, the issues encountered were also more complex. Although the Legal Assistance and Claims Division consisted of only two judge advocates, as a practical matter every SUPCOM attorney provided legal assistance, from Colonel Hagan and the deputy staff judge advocate, Lt. Col. Kevin V. Murphy, to Majs. Shelby L. Starling and Thomas G. Robisch, both of whom worked in the Civil and Operational Law Division.[38]

Until Congress granted tax relief, the legal assistance attorneys had established a structure for the command tax program, focusing on how to provide tax forms and advice on completing these forms to every soldier in the command. These judge advocates also advised on a variety of legal matters, including filing for divorce and legal separation and responding to creditors. The Soldiers' and Sailors' Civil Relief Act was particularly important for mobilized reservists, as it reduced the interest rate on existing loans to 6 percent per year. Thus, a soldier with a high-paying civilian job who after deploying earned significantly less in uniform benefited from lower credit card and mortgage payments. As this provision of the act was not widely known, SUPCOM lawyers spent a considerable amount of time informing creditors that reserve component members of the command not only had the right to lower payments, but that this benefit continued for the duration of their active service. Additionally, as any civil proceeding involving reservists called to active duty could be stayed under the Soldiers' and Sailors' Civil Relief Act, judge advocates assisted soldiers in using the act to delay any suit filed by a creditor or other litigant who either did not know of the act's benefits or was resistant to its protections.

XVIII Airborne Corps

The XVIII Airborne Corps had more military lawyers experienced in operational deployments than any other Army legal office. Its recent major deployments had been to the U.S. Virgin Islands in Operation HAWKEYE (a relief effort) and to Panama in Operation JUST CAUSE. Annual engineer unit rotations to Honduras also had produced a large number of judge advocates capable of operating on their own at great distances from their home office. Corps attorneys had also participated in INTERNAL LOOK, a major training exercise held in July 1990. It coincidentally involved a Southwest Asia scenario focusing on the same countries to which the corps would deploy the following month. Additionally, a large number of XVIII Airborne Corps lawyers had received training in operational law and had been provided recently produced operational law reference materials.

The corps' top lawyer, Colonel Bozeman, was not only a veteran of combat in Vietnam but also had been tested the previous year as the joint task force staff judge advocate for Operation JUST CAUSE. His deputy, Lt. Col. Patrick Finnegan, likewise had operational experience, having served as the task force legal adviser for the humanitarian relief mission to the Virgin Islands. Since legal and nonlegal issues encountered in Saudi Arabia often were similar to those they had dealt with on earlier missions, both Bozeman and Finnegan—and other corps judge advocates with experience in JUST CAUSE or HAWKEYE—were ideally suited for early deployment in DESERT SHIELD.[39]

Although there was little time for planning prior to the departure of the first of the corps' units to Saudi Arabia, the pace of deployment, combined with the command decision to send large numbers of combat forces first, with combat support and combat service support personnel to follow, made it somewhat easier for Colonel Bozeman and his legal staff to support both the deployed units and the continuing activity at Fort Bragg. Earlier deployments had made all corps units well aware of the importance of overseas movement processing. Consequently, many troops who had recently deployed to St. Croix or Panama needed little or no updating of their legal documents. DESERT SHIELD, however, required significantly more soldiers than had earlier deployments, including many from the reserve components. Active duty judge advocates at Fort Bragg, assisted by the recently mobilized 204th JAG Detachment, prepared more than 17,000 wills for departing personnel and provided round-the-clock legal assistance to family members.[40]

Army judge advocates also briefed all soldiers on the rules of engagement (ROE) for DESERT SHIELD. The CENTCOM judge advocates had developed rules of engagement based on the Joint Chiefs of Staff peacetime rules on the use of force, and the corps lawyers used these as a basis for drafting XVIII Airborne Corps rules of engagement. Before flying to Southwest Asia all soldiers were briefed on the corps rules governing the use of force. Refining the use of wallet-size ROE cards pioneered by the 82d Airborne Division in JUST CAUSE, the corps judge advocates developed two separate, different-colored cards. More than 100,000 white cards were printed with the peacetime rules of engagement applicable during DESERT SHIELD, and a like number of blue cards were printed with the rules that would be applicable during hostilities. The reverse of the peacetime card also contained cultural tips. These cards, written in easily understood English and distributed to every soldier, provided a ready reference for questions concerning the lawful use of force.[41]

Bozeman faced a nearly total deployment of the corps units, but he retained a substantial continuing requirement to provide legal services for Fort Bragg, a major mobilization station. Which of his personnel should he send to Saudi Arabia? Most judge advocates and legal specialists would be deploying, but some would have to stay behind for the garrison mission. For most, staying behind meant being left out, for while the importance of the garrison mission was clear to Bozeman and Finnegan, it was not as clear to some of the judge advocates concerned. Colonels Bozeman and Finnegan had to make, and explain, deployment decisions carefully. Bozeman's Panama experience had left him with an acute awareness that the corps might face many issues involving the leasing and purchasing of supplies and equipment, so he decided to first deploy his operational law specialist, Capt. Mark C. Prugh, and his contract law chief, Capt. Michael L. Larson. They arrived with the assault command post in Dhahran in the searing heat of early August. The third corps lawyer to deploy was Capt. Patricia A. Martindale, a legal assistance officer. Arriving in Dhahran on 21 August, she was the first woman judge advocate in Southwest Asia.[42]

Colonel Bozeman arrived in Dhahran with Lt. Gen. Gary E. Luck, the corps commander, on 28 August 1990. Over the next few months, as corps units arrived in theater, Bozeman selectively drew down his Fort Bragg operation while systematically building up his forward-deployed organization. Coordinating with Colonel Finnegan, who had remained at Fort Bragg as the installation staff judge advocate, Bozeman timed the deployment of lawyers to precede or coincide with the arrival of the brigades they supported. Rather than immediately detailing these judge advocates to their brigades and groups, however, Colonel Bozeman first brought each incoming corps lawyer to the main command post in Dhahran. This served several purposes. First, deploying all judge advocates to the forward staff judge advocate office meant that in-theater legal operations were more easily organized, avoiding gaps in both expertise and personnel. Second, it oriented everyone to the legal requirements in theater. Both mid-level and junior officers learned how legal services were organized and what legal policies were in effect at both corps and theater levels, thus promoting efficiency as well as uniformity of legal advice.[43]

As DESERT SHIELD progressed and Bozeman's legal operation matured, attorneys were detailed to brigades and groups and were eventually attached on orders for administration and logistics. Most were responsible for two or three special court-martial convening authorities. As a condition of the detail of these attorneys, the units agreed to provide office space, a sleeping area, access to telephone or radio communication, and transportation. Many units provided a private "office" in the tent area, since legal assistance clients were seen at the unit. Transportation originally posed a major concern, as corps lawyers had no access to organic vehicles. However, rental cars and vehicles contributed by other countries eventually alleviated most of these transportation problems.[44]

With individual lawyers detailed to brigades and groups in the corps, the supervision of their legal work became difficult, as most brigade-level units were isolated by distances amounting to more than a hundred miles or by difficulties with communications. Colonels Bozeman and Finnegan used several methods to monitor distant legal operations to assure support for the brigade-level legal officers and legal specialists. First, senior legal staff members visited the brigade lawyers as frequently as possible. These visits were usually accomplished on a circuit trip of slightly over three hundred miles. Because of high maintenance demands on helicopters in the desert, the circuit had to be made by car or truck and usually involved at least one overnight stay.

These visits became even more important after the corps headquarters moved to Rafha, a few miles from Iraq. This move split the corps lawyers into three groups: the principal operation at Rafha, an administrative element with the corps command post at King Khalid Military City, and another group of lawyers with the corps rear element at Dhahran.

Second, XVIII Airborne Corps kept its judge advocates fully informed through a series of information papers on recurring or significant issues. Later, these papers served as good primers on theater legal issues for VII Corps units deploying to Saudi Arabia. Finally, two sep-

Office of the Staff Judge Advocate, XVIII Airborne Corps, in Rafha, near the Iraqi border, February 1991. Standing, left to right: *Col. John R. Bozeman, Capt. Kevin A. Ohlson, Lt. Col. Richard H. McCall, Jr., Capt. David W. Francis, Capt. Gary A. Khalil, Capt. Michael D. Watson, Capt. David Jacquot, Capt. Joseph J. Frattallone, Capt. John F. Yaninek, Capt. Allen K. Goshi, Capt. Christopher M. Garcia.* Kneeling, left to right: *Lt. Col. Patrick Finnegan, Capt. Michael T. Rudisill, Capt. Karen R. O'Brien, Capt. Maritza O. Ryan, Capt. Patricia A. Martindale, Capt. Amaury R. Colon-Burgos, Capt. Mark C. Prugh.*

arate legal conferences were held for all corps attorneys, one during DESERT SHIELD and one during conduct of the air war. While issues facing the XVIII Airborne Corps lawyers were similar to those confronting advisers in most major deployments, DESERT SHIELD produced a number of unique issues.

Contract Issues

Arriving in Dhahran at 4:30 P.M. on 9 August, Captain Larson was the third Army lawyer in theater.[45] Within an hour he was advising a corps contracting officer who was negotiating the purchase of bottled water from a Saudi merchant. A contract law expert with prior judge advocate assignments at White Sands Missile Range and on Okinawa, Larson was for several weeks the only Army contract attorney in Saudi Arabia. Rather than simply conducting legal reviews, he found himself primarily assisting XVIII Airborne Corps contracting officers in organizing their operations. Early on, Larson also assisted the Saudis in providing host nation food support, one result of which was the provision of 1,600 meals to hungry American soldiers by the Hardee's hamburger franchise in Dhahran.[46]

In the first days of Desert Shield the corps' biggest need was for transportation. Buses were needed for troops and rail cars and heavy equipment transporters for materiel. Captain Larson's contracting officers, however, possessed only limited contracting authority. Only one officer could negotiate contracts up to $25,000; the remaining officers were authorized to purchase goods and services worth up to only $2,500. Yet transportation requirements alone greatly exceeded these amounts. For example, buses ordinarily would have been leased for thirty days, but such a contract would have exceeded the $25,000 limit. The solution was to lease buses for a three-day period, with the understanding that the lease would be renewed again and again. This unorthodox solution may have violated the prohibition on splitting contracts to avoid dollar limits, but the combat tempo of the mission offered no alternative.[47]

Over the ensuing weeks Larson handled hundreds of contract matters. At first, all contracts were negotiated without soliciting bids, but within a month a solicitation for laundry services was advertised. Later, an invitation to bid on a contract to operate the headquarters troop dining facility also was announced. Both solicitations resulted in competitive contract awards.

Initially, XVIII Airborne Corps contracts did not address the issue of performance during combat. This was a matter of concern to Capt. Peter L. Brown, who replaced Captain Larson as the corps contract attorney in October 1990 and who worked closely with his 22d SUPCOM contract law counterparts. As a result, by January most contracts for transportation, supplies, and services critical to the future support of Operation Desert Storm contained a "Continued Performance During Hostilities" clause. Contracts for heavy equipment transportation, for example, gave the United States, through the contracting officer, the right to direct that a contractor continue performance during combat and, upon demand, substitute U.S. employees for contractor employees in the operation of the equipment. Although the clause did not confer upon the United States legal authority to use force against a contractor or his employees to obtain contract performance or permit the seizure of the property, the inclusion of this hostilities clause in contracts served notice that not only would breach of contract occur if civilian operators refused to drive or otherwise operate equipment, but also that a possible remedy was available.[48]

As DESERT SHIELD progressed and as the corps contracting process matured, a thorny issue developed regarding the offer of gifts to American contracting officers. Such gifts were the custom in Saudi Arabia, and U.S. officials were concerned that refusal to accept a proffered item would be viewed as offensive. Judge advocates took a firm position on the matter. For example, when a contracting officer was offered a gift by a contractor who had been awarded a small-purchase contract and was competing for a larger one, the corps judge advocates prepared a memorandum for the signature of the corps chief of staff. The letter, signed on 30 October 1990, reminded all corps contracting officers that, regardless of custom, the law required that they avoid both actual conflicts of interest and those situations that might give the appearance of such a conflict. No item of value could be accepted from any business entity, and any offer of a gift or gratuity was to be reported immediately.[49]

Legal Assistance

Captain Martindale arrived in Dhahran by C–141 aircraft on 21 August. Although a judge advocate for less than a year, Martindale was made responsible for all of XVIII Airborne Corps' legal assistance requirements. Having brought with her a laptop computer and a dot-matrix printer, Martindale began almost immediately to write and execute wills. Over the next few weeks she saw an average of fifty troops daily. Together with Capt. John F. Yaninek, who arrived from Fort Bragg a few weeks later, she delivered a wide range of legal assistance services. A key to their success was the satellite telephone system provided by the corps signal office. This asset allowed those providing legal assistance to call any business or attorney in the United States at any time of the day or night. With the time in Saudi Arabia eight hours later than that in North Carolina, Martindale and Yaninek spent every evening on the telephone to creditors, merchants, and attorneys in Fayetteville, representing their soldier-clients. Many a crisis was avoided or solved without resorting to mail correspondence. Moreover, the satellite hookup was configured so that the cost of the telephone calls was billed to the corps staff judge advocate office at Fort Bragg. This system eliminated any need to work through the Saudi telephone system to pay for calls.

Over the next several months, both routine and unusual legal assistance issues required great effort by Captains Martindale and Yaninek and other corps attorneys. These tasks included assisting non-U.S. citizens serving in the Army who desired American citizenship, advising activated reservists on their veterans' reemployment rights, and explaining rights and obligations under the Soldiers' and Sailors' Civil Relief Act. They also counseled U.S. soldiers wishing to marry concerning the applicable foreign law and procedure or, for those interested in an alternative, those U.S. states that permitted marriage by proxy. Additionally, in concert with CENTCOM and ARCENT judge advocates, corps lawyers prepared to help soldiers in filing tax returns and coordinated with different states regarding their requirements. Each brigade and battalion designated unit tax advisers, and judge advocates made arrangements for these individuals to receive training in Dhahran from representatives of the Internal Revenue Service. At the same time, soldiers were encouraged to allow a spouse or another designated person to file their state and federal taxes for them through a general power of attorney. As noted earlier, legislative tax relief ultimately obviated the need for a corps-wide tax program, but the attorneys in Saudi Arabia were prepared for the worst.[50]

Administrative Law

Major administrative law issues involved female soldiers, religion, flags, and civilians accompanying the forces. Under then-existing Army policy, if a soldier's pregnancy occurred or was discovered after her deployment, she would ordinarily remain in the overseas environment until her seventh month of pregnancy. Given the combat tempo of Desert Shield, however, this was unrealistic. Consequently, Army Central Command directed that all pregnant soldiers be immediately returned to home station. This ruling led to complaints that female soldiers had found a convenient way out of the desert, and some commanders wanted to issue "no sex" orders or to punish soldiers who became pregnant. Recognizing the difficulty of enforcing such orders, as well as the likelihood of adverse publicity, Colonel Bozeman recommended that commanders and supervisors avoid issuing any orders dealing with sexual relations that would not be imposed in the United States. Instead, Bozeman urged commanders to consider establishing single-sex sleeping areas and prohibiting visits to those areas by members of the opposite sex. He also advised leaders to appeal to their soldiers' sense of duty by emphasizing that a pregnancy meant a loss to the unit that was not easily absorbed.[51]

Other issues involving female soldiers resulted from the basic conservatism of Saudi society. CENTCOM policy emphasized that good host-nation relations required "cultural awareness," but, since Saudi law made certain activities illegal if performed by women, the issue went beyond simply a need to respect Islamic culture. Thus, Colonel Bozeman advised, as did all lawyers in the CENTCOM area of operations, that XVIII Airborne Corps not discriminate against women in order to satisfy local culture.

For example, it was illegal for women to drive motor vehicles in Saudi Arabia. Compliance with Saudi law would not only have discriminated against female soldiers but, because a large number of women served in transportation units, would have hurt readiness by seriously disrupting the corps' ability to move and supply its combat forces. Consequently, the corps dealt with this matter by issuing guidance stating that both male and female soldiers "may drive vehicles (military and civilian) . . . when on-duty, and in uniform (including headgear)." As a practical matter, a woman driver often was in the company of a male soldier. But Saudi officials were informed that the U.S. military did not have "male" and "female" drivers, but "soldier" drivers and that, by requesting U.S. assistance in defending their country, they accepted the Army's force structure and its large number of women soldiers.[52]

Similarly, Saudi culture required conservative dress in public and, while the restrictions on men were relatively few, American female soldiers discovered that the local religious police or "muttawwa" frowned on immodest dress. As a result, corps lawyers advised the command group on the development of a "dress code," prohibiting female soldiers from wearing revealing t-shirts, earrings, and shorts in places where they might have contact with the Saudi public. Furthermore, women in uniform were not to visit Saudi communities without a male escort, and they were cautioned, when wearing civilian clothes in Saudi communities, to close their blouses up to the neck, to have sleeves covering their elbows, and to wear dresses or slacks close to ankle-length. Although women soldiers generally understood the necessity to respect Saudi culture, they disliked what they saw as unequal treatment, particularly since the desert heat made conservative dress uncomfortable.[53]

As these administrative law questions arose outside the United States, many took on an international law flavor. When the press reported that the Saudi government had objected to the display of the American flag on U.S. uniforms and the use of the "flag stamp" on mail, Army lawyers were asked to review both of these issues. Under international law, flying the flag of one sovereign in the state of another may occur only with the permission of the host state. As the Department of Defense did not intend to request such permission from the Saudis, the American flag was not flown in areas assigned to U.S. forces. Additionally, security concerns dictated that the U.S. flag not be flown in those areas. Thus, while American forces later flew the U.S. flag in Iraq, it was not raised over Saudi soil. As to the issue of uniforms, since the American flag patch being worn on U.S. uniforms was lawful under Army regulations and since the Saudis never officially raised the matter, there was no reason to prohibit this practice. Finally, the Saudis never objected to the use of flag postage stamps.[54]

The practice of any religion other than Islam is illegal in Saudi Arabia. Since this restriction posed a major challenge to the command's obligation to provide for the religious needs of soldiers, the advice of Army lawyers was sought. At the XVIII Airborne Corps, worship services were referred to as "morale services (C), (P), or (J)" for Catholic, Protestant, or Jewish soldiers, respectively. Chaplains were directed to hold services out of public view, and religious materials were provided only to U.S. personnel and returned after the services. Chaplains wore their distinctive branch insignia only in U.S.-controlled areas; the cross or torah was removed when contact with the local populace was likely. Similarly, Bibles or other religious literature were identified as "morale publications." These measures struck a balance between the need for regular religious services for all corps troops and an obligation to respect host country law. Rumors to the contrary, Saudi officials never objected to the command's policy concerning religious activities in their country.[55]

Finally, as the likelihood of war—and casualties—increased, a new administrative law question arose. Should Army civilian employees and civilian contractors accompanying U.S. forces carry Army-issued weapons, protective gear, and Geneva Convention identification cards? Could they also wear military uniforms? The corps' judge advocates, after discussing these issues with higher headquarters, advised that civilians could be provided most equipment, including uniforms, and, as the civilians at issue were "accompanying the force," they should be given identification cards reflecting their status under the 1949 Geneva Convention. Government weapons generally would not be issued; however, firearms could be hand-receipted to Department of the Army civilians if the theater commander determined that they were required for self-defense.[56]

VII Corps

Lawyers at VII Corps first learned of a possible deployment to the Persian Gulf on 29 October 1990, when Lt. Gen. Frederick M. Franks, Jr., the commanding general, included the corps staff judge advocate in a limited-access planning cell. Judge advocate operations increased in tempo after the public announcement of VII Corps' deployment on 8 November 1990. Over the next six weeks, as they prepared for their departure for the Arabian peninsula, Colonel Huffman; his deputy, Lt. Col. Edward I. Frothingham; and all the VII Corps' judge advocates worked twelve to fifteen hours a day, seven days a week. It was a grueling pace, for not only were some 6,800 wills and 13,800 powers of attorney prepared for over 40,000 departing troops, but the judge advocates also had to prepare for their own deployment. This process included M16 rifle and .45-caliber pistol requalification and reviews of combat first aid procedures, land navigation, and the use of chemical protective masks.[57]

Determining the organization of VII Corps legal services in Southwest Asia was the first major decision for Colonel Huffman. In Europe, legal services were organized by geographic area rather than along command lines. Since the early 1970s, the nearest higher headquarters provided all units in a particular location with military justice, administrative law, and claims and legal assistance support, regardless of whether they were part of its military chain of command. A subsequent refinement of this area jurisdiction concept established consolidated legal centers at which unit legal specialists, detached from their parent units, were attached to a legal office for duty, training, and administration. Now, as VII Corps moved from Germany to Saudi Arabia—and from peace to war—its rear area–based organization disappeared, along with its eight full-service legal offices located from Heilbronn to Munich. Only a minimum amount of legal support remained behind to care for the family members and civilian component of the force that remained in Germany.

In Saudi Arabia, VII Corps legal services were structured along command lines. *(Chart 9)* In reorganizing, Colonel Huffman had to decide how many judge advocates to take to the Persian Gulf. The corps' table of organization and equipment authorized seventeen lawyers at the corps headquarters. Another eleven lawyers were authorized by the corps' table of distribution and allowances, resulting in a total of twenty-eight judge advocates available for deployment. After coordinating with Col. Thomas M. Crean, Judge Advocate, U.S. Army, Europe, Colonel Huffman decided to take his full table of organization and equipment, plus two judge advocates from the table of distribution and allowances. Nine lawyers remained in Germany. He selected the nineteen deploying judge advocates on the basis of both soldier skills and technical expertise. Recognizing that the corps' mission in Europe still required first-class legal support, however, Huffman also left some of his brightest and most experienced people in Germany. As for those men and women going to the desert, Colonel Huffman, working with Lt. Col. Charles E. Trant, the corps' chief of justice who would

CHART 9—OFFICE OF THE STAFF JUDGE ADVOCATE, VII CORPS, JANUARY 1991

serve during the deployment as his deputy, decided that eleven attorneys would be located at corps headquarters. The remainder would be allocated in such a way that the engineer brigade, engineer group, two field artillery brigades, military police brigade, and armored cavalry regiment would each have at least one judge advocate. This distribution of assets would ensure the provision of adequate legal support to the six brigade-size units under the corps' control, regardless of how far they were located from headquarters.[58]

Personnel decisions were based only on performance; gender was not considered. Consequently, VII Corps women attorneys performed the same duties—from filling sandbags and pulling guard duty to prosecuting courts-martial and advising commanders—as their male colleagues. In selecting a lawyer for the job of chief of claims, for example, some counseled against choosing a woman "because the Saudis would not want to deal with a female in foreign tort claims." But Colonel Huffman picked Capt. Nancy A. Lorenzo for the job, commenting that any Saudi claimant who objected to dealing with Captain Lorenzo did not have to file a claim. Events demonstrated that Lorenzo was both highly effective in processing foreign claims and admired by those filing them; a Saudi male claimant even offered her 750,000 riyals (about $200,000) to be his third wife.[59]

A final decision was that of determining what to take. Colonels Huffman and Trant realized that materiel authorized for the staff judge advocate under the corps' table of organization and equipment was inadequate. It did not include motor vehicles, tentage, or other major field equipment items. Trucks, vans, and tents were immediately requested from the headquarters commandant, but critical equipment, especially transportation assets, remained in short supply. Hauling computers, printers, and people from one to location to another remained problematic throughout DESERT SHIELD.[60]

After the XVIII Airborne Corps staff judge advocate advised that VII Corps could anticipate purchasing significant quantities of goods and services in theater, Colonel Huffman selected Maj. George B. Thomson, Jr., as the first lawyer for deployment. Thomson's experience in contract law made him the logical choice, and he left for Dhahran on 23 November, becoming the first VII Corps judge advocate in Southwest Asia. Capt. Jorge A. Lorenzo, chief of the Operational Law Division, followed on 7 December. Huffman deployed with General Franks on 13 December, with Colonel Trant and the rest of the corps lawyers departing Germany with the main body of corps headquarters the day after Christmas.[61]

Administrative Law

Even before departing for Saudi Arabia, VII Corps judge advocates conferred with their counterparts at U.S. Central Command and XVIII Airborne Corps. Relying heavily on work already done, particularly by their colleagues from Fort Bragg, VII Corps lawyers drafted policy memorandums on flying the U.S. flag, visiting towns and cities, vehicle accident procedures, and foreign criminal jurisdiction. After their arrival in the desert, administrative law attorneys assisted in the formulation of other corps-wide policies, including the relief of officers for cause, accidental weapon discharges, "no-sex" orders, and the acceptance of gifts from Saudi contractors.

One of the thorniest issues involved conscientious objectors. In most units deploying to the Persian Gulf, at least one soldier claimed to oppose participating in war in any form, due to religious training or belief. Such filings came as a surprise to some in the chain of command, who assumed that an all-volunteer force would not include soldiers who objected to combat. The complex procedures in the regulation on conscientious objection and a lack of command familiarity with the process added an element of confusion to the two problems requiring resolution: applications for conscientious objector status were not being processed in a timely manner, and those filing applications were refusing to deploy to Southwest Asia. Army lawyers advised unsympathetic, if not hostile, unit commanders that any application had to be forwarded to the special court-martial convening authority for investigation without delay. The judge advocates also informed commanders that application for conscientious objector status would not halt the applicant's deployment; the decision regarding status would be made in the desert. Problems associated with conscientious objectors were never completely resolved, primarily because units were consistently slow in processing applications.[62]

Rules of Engagement and Law of War Training

CENTCOM lawyers had disseminated the Joint Chiefs of Staff Peacetime Rules of Engagement and, like XVIII Airborne Corps, VII Corps worked to tailor these to its mission. Thousands of yellow cards were distributed, with one side printed with "Peace Time Rules of Engagement" and the other side containing "War Time Rules of Engagement." The VII Corps' ten peacetime rules were concise and easy to understand. The first stated simply that the United States "is not at war"; however, the second rule emphasized that a soldier had "the right to use force to defend . . . against attacks or threats of attack." The remaining rules addressed a number of issues, including the need to treat all people and property with dignity and respect.[63]

The VII Corps "War Time Rules of Engagement" began with the phrase "ATTACK all enemy soldiers, vehicles, positions, supplies, and equipment using these rules" (emphasis in original). Eleven restrictions on the use of force employed during hostilities, including prohibitions on looting, taking war trophies, and using booby traps that might kill or harm innocent persons, then followed. One rule required that soldiers use only the "minimum amount and type of ammunition necessary to accomplish the mission." Another instructed troops to "prevent unnecessary suffering." Judge advocates arranged for training on the peacetime rules of engagement while units were at port in Ad Dammam and Al Jubayl waiting for the arrival of their equipment. Training shifted to the wartime rules as DESERT STORM neared. Rules of engagement training ensured that soldiers understood and were comfortable with command restrictions on using weapons. The fact that there were no reported VII Corps war crimes during DESERT STORM also reflected the effectiveness of the training provided by judge advocates.[64]

Although General Franks had decided that the rules of engagement would be uniform throughout VII Corps, the issue caused certain problems. The British 1st Armoured Division was an integral part of VII Corps' battle formation; however, the British rules of engagement were written for an urban scenario akin to that in Northern Ireland. Moreover, the British forces were not inclined to adopt American rules of engagement, as they were familiar and comfortable with their existing rules. Since the intent of the British rules was to prevent an escalation of violence, they were more restrictive than VII Corps' "War Time Rules of Engagement." Concerned that troops operating side-by-side but acting under dissimilar rules of engagement would react differently to situations arising during hostilities, Huffman suggested to the British legal adviser that his division adopt the VII Corps rules. The extent, if any, to which the British 1st Armoured Division subsequently revised its rules of engagement is not known. But when the 2d Armored Cavalry Regiment, 1st Infantry Division, and 1st and 3d Armored Divisions attacked the Iraqis on 24 February, these units operated under uniform rules of engagement.[65]

Military Justice and Discipline

With prior experience as a military trial judge, Colonel Trant, now serving as the VII Corps deputy staff judge advocate, worked closely with the corps chief of military justice, Maj. Calvin L. Lewis, in overseeing disciplinary matters in VII Corps. While trial counsel spent much of their time advising commanders on nonjudicial punishment and court-martial options, Trant and Lewis focused on writing a new VII Corps supplement to Army Regulation 27–10, *Military Justice*. The existing corps supplement reflected the area jurisdiction structure used in Germany. As the corps now operated along command lines, a new regulatory guide was needed, particularly since the nature of some corps units had changed dramatically since their arrival in Saudi Arabia. The corps support command, for example, had numbered some 6,000 personnel in Germany. Now it was 2d Corps Support Command and, with roughly 36,000 troops, was a separate general court-martial jurisdiction. The new corps supplement, which General Franks approved in February, addressed this and other similar changes.

Colonel Trant, working with Colonel Finnegan, his counterpart at the XVIII Airborne Corps, also drafted a new prohibited activities regulation for Southwest Asia. The guidance from the two corps staff judge advocates was to avoid the situation in which one corps prohibited an activity that the other allowed. Trant and Finnegan, using U.S. Army, Europe, Regulation 600–1, *Prohibited Activities*, as a model, drafted two corps regulations that were mirror images. Some Germany-specific prohibitions (like firearms in barracks and the length of knife blades) were eliminated or modified, and new items were included. For example, both VII Corps Regulation 600–1 and its XVIII Airborne Corps counterpart established an enlisted fraternization prohibition governing relations between noncommissioned officers and junior enlisted personnel.[66]

On the Eve of the Ground War

On the eve of the ground war, some 200 Army lawyers were serving nearly 300,000 soldiers in Southwest Asia. The challenges facing these judge advocates varied. For Lt. Col. James J. Smith, the 82d Airborne Division staff judge advocate, providing his command group with good legal support was complicated by distance; part of the division was with the French armored force in western Saudi Arabia, while the rest of the division was in the eastern part of the country.[67] Similarly, Maj. Robert L. Swann, the 3d Armored Division deputy staff judge advocate, supervised the delivery of legal services to widely dispersed armored units. Both Smith and Swann worked long hours under adverse conditions. They, like all judge advocates in tactical assembly areas, combined their practice of law with building fighting positions, manning the perimeter, and supervising soldiers pulling guard duty.

But not all judge advocates on the eve of the ground war were practicing military law. Mid-January found Capt. Tarek Sawi, who had been responsible for legal assistance and operational law at 22d Support Command, in an Egyptian Army tank transport battalion. This unit, tasked with transporting tanks and fighting vehicles to American armored units on the front lines, needed a liaison officer to accomplish its critical mission, and Sawi's fluent Arabic made him the best person for the job. He did not address a single legal issue, but the food, lodging, tires, and spare parts Captain Sawi obtained for the Egyptians were important to their—and the U.S. Army's—success.[68]

In the final analysis, commanders in Southwest Asia sought the legal advice of judge advocates at every stage of operational planning. The Judge Advocate General's Corps' development of and focus on a comprehensive operational law program had paid tremendous dividends. And, as an essential part of this operational law effort, the Army's Law of War training program had achieved its desired effect. Commanders at all levels insisted that their operations be conducted in strict compliance with the requirements of the law. Thus, judge advocates worked more closely with commanders and operators than ever before. DESERT SHIELD dramatically demonstrated that an ability to address the wide range of legal issues that arise in an operational environment had become an integral part of judge advocate practice. DESERT STORM would reveal that new operational law challenges lay ahead, particularly at the end of the 100-hour war.

Notes

[1] Interv, author with Capt Patricia A. Martindale, 13 Jan 97, Historians files, OTJAG.

[2] Frank N. Schubert and Theresa L. Kraus, eds., *The Whirlwind War* (Washington, D.C.: Government Printing Office, 1995), pp. 49, 157.

[3] Ibid., p. 130.

[4] Ibid., p. 157.

[5] Ibid., pp. 34, 251–69.

[6] Interv, author with Col Raymond C. Ruppert, 2 Dec 96, Historians files, OTJAG.

[7] Interv, author with Maj Robert A. Burrell, 18 Dec 96, Historians files, OTJAG.

[8] Interv, author with Ruppert, 2 and 3 Dec 96.

[9] CENTCOM General Order (GO) no. 1, *Prohibited Activities for U.S. Personnel Serving in the USCENTCOM AOR*, 30 Aug 90, Historians files, OTJAG.

[10] Interv, author with Ruppert, 3 Dec 96.

[11] CENTCOM GO 1, pars. d and e; interv, author with Ruppert, 3 Dec 96.

[12] Interv, author with Ruppert, 3 Dec 96.

[13] Ibid., 2 Dec 96; memo, SJA, XVIII Abn Corps, for Corps Provost Marshal, 25 Sep 90, sub: Confinement Facility—Operation DESERT SHIELD, Historians files, OTJAG; memo, SJA, USCENTCOM, for Staff Judge Advocates, 17 Nov 90, sub: Confinement Facility—Operation DESERT SHIELD, Historians files, OTJAG.

[14] Interv, author with Ruppert, 2 Dec 96.

[15] Chairman, Joint Chiefs of Staff, Memorandum 846–88, *Peacetime Rules of Engagement for U.S. Forces*, 28 Oct 88.

[16] Schubert and Kraus, *Whirlwind War*, p. 161.

[17] Interv, author with Lt Col Douglas K. Fletcher, 22 Jan 97, Historians files, OTJAG.

[18] Joint Universal Lessons Learned System (JULLS) Long Rpt 51469–80969, 16 May 91, sub: Donations of Money and Property in Support of the War Effort, Historians files, OTJAG.

[19] Schubert and Kraus, *Whirlwind War*, pp. 60–63; interv, author with Fletcher, 22 Jan 97. See also Janet McDonnell, *Supporting the Troops* (Washington, D.C.: Corps of Engineers, 1996), pp. 69–72.

[20] U.S. Department of Defense, *Conduct of the Persian Gulf War: Final Report to Congress* (Washington, D.C.: Government Printing Office, 1992), pp. 622–23.

[21] After Action Rpt, Cdr, 199th JAG Det, to Ch, Intl Affairs Div, OTJAG, n.d., sub: Comments and Recommendations Concerning Investigation/Documentation of Enemy War Crimes; interv, author with Maj Mark P. Brewer, 10 Dec 98, both in Historians files, OTJAG.

[22] After Action Rpt, Cdr, 208th JAG Det, to Ch, Intl Affairs Div, OTJAG, 1991, sub: War Crimes Documentation Center, Historians files, OTJAG.

[23] JULLS Long Rpt 62159–45948, 2 Jul 91, sub: International Agreements for EPW [Enemy Prisoners of War] Transfer, Historians files, OTJAG.

[24] Interv, author with Ruppert, 2 Dec 96.

[25] Protocol, CENTCOM and British Forces Middle East, 31 Jan 91, sub: An Arrangement for the Transfer of Enemy Prisoners of War and Civilian Internees From the Custody of the British Forces to the Custody of the American Forces, Historians files, OTJAG.

[26] Intervs, author with Burrell, 18 Dec 96, and with Ruppert, 2 Dec 96.

[27] Interv, author with Burrell, 18 Dec 96.

[28] Interv, author with Ruppert, 3 Dec 96.

[29] Interv, author with Col Bernard A. Pfeiffer, 9 Dec 96; JULLS Long Rpt 32433-53212, 24 Mar 91, sub: Filing Deadlines for SF 278, Financial Disclosure Reports, both in Historians files, OTJAG.

[30] Interv, author with Daniel L. Winand, 22 Jan 97, Historians files, OTJAG.

[31] Ibid.; Memo for Record, SJA, XVIII Abn Corps, 12 Mar 91, sub: SJA Significant Activities During the Ground Campaign, p. 1, Historians files, OTJAG.

[32] Interv, author with Winard, 22 Jan 97.

[33] Interv, author with Lt Col David H. Brunjes, 28 Jan 97, Historians files, OTJAG.

[34] *United States* v. *David S. Wiggins*, Army Court of Military Review (ACMR) 9100982, 30 Jul 93.

[35] Interv, author with Col William R. Hagan, 27 Jan 97, Historians files, OTJAG.

[36] Ibid.; William R. Hagan, "The Judge Advocate in Desert Storm/Shield," *Military Advocate* (Spring 1991):3.

[37] OTJAG, Desert Storm Assessment Team Rpt to TJAG of the Army [DSAT Rpt], Contact Law, 22 Apr 92, pp. 3–4; interv, author with Hagan, 27 Jan 97.

[38] Office Organization Diagram, Office of the Staff Judge Advocate, 22d Support Command, n.d., Historians files, OTJAG.

[39] Interv, author with Col John R. Bozeman, 19 Feb 97, Historians files, OTJAG.

[40] Patrick Finnegan, "Operational Law: Plan and Execute," *Military Review* 76 (March–April 1996):33.

[41] Ibid.

[42] Intervs, author with Maj Mark C. Prugh, 9 Jan 97; with Michael L. Larson, 3 Feb 97; and with Martindale, 13 Jan 97, all in Historians files, OTJAG.

[43] Interv, author with Col Patrick Finnegan, 16 Jul 96, Historians files, OTJAG.

[44] Finnegan, "Operational Law," p. 34; interv, author with Finnegan, 16 Jul 96.

[45] The second lawyer in Saudi Arabia for Desert Shield was Lt. Col. James J. Smith, the staff judge advocate of the 82d Airborne Division. He arrived in Dhahran on 9 August 1990, a few minutes after Captain Prugh.

[46] Interv, author with Larson, 3 Feb 97.

[47] Ibid.

[48] Information Paper, SJA, XVIII Abn Corps, 20 Jan 91, sub: Use of Contractor Transportation Equipment and Other Critical Supplies/Services in Support of Operation Desert Storm, Historians files, OTJAG; interv, author with Larson, 3 Feb 97.

[49] Memo, Brig Gen Edison E. Scholes for Contracting Officers, XVIII Abn Corps Acquisition Section, 30 Oct 90, sub: Standards of Conduct, Historians files, OTJAG.

[50] Information Paper, SJA, XVIII Abn Corps, 11 Nov 90, sub: Immigration Issues—Operation Desert Shield; Memo, Hq, XVIII Abn Corps, 20 Nov 91, sub: 1991 Tax Assistance Program—Operation Desert Shield, both in Historians files, OTJAG; interv, author with Martindale, 9 Jan 96; Finnegan, "Operational Law," p. 35.

[51] Information Paper, SJA, XVIII Abn Corps, 12 Dec 90, sub: No Sex Orders—Operation Desert Shield, Historians files, OTJAG.

[52] JULLS Rpt, unnumbered, Mar 91, sub: Women Driving in Saudi Arabia, Historians files, OTJAG; Finnegan, "Operational Law," p. 36; interv, author with Ruppert, 3 Dec 96.

[53] DSAT Rpt, sub: Personnel, p. 1; Finnegan, "Operational Law," p. 35; JULLS Rpt, unnumbered, Mar 91, sub: Dress Codes, Historians files, OTJAG.

[54] Finnegan, "Operational Law," p. 36.

[55] Interv, author with Ruppert, 3 Dec 96.

[56] Memo, SJA, XVIII Abn Corps, for Staff Judge Advocates, 17 Sep 90, sub: Civilians Supporting Operation DESERT SHIELD, Historians files, OTJAG; Finnegan, "Operational Law," p. 36.

[57] Memo, VII Corps Base SJA for Chief of Staff VII Corps, Jan 91, sub: Deployment After Action Report, p. 1; VII Corps Cmd Rpt, Operation DESERT SHIELD/DESERT STORM, n.d., Tab L (Legal), pp. 1–3, 13; intervs, author with Brig Gen Walter B. Huffman, 4 Nov 96, and with Col Charles E. Trant, 10 Jan 97, all in Historians files, OTJAG.

[58] Interv, author with Huffman, 4 Nov 96; VII Corps Cmd Rpt, Tab L, p. 3.

[59] DSAT Rpt, "Personnel," p. 9; interv, author with Huffman, 4 Nov 96.

[60] VII Corps Cmd Rpt, Tab L, pp. 4–5.

[61] Interv, author with Trant, 10 Jan 97.

[62] DSAT Rpt, "Administrative Law," 22 Apr 92, p. 5.

[63] "Peace Time Rules of Engagement," VII Corps, 1990, Historians files, OTJAG. See Appendix B for the text of the "Peace Time Rules of Engagement," "War Time Rules of Engagement," and "Customs and Courtesies" cards issued to VII Corps troops serving in the Gulf.

[64] DSAT Rpt, p. 12; interv, author with Huffman, 4 Nov 96.

[65] Interv, author with Huffman, 4 Nov 96.

[66] Interv, author with Trant, 10 Jan 97.

[67] Interv, author with Col James J. Smith, 30 May 96, Historians files, OTJAG.

[68] Interv, author with Maj Tarek Sawi, 18 Apr 96, Historians files, OTJAG.

5

Operation Desert Storm, 1991

To give the right legal advice, you must be at the right place at the right time. But being a good lawyer is not enough to put you there; you must also be a good staff officer, able to do any mission supporting the operation.[1]

—Maj. Mark Cremin
Judge Advocate, 1st Armored Division

On 16 January 1991, United States and coalition forces launched their air war against Iraqi forces. While aircraft and missiles pounded enemy positions in Iraq and Kuwait, General Schwarzkopf and the coalition prepared for the ground campaign. On the left flank, near the intersection of the Kuwaiti, Saudi, and Iraqi borders, were Egyptian and Syrian heavy forces and the 1st Cavalry Division from Fort Hood, Texas. In the center were the U.S. Marine 1st and 2d Divisions and a brigade of the Army's 2d Armored Division. On the right flank, along the coast, were Saudi, Qatari, Bahraini, and other Persian Gulf forces. Behind these units was the bulk of the allied fighting power: the XVIII Airborne and VII Corps.

The XVIII Airborne Corps, stationed at Fort Bragg, consisted of the 82d Airborne, 101st Airborne (Air Assault), and 24th Infantry Divisions, as well as the 3d Armored Cavalry Regiment. It also exercised operational control of the French 6th Light Armored Division. The VII Corps, headquartered at Stuttgart, consisted of two Germany-based divisions, the 1st Armored from Ansbach and the 3d Armored from Frankfurt, and the 2d Armored Cavalry Regiment from Nuremberg. These "heavy" units were further supplemented by the 1st Infantry Division (Mechanized), the 1st Cavalry Division, and two brigades of the British 1st Armoured Division. Of the roughly 296,000 soldiers in theater, the two corps had 258,000. The VII Corps alone had 142,000 soldiers, 1,500 main battle tanks, 1,500 Bradleys and other armored personnel carriers, 669 artillery pieces, and 223 attack helicopters.[2]

In late January and early February, the two corps would meet on the left of the allied front, along the Saudi-Iraqi border. Thus, on 23 February 1991, as President Bush demanded for the last time that Saddam Hussein comply with all UN resolutions imposed since the 2 August invasion, Schwarzkopf's forces stretched inland some 300 miles from the Persian Gulf. Consequently, when the Iraqi dictator refused Bush's final ultimatum on 24 February, the all-out ground attack that followed took place along a broad front. Badly mauled by the air attacks of the preceding thirty days and surprised by the speed of the advancing Americans, Iraqi forces offered little resistance. By day four of the battle, American and allied ground forces had driven enemy forces from Kuwait and had pushed far into Iraq, either crushing enemy units in place or taking their surrender. After nearly 100 hours of continuous movement and combat, the war was over. On 28 February, when the cease-fire ordered by President Bush went into effect, XVIII Airborne Corps was in Iraq's Euphrates River valley and VII Corps was in Iraq at Al Busayyah. Coalition forces, including two Marine divisions and other soldiers of Marine Central Command, had liberated Kuwait. Thirty-five of Iraq's forty-three combat divisions were decimated, and the allies had destroyed 3,800 of the enemy's 4,300 tanks and taken 60,000 prisoners. It was the fastest and most complete victory in American military history.[3]

United States Central Command (CENTCOM)

With the start of DESERT STORM, CENTCOM judge advocates focused almost exclusively on providing timely and accurate advice in support of combat operations. To guarantee that a lawyer was always available, two operational law attorneys worked twelve-hour shifts, seven days a week, in the CENTCOM operations center.[4] This round-the-clock presence reflected the fact that judge advocates were now completely integrated into military operations. The experiences of commanders, planners, and lawyers in URGENT FURY and JUST CAUSE had demonstrated that judge advocate participation in all aspects of planning and execution contributed substantially to mission success. And the Judge Advocate General's Corps had now developed an extensive program for training and supporting those judge advocates who practiced operational law in the field. *(Map 9)*

Law of War, Rules of Engagement, Targeting Issues, and Prisoners of War

During the course of DESERT STORM, commanders frequently called upon judge advocates to address the legality of attacking a particular building, bridge, road, railroad, or other structure. In advising commanders on such issues, CENTCOM judge advocates consistently, on a target-by-target basis, balanced the military necessity for attacking a target against any negative effects, such as collateral civilian damage or casualties.

At General Schwarzkopf's direction, the CENTCOM staff planned the military campaign with a view toward minimizing collateral civilian casualties and damage to civilian objects. Consequently, some targets were avoided because the value of destroying these targets was outweighed by the potential risk to nearby civilians or, in the case of certain archeological and religious sites, to civilian structures. For example, the government of Iraq parked two jet fighters next to the ancient temple of Ur, believing that the coalition's respect for cultural property would preclude any attack on these planes. While U.S. Central Command had previously identified the temple of Ur as an "Off Limits Target," the Law of War permitted an attack on the aircraft, with Iraq, having intentionally positioned the planes near the temple, bearing responsibility for any collateral damage to the structure. General Schwarzkopf, however, upon learning that there was no runway or servicing equipment for the fighters, decided that the limited value of destroying the planes was outweighed by the risk of incidental damage to the temple. The Iraqi fighters were not targeted.[5]

While a policy of minimizing collateral civilian casualties and protecting civilian property furthered compliance with the Law of War and was a touchstone of CENTCOM's rules of engagement (ROE), CENTCOM judge advocates discovered that the policy and rules were sometimes misinterpreted. During the air war, for example, Colonel Ruppert learned that some "pilots were pulling off legitimate targets. They thought they could not engage them because they were in civilian neighborhoods . . . Pilots had to risk their lives [again] in flying the mission a second time." Ruppert and the other military lawyers emphasized that this unduly restrictive interpretation of the rules of engagement was not only erroneous but, as it increased the risk to American pilots, was a threat to the overall allied air campaign.[6]

Other situations also illustrate how CENTCOM judge advocates considered the legal principles of military necessity, proportionality, and unnecessary suffering in advising on targeting. In one instance, the destruction of bridges over a river caused the Iraqis to convert a nearby

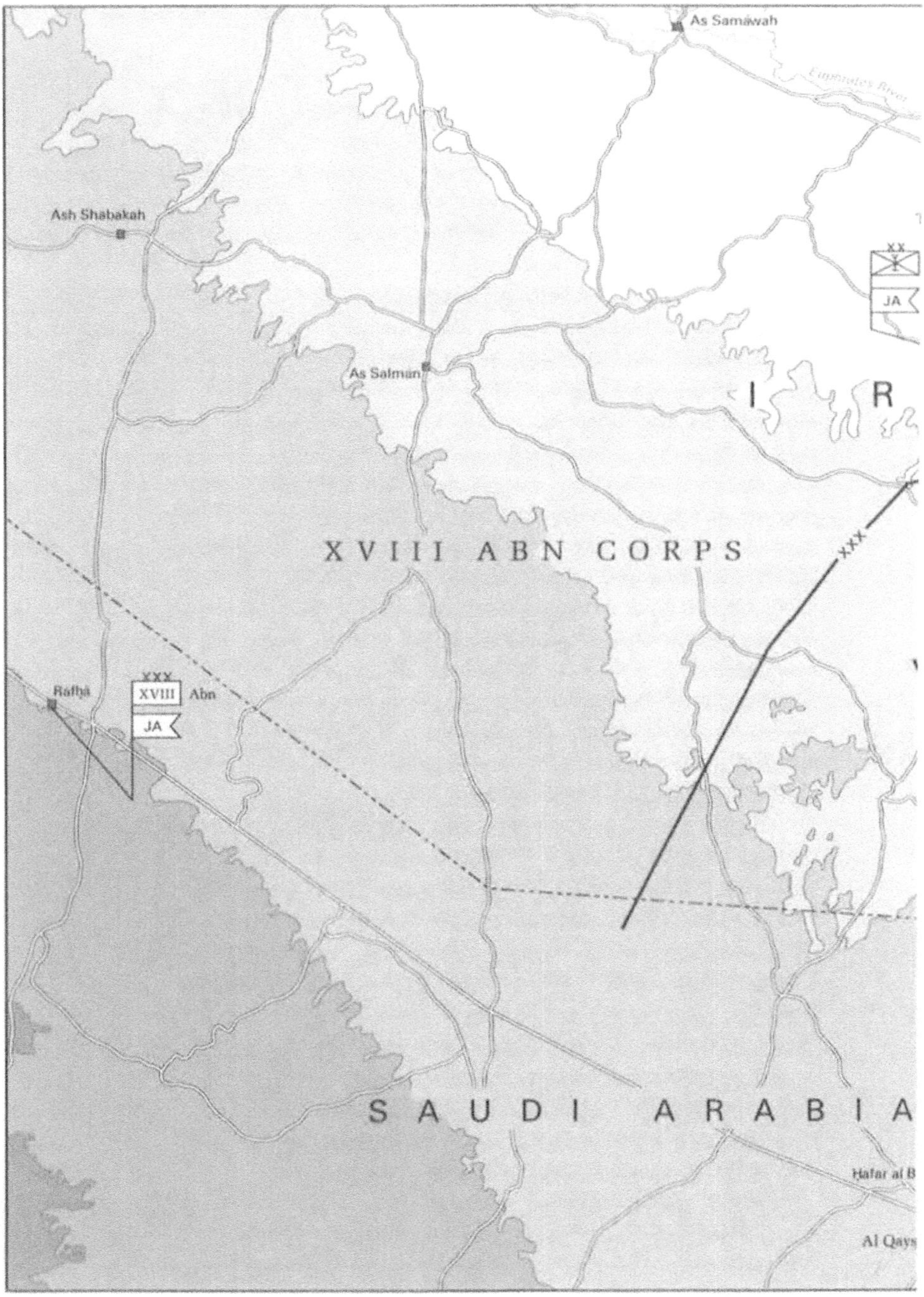
As Samawah
Euphrates River
Ash Shabakah
As Salman
I R
XVIII ABN CORPS
XXX
XXX
XVIII Abn
JA
Rafha
XX
JA
SAUDI ARABIA
Hafar al B
Al Qays

Map 9

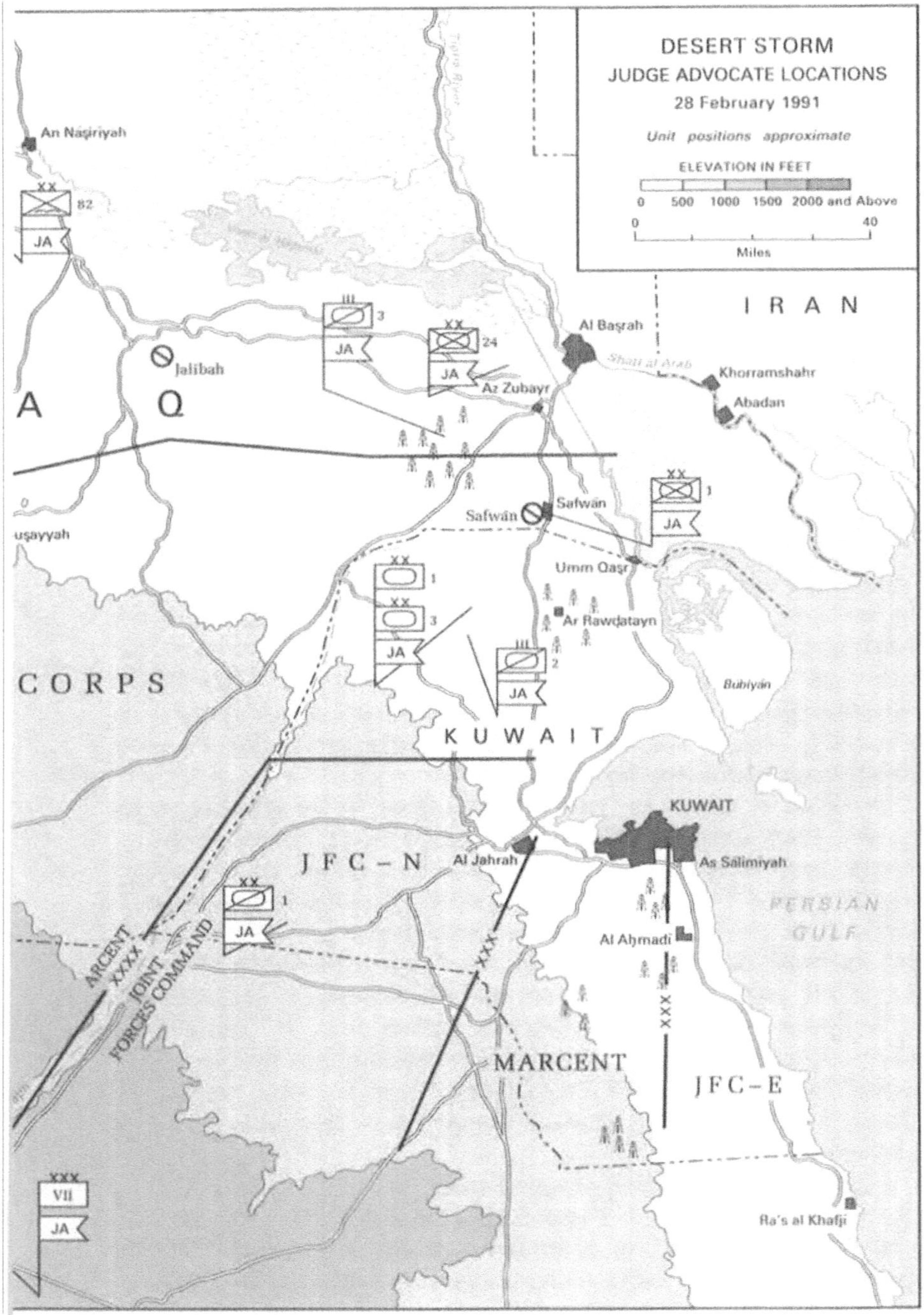

DESERT STORM
JUDGE ADVOCATE LOCATIONS
28 February 1991
Unit positions approximate
ELEVATION IN FEET
0 500 1000 1500 2000 and Above
0 40
Miles
IRAN
An Naşiriyah
Al Başrah
Jalibah
Az Zubayr
Khorramshahr
Abadan
Safwān
Umm Qaşr
Ar Rawḍatayn
Bubiyān
KUWAIT
Al Jahrah
As Salimiyah
Al Aḥmadi
Ra's al Khafji
PERSIAN GULF
CORPS
JFC-N
JFC-E
MARCENT
ARCENT
JOINT FORCES COMMAND
JA
VII

dike into a major supply route for transporting soldiers and war materiel. The targeting committee wished to destroy the dike but, recognizing that it protected the local population from flooding, asked if such an attack was lawful. Similarly, several Iraqi vehicles, identified as possibly containing chemical weapons components, were parked next to a hospital. Was it lawful to target these military vehicles despite the risk of collateral damage to the medical facility? Using the principles of military necessity, proportionality, and unnecessary suffering, Colonel Ruppert concluded that both the dike and the trucks were legitimate targets under the Law of War.[7]

Giving the correct legal advice on targeting was complicated by Iraq's intentional commingling of military targets with the civilian population. Just as the Iraqis had used the temple of Ur to shield aircraft from attack, they also used civilian structures to protect other legitimate targets. The most tragic result of this Iraqi policy was an incident involving the Al-Firdus bunker in Baghdad. Originally constructed as an air raid shelter for exclusive use by civilians, a portion of the bunker had been converted into a military command and control center. Access to the bunker was controlled, and it also was camouflaged. Iraqi authorities, however, permitted selected civilians to continue using part of the bunker as an air raid shelter. CENTCOM planners knew only that the bunker was an Iraqi command and control center. The CENTCOM judge advocates, applying the principles of military necessity and proportionality, advised that an attack on the bunker was legitimate, as it was being used to further Iraq's war effort.

Shortly after the bunker was destroyed, however, the government of Iraq and some members of the news media loudly protested that a "civilian bomb shelter" had been attacked in violation of the Law of War. Since the CENTCOM lawyers were manning the operations center around the clock, they were immediately able to provide the correct legal rationale for the targeting of the bunker. Furthermore, while CENTCOM planners were unaware that civilians had taken refuge in the Al-Firdus bunker, judge advocates advised that the presence of these civilians would not have precluded an attack on the structure. U.S. Central Command may well have refrained from targeting the bunker until the civilians had been removed, but the Law of War did not require such restraint.[8]

Although 22d Support Command would have overall responsibility for processing prisoners of war during DESERT STORM, that unit did not play any significant role in the development of a theater policy on prisoners. CENTCOM judge advocates provided theaterwide guidance, with the implementing details left to the service components. Thus,

General Schwarzkopf determined that U.S. forces would not operate long-term prisoner of war camps. Instead, enemy soldiers captured by American, British, and French forces would be transferred to Saudi control. The details of the processing and treatment of Iraqi prisoners between the time of capture by U.S. forces and transfer to Saudi-run camps, however, were to be left to Army Forces Central Command and its Navy, Air Force, and Marine Corps counterparts. This, in turn, meant that, given the Army's status as the Defense Department executive agent for all prisoner of war operations and the reality that the vast majority of enemy soldiers would be captured by ARCENT units, judge advocates at XVIII Airborne Corps, VII Corps, and the 22d Support Command would take the lead in developing policies on prisoner processing and treatment. The XVIII Airborne Corps lawyers began developing policies on prisoners first, using the guidance issued by CENTCOM. After their arrival in Southwest Asia, VII Corps judge advocates looked to the XVIII Airborne Corps for assistance in establishing VII Corps rules on enemy prisoner processing. In any event, a theaterwide policy on prisoners of war was in place long before DESERT STORM began, affording commanders with sufficient time to provide appropriate training to their soldiers.

Occupation Law Issues

On 3 March 1991, General Schwarzkopf, accompanied by Colonel Ruppert, met his Iraqi counterpart in a tent in Safwan, Iraq. The day prior, Schwarzkopf had handed Ruppert a document titled "Unilateral Cessation of Offensive Operations," which Ruppert reviewed and returned to his boss. The document, which dictated the terms of the cease-fire, was signed by the Iraqi senior military commander.[9] However, for CENTCOM judge advocates, the end of the war brought new legal issues to the fore.

U.S. forces now occupied most of southeastern Iraq. For political reasons, the U.S. Central Command took the position that the United States was not an "occupying force." But, as the physical seizure and control of Iraqi territory triggered the application of the 1949 Geneva Convention Relative to the Protection of Civilian Persons in Time of War (Geneva Civilians Convention), CENTCOM also pledged that its forces would comply with the duties and responsibilities of an occupying force set out in that convention.[10] Providing legal advice on the duties of an occupier, however, proved to be a difficult task: The American Army had not occupied enemy territory since the Korean War, and, as a result, no commanders, general staff officers, or judge advocates had any experience dealing with occupation law.

Under international law, military occupation does not transfer the sovereignty of an occupied territory to the occupier. Rather, it provides the occupying power with some of the rights of sovereignty, particularly those related to the maintenance of law and order. Thus, American forces were required to perform police functions and to establish security for Iraqi civilians in the geographic area under U.S. control. As Iraq technically retained sovereignty over the occupied area, however, restoring public order and safety did not mean a substitution of U.S. law for Iraqi law. Rather, Iraqi civil and penal laws continued in force, at least in theory. Additionally, CENTCOM judge advocates advised that private property could not be confiscated by the occupying force, but that, if needed for direct military use, it could be requisitioned, with payment made to the property's owner. Probably the greatest concern to U.S. commanders in Iraq, however, was the legal requirement that civilians in the occupied area have access to sufficient food and medical supplies, particularly since the Geneva Civilians Convention required that adequate food and medical supplies be brought to the occupied territory if local Iraqi resources proved inadequate.

Initially, except for the Safwan region in the VII Corps area, the mostly unpopulated desert presented few issues concerning Iraqi civilians. Nonetheless, General Franks moved swiftly to meet his Geneva Convention responsibilities by forming a civil affairs (G–5)/staff judge advocate operational cell, which provided operational guidance to VII Corps units during the nine-week "occupation."[11] Popular uprising in parts of Iraq led thousands of civilians to flee the fighting between Iraqi military units and Shi'ite forces and to enter territory held by coalition forces. In accordance with the law, U.S. and allied forces then provided food, water, and medical care for these refugees. Additionally, this influx required U.S. forces to give additional attention to other Geneva Convention requirements such as protecting the local populace from criminal acts. This relief effort continued as long as CENTCOM forces were in southern Iraq. Even after the departure of these units, American and allied forces participating in Operation PROVIDE COMFORT continued aiding Kurdish refugees in northern Iraq.[12]

The Geneva Civilians Convention contemplated that impartial humanitarian organizations such as the International Committee of the Red Cross would assume the responsibilities of the occupying power when that power withdrew from enemy territory. As coalition forces prepared to withdraw from Iraq, however, no international relief agency was prepared to assume the ongoing relief effort. As this meant that American legal responsibility for the refugees continued, the Iraqis were offered the opportunity to move to a refugee camp at Rafha, Saudi

Arabia. About 20,000 refugees (including more than 8,000 from the Safwan area) accepted this offer.[13]

Just as Army lawyers had not advised on occupation law since the defeat of the Axis powers in 1945, they also had not faced issues related to liberated territory since that time. The liberation of Kuwait revealed that major rebuilding of that nation was required. CENTCOM's Task Force Freedom, as part of its mission to assist in the restoration of Kuwait's legitimate government, wished to participate. But to what extent could Department of Defense manpower, equipment, and supplies be used to provide relief to Kuwait's civilian population following their liberation from enemy forces? Could soldiers repair electrical power plants, water distribution systems, telephone exchanges, and highways? Could they engage in firefighting and control air traffic? After conferring with the CENTCOM judge advocates, Army Lt. Col. Charles Criss, the Task Force Freedom legal adviser, advised that a commander might properly provide life-sustaining services on a temporary basis and otherwise maintain order until the Kuwaiti government could be reestablished. This opinion became the basis for Task Force Freedom's initial nation-building efforts in Kuwait.[14]

New War Crimes Issues

Iraqi forces retreating from Kuwait intentionally dynamited 732 producing oil wells. Over 650 of these caught fire, causing oil-laden clouds that rose as high as 22,000 feet. At the peak of destruction, the fires daily consumed approximately five million barrels of oil, valued at about $100 million, and generated more than one-half million tons of airborne pollutants. The Iraqis also intentionally polluted Kuwaiti waters by dumping great volumes of unrefined or crude oil into the Arabian Gulf. Were these wanton acts of destruction tantamount to "environmental" war crimes? CENTCOM judge advocates, after conferring with their counterparts at the Joint Chiefs of Staff and the Army Judge Advocate General's International and Operational Law Division, concluded that they were, as Article 147 of the Geneva Civilians Conventions prohibits the "extensive destruction . . . of property, not justified by military necessity and carried out unlawfully and wantonly."[15] The retreat of Iraqi forces meant that the persons responsible for these new environmental war crimes were unavailable for prosecution. Under international law, however, as reaffirmed by UN Security Council Resolution 687, Iraq, as a nation, was liable for this environmental damage and the depletion of Kuwait's natural resources.[16]

The 22d Support Command

After the air offensive began on 16 January, ground forces, including those belonging to the 22d Support Command (SUPCOM), began repositioning for the upcoming attack on Iraqi forces. Although the organization of legal services remained unaltered, the location of the staff judge advocate did change; after the "center of mass" of SUPCOM operations shifted from Dhahran to King Khalid Military City, Colonel Hagan moved there as well. With the start of the ground war, SUPCOM judge advocates now made their greatest contribution to military operations—in the areas of battlefield acquisition, contract performance during combat, and enemy prisoners of war.

Acquisition and Contract Law Issues

Following deployment, the Army supply system was the primary source of each unit's war-fighting equipment and supplies. Items unavailable or in short supply were acquired by ordering officers and Class A agents. The acquisition system worked in most cases, so that items such as fresh food, water, or clothing could be obtained through the contracting process. On the modern battlefield, however, the contracting system revealed serious shortcomings, particularly in the early stages of the offensive. Ordering officer monetary authority was limited, and ordering officers and contracting officers were often unavailable to acquire urgently needed property or services. As a result, private property sometimes had to be seized during hostilities when indispensable for defeating the enemy. Under the Law of War, such seizures are permitted. The law further provides that any private property seized during combat must later be returned to the owner and compensation paid for the use of that property and for any damage to or loss of this property.

As planning for DESERT STORM continued, American forces anticipated that they would seize private property after crossing into Iraq and Kuwait. Colonel Huffman, the VII Corps staff judge advocate, queried Colonel Hagan about record keeping for such battlefield seizures. Documentation was required for property accountability and to serve as a record in the event that subsequent claims were made against the United States for loss or damage arising out of such seizures. Huffman also conferred with Colonel Bozeman, his counterpart at XVIII Airborne Corps.

Bozeman, familiar with the chaos resulting from battlefield acquisition during URGENT FURY and JUST CAUSE, had already been considering ways to avoid a similar situation in DESERT STORM. Seizures in Grenada and Panama, particularly of privately owned automobiles and trucks, had been haphazard, hampering later efforts to pay claims for the loss of or damage to these vehicles and resulting in unnecessary investigations and overpayments.

Huffman and Bozeman agreed that the ground war would require battlefield seizures as a supplement to the contracting process, as it was impractical to contract for control of such property. But they wished to avoid uncontrolled, undocumented seizures, and they wanted the owner of the property to know why it had been seized and that compensation would later be paid. Determining that only a uniform system could inject discipline into the seizure process, and recognizing that any system must apply theaterwide, Colonels Bozeman and Huffman looked to the 22d Support Command. That command's role in theater-level logistics made it the best organization for developing guidelines for battlefield seizures. Consequently, Colonel Huffman met with Colonel Hagan and tasked him with creation of a battlefield acquisition system. Hagan in turn assigned Colonel Lowell, in his Contracts Division, this mission. Lowell immediately assembled a team of lawyers, and they produced SUPCOM SJA Form 27–1, the Property Control Record Book.[17]

This battlefield seizure book conferred no authority on commanders to seize property. It simply provided them with an easy way to document seizures made under the Law of War. Each Property Control Record Book was numbered and contained ten sets of serial-numbered property control record forms. The cover of the book stated that its use was prohibited in Saudi Arabia; only in Iraq and Kuwait were seizures authorized. A short explanation of the legal basis for battlefield seizures and instructions on the use of the forms were printed on the inside cover. Commanders were reminded that private property could not be seized for mere convenience and that civilians could not be left without adequate food, shelter, clothing, or medical supplies. The instructions also detailed the appropriate distribution of the four multicolored property control record forms. The Property Control Record Book required a soldier seizing property to enter pertinent information concerning the seized property and also contained a receipt, written in both English and Arabic, to be signed by the property owner, if available. The inside back cover of the book contained a seizure record.[18]

After General Pagonis approved the Property Control Record Book, Colonel Lowell contracted with a Saudi printer for its publication, and a few days later ARCENT combat units received copies of the book. Due to the short duration of the ground war, none of these books were used in making battlefield seizures. However, Army judge advo-

RECEIPT

This is a receipt for your property that has been used or taken by the Armed Forces of the United States of America. The unit commander determined that this property was essential to ensure the success of the mission or to protect the safety of the soldiers of his command. This receipt may be used to redeem your property or to document any claim.

I acknowledge receipt of this document.

Name ____________

Address ____________

ايصال

هذا ايصال باستلام ممتلكاتكم التي استخدمتها القوات الأمريكية وأن قائد الوحدة يرى بأنها ضرورية لضمان نجاح المهمة أو ضمان سلامة جنوده ، يمكن استخدام هذا الايصال لاسترداد ممتلكاتكم أو لتوثيق أي مطالبة .

أقر باستلام هذه الوثيقة .

الاسم ____________

العنوان ____________

DOCUMENT NO. 000481

PROPERTY CONTROL RECORD

COUNTRY ____________

GRID ____________ DATE ____________

1. Owner's Name ____________
2. Owner's Address ____________

3. Reason for military necessity ____________

4. Description of property ____________

5. Condition of property ____________

6. Remarks ____________

Signature ____________
Printed Name ____________
Rank ____________ SSN ____________
Unit ____________

The four carbonless receipts in the Property Control Record Book were printed in both Arabic and English; below, *the accompanying Instructions to Commanders emphasized that the Law of War did not create any "license to loot."*

INSTRUCTIONS TO COMMANDERS

1. You must accomplish your mission and ensure the safety of the lives and equipment entrusted to you. You must also obey the law and respect the lives and property of the local population.

2. During combat operations, international law allows you to seize property if you have valid military necessity. Seizing private or public property for mere convenience is unlawful. You may not leave civilians without adequate food, clothing, shelter, or medical supplies. **COMBAT OPERATIONS DO NOT GIVE YOU A LICENSE TO LOOT.** Improper seizure of property may result in personal liability.

3. This Property Control Record is used to document seizure of property on the battlefield by US Armed Forces. It is very important that the form be filled out completely, legibly, and accurately. Property should be described in as much detail as possible. Get photographs if you can!

4. After you have completed this form, give Copy 1 (white) to the property owner, if available; forward Copy 2 (blue) to your battalion S-4. Copy 3 (pink) stays with the property that was seized and Copy 4 (green) remains attached to this book. Fill in the Seizure Record inside the back cover.

5. Direct questions about use of this form to the nearest judge advocate.

ARCENT SUPCOM SJA FORM 27-1 (Jan 1991)

cates had anticipated an operational problem and had provided a practical solution. The Property Control Record Book's value as a tool for commanders was later demonstrated when it became the model for a similar publication issued by U.S. forces in Haiti during Operation Restore Democracy.[19]

Contractor vehicles represented a special problem. A number of owner-operators feared that their heavy equipment transporters would be damaged. Consequently, they refused to relinquish them to U.S. military drivers without an assurance that any loss or damage to their vehicles would be compensated. The SUPCOM Contract Law Branch immediately prepared a document entitled "U.S. Government Liability for Loss of or Damage to Leased Vehicles." It was translated into Arabic and signed by the commander of the ARCENT Contracting Command. The document recited a Federal Acquisition Regulation clause, providing for liability insurance in all government contracts, and stated that the protection of the clause extended to loss of or damage to vehicles leased from their owner-operators. Significantly, the document did not commit the government to pay directly for any loss or damage, but its insurance provisions satisfied the Saudi owner-operators, and the trucks rolled.[20]

The possible refusal of Saudi or third-country civilian drivers to operate their contract vehicles during war also troubled the Americans. If these civilians refused to drive into an area of imminent or ongoing hostilities, the provision of fuel and ammunition to combat forces would be severely reduced. A practical solution was to have a U.S. soldier "ride shotgun" in each contracted vehicle headed north to Iraq and Kuwait. This encouraged the contract driver to stay with his vehicle but, if this individual refused to continue driving, the soldier could do so. Colonels Murphy and Lowell of the 22d Support Command, working with Colonel Hoskey, the head of the ARCENT Contract Law Division, also proposed a second, complementary solution. The core of their idea was a printed card explaining in both Arabic and English that the driver was obligated to transport war materiel, as contracted. The card further notified the driver that if he refused to proceed as dispatched, a U.S. driver would be substituted and the vehicle would be used for the duration of the contract, with or without the presence of the contract driver. A second card, also in Arabic and English, was to be used by the U.S. assistant driver in requesting Saudi police aid in completing the mission. In Desert Storm, the heavy reliance on contractor-furnished drivers called for unusual approaches, and the production of these two cards clearly demonstrated how contract law could be used to support the combat mission.[21]

Prisoner of War Issues

While basic policies were formulated by CENTCOM, VII Corps, and XVIII Airborne Corps, the 22d Support Command had overall responsibility for the confinement and treatment of Iraqis captured by American, British, and French forces. This included building and operating facilities; processing prisoners from the front line to U.S.-run camps; providing food, medical care, and transportation; and then arranging for the transfer of prisoners to Saudi-run camps. Actual day-to-day operations fell to the 800th Military Police (Enemy Prisoner of War) Brigade, a major subordinate element of SUPCOM. This brigade, with over seven thousand soldiers, was a composite Army National Guard and Army Reserve unit from the northeastern United States specializing in prisoner of war operations.[22]

A major and a captain judge advocate comprised the legal staff of the 800th. Their mission was to provide a full range of legal services, in addition to handling legal issues surrounding the transfer of prisoners from American to Saudi control. They also were to identify Iraqi prisoners alleged to have committed war crimes. As thousands of Iraqi prisoners flooded into camps operated by the military police, it quickly became evident that the unit was inadequately staffed in both numbers and grades of judge advocates for them to perform their legal support mission to their headquarters, much less to the four prisoner of war camps, two of which were several hundred miles from the brigade headquarters in Dhahran. At the urging of the CENTCOM and ARCENT staff judge advocates, the 22d SUPCOM staff judge advocate provided three additional military lawyers to support missions of the 800th Brigade. Colonel Murphy, the SUPCOM deputy staff judge advocate, traveled from camp to camp providing oversight and advice on an ad hoc basis. Maj. Michael Carmin, who had been working in the SUPCOM military justice operation, and Capt. (later U.S. Congressman) Stephen E. Buyer, who had been serving in SUPCOM's King Khalid Military City legal office, were assigned as command judge advocates in two collocated prisoner of war camps. Their work, Colonel Hagan reported, "prevented the camps from turning into legal and public relations nightmares."[23]

Many unusual questions surfaced, all of which came to judge advocates for opinions. Could commanders accept offers from enemy prisoners and civilian detainees to spy for U.S. forces? Could they be utilized in psychological operations? Could captured soldiers be used as gravediggers for the enemy dead? After coordinating with other judge advocates in the theater, the SUPCOM attorneys advised that Articles

انت سجين حرب ، لن تؤذى او تجرح الا اذا حاولت الهرب
يجب ان تبقى صامتاً وتفعل ما يطلب منك . ستعامل باحترام
وعدل . سيقام بتفتيشك ، من المحتمل ان تؤخذ منك اغراضك
الشخصية مؤقتاً ولكنها ستعاد اليك .

You are a prisoner of war. You will not be hurt or injured unless you try to escape. You must remain quiet and do what you are told. You will be respected and treated fairly. You will be searched. You may be temporarily deprived of your personal property, but it will be returned to you

Iraqis taken prisoner by VII Corps units received this card, printed in both English and Arabic.

49 and 52 of the 1949 Geneva Prisoners of War Convention permitted prisoner labor as long as the work was not unhealthy, dangerous, or humiliating. Consequently, using enemy prisoner volunteers as intelligence collectors, translators, and interpreters was lawful, and such service also entitled them to compensation. Furthermore, the use of prisoners for burial details was also lawful. The expeditious burial of enemy dead helped preserve a healthful environment, was not contrary to the Islamic faith, and ensured that the burial was in accordance with Islamic religious beliefs.[24]

Even military criminal law questions arose in the prison camps, as enemy captives were subject to the Uniform Code of Military Justice and could be tried by courts-martial. After an Iraqi prisoner allegedly set fire to more than sixty tents in his prison camp, the command directed an inquiry into the incident. This investigation, conducted pursuant to the *Manual for Courts-Martial*, found insufficient evidence to warrant further criminal proceedings, and the prisoner was transferred to Saudi control. However, judge advocates had been fully prepared to court-martial the prisoner.[25]

From 18 January 1991, when the first Iraqi soldier was captured, to 2 May 1991, when the last prisoner in U.S. custody was transferred to Saudi Arabian control, the Army processed 69,822 Iraqi prisoners of war. Relative to the length of the campaign, this was the most extensive U.S. prisoner of war operation since World War II.[26] Moreover, when the U.S. custody of Iraqi captives ended, International Red Cross officials reported that the treatment of these prisoners by the American forces had complied more fully with the Geneva Conventions than the treatment afforded by any nation in any previous conflict in history. Army lawyers deserved much of the credit, particularly since, in advising on the policies and procedures for handling prisoners of war, judge advocates ensured that there was no adverse effect on the planning and execution of military operations.[27]

The activities of VII Corps and 1st Armored Division judge advocates typify those of the military lawyers who deployed during DESERT STORM. Accordingly, discussion will focus on the organization and actions of these two units.

VII Corps

At the start of DESERT STORM, Colonel Huffman and most of the VII Corps judge advocates were located at the corps rear command post in Tactical Assembly Area THOMPSON. An utterly isolated desert outpost, it had been newly created when an engineer unit bulldozed sand into a berm. Along with other senior officers in the assembly area, Huffman became a perimeter sector commander, responsible for some 250 meters of the roughly five-kilometer-long perimeter. Assisted by Colonel Trant, he was in charge of constructing bunkers and fighting positions, as well as guarding his sector. The need for security meant that every judge advocate captain pulled guard duty at night and filled sandbags. Major Lewis, the head of the Military Justice Division; Major Thomson, the chief of operational law; and Maj. Carlton L. Jackson, the senior defense counsel attached to VII Corps for duty, served as sergeants of the guard. Thus, in addition to their busy legal practice, judge advocates assumed their share of force protection. As another nonlegal duty, Colonel

Huffman's office also developed the officer rating scheme for General Franks' corps in Southwest Asia. No other office, the corps commander believed, could do it faster or better.[28]

At the beginning of February, the corps moved its rear command post to the Saudi air base at Al Quaysumah, some thirty kilometers east of Hafar al Batin. The VII Corps lawyers went with it, continuing to practice law, fill sandbags, and perform perimeter guard duty. Much of the time the judge advocates dressed in their chemical protective suits, for, as the start of the U.S. ground offensive neared, the Americans anticipated an Iraqi chemical weapons attack. The VII Corps judge advocates remained at Al Quaysumah through the ground war, leaving only when VII Corps units began their withdrawal from Iraq at the end of March. They then moved to King Khalid Military City, referred to as "KKMC" by some, but more popularly known as "The Emerald City" because of the green roofs of its mosques. In May, VII Corps attorneys returned to Germany with the corps headquarters.[29]

Operational Law Issues

With the start of the ground offensive, Colonel Huffman moved to the corps main command post, located west of Hafar al Batin and north of Tapline Road, a main supply route running parallel to the Trans-Arabian Pipeline. This placed him in the tactical operations center and close to General Franks.

The VII Corps main effort in the attack was the initial breakthrough operation. This crucial mission, assigned to the 1st Infantry Division (Mechanized), required a deliberate breach of Iraqi defensive fortifications as quickly as possible. After expanding and securing the breach site, the British 1st Armoured Division was to pass through the lines and continue the attack against the enemy forces.

To accomplish this breaching mission, 1st Infantry moved forward and plowed through the berms and minefields erected by the Iraqis. Many enemy soldiers surrendered during this phase of the attack and were taken prisoner. The division then assaulted the trenches containing other Iraqi soldiers. Once astride the trench lines, the division turned the plow blades of its tanks and combat earthmovers along the enemy defense line and, covered by fire from its Bradley armored infantry fighting vehicles, began filling in the trench line and its heavily bunkered, mutually supporting fighting positions.[30]

In the process, many more Iraqi soldiers surrendered to division personnel; others died in the course of the attack and the destruction or bulldozing of their defensive positions. During the bulldozing operation, General Franks radioed to Colonel Huffman, asking if burying the

Corps-level staff judge advocates. Left to right: *Col. John R. Bozeman, XVIII Airborne Corps; Col. Walter B. Huffman, VII Corps; and Lt. Col. William R. Hagan, 22d Support Command.*

enemy alive in his own trenches was permitted under the Law of War. If not, said Franks, he would "stop it now." Colonel Huffman assured him that the breaching operations were lawful. He advised, however, that the location where Iraqi defenders were being buried should be marked for later reporting to the International Committee of the Red Cross.[31]

As VII Corps units continued their rapid advance, crushing Iraqi units, thousands of enemy soldiers were captured. The volume of prisoners was so great that it threatened to slow the American advance; not only was transportation unavailable for moving the prisoners to rear holding areas, but there were too few soldiers to guard them. Could Iraqi prisoners simply be provided food and water and instructed to start walking south? Colonel Huffman, over the protest of some of the corps' commanders, answered that this was not possible. Under the Law of War, VII Corps was responsible for the safety of these prisoners. There were minefields to the south, and Iraqi prisoners might be killed or injured if they inadvertently walked through such fields. Additionally, groups of Iraqi soldiers traveling on foot behind the front lines might be targeted by U.S. soldiers or aircraft. Were they to be killed or injured, VII Corps would bear a responsibility for violating the Law of War. Huffman realized that large numbers of prisoners were hampering the forward progress of VII Corps units, but advised that the requirements of the law left no alternative: prisoners must be safeguarded and moved, under escort, to the rear.[32]

As discussed earlier, VII Corps judge advocates also aided the corps in meeting its Geneva Convention responsibilities regarding Iraqi civilians, providing operational guidance to VII Corps units during the nine weeks of "occupation."

Administrative Law

Administrative law problems ranged from conscientious objectors and the acceptance of gifts to the inadequacy of soldier family care plans and disharmony among active component, National Guard, and Army Reserve regulations. As the fighting ended, however, the toughest administrative law issues concerned investigating friendly fire incidents, classifying U.S. casualties, and war trophies.

The accidental killing and wounding of American soldiers from friendly, rather than hostile, fire are terrible concomitants of war. The emotional turmoil accompanying a fratricide makes its investigation more difficult than that of other incidents. Additionally, the tempo of combined arms operations on the modern battlefield made the DESERT STORM friendly fire investigations more complex than those conducted in previous conflicts. It was essential to investigate fratricides, however, in order to learn the causes of the incidents and to fix responsibility. Only then could safety be improved and measures undertaken to prevent future similar incidents.

Perhaps the most infamous fratricide involving VII Corps soldiers was the accidental killing of Cpl. Douglas L. Fielder, a 1st Armored Division combat engineer, by soldiers from the 3d Armored Cavalry Regiment, XVIII Airborne Corps. On 27 February 1991, the third day of fighting, Fielder and four other soldiers were stranded at an Iraqi airfield, awaiting a recovery vehicle for their disabled ammunition and explosives carrier. Mistaking this group for the enemy, a cavalry unit commanded by Capt. Bodo Friesen opened fire, wounding one of Fielder's comrades before a cease-fire was called. When the squadron commander, Lt. Col. John H. Daly, Jr., arrived on the scene a few minutes later, however, he granted permission for his gunner to open fire. Those shots killed Fielder.[33]

Within hours of the incident, the judge advocate captain with the 3d Armored Cavalry Regiment began a formal investigation of the incident. At the end of his inquiry, he concluded that "all personnel involved acted responsibly" and recommended that they be absolved of any liability or responsibility for the fratricide. This same judge advocate later reopened the investigation to answer some additional questions posed by the VII Corps commander, but he did not alter his earlier conclusion or recommendation.[34]

A review of this investigation by Fielder's battalion commander and the VII Corps staff judge advocate raised significant questions concerning the adequacy of the inquiry. These concerns resulted in XVIII Airborne Corps' conducting a second investigation in October 1991. The second investigating officer also concluded that all involved had acted responsibly. However, a later review, conducted at the direction of the commander in chief of Forces Command, disagreed with the findings and recommendations of both investigations. The FORSCOM staff judge advocate concluded that four of the officers involved in the fratricide were negligent and derelict in performing their duties and recommended reversing the two investigating officers' findings.

Continued dissatisfaction with the Army's handling of this fratricide incident four years later resulted in Senator Fred Thompson of Tennessee's call for an examination of the matter by the General Accounting Office. In April 1995 that office published a report that criticized the Army's investigations as "incomplete and inaccurate."[35] Even before the involvement of the Army Forces Command staff judge advocate and the General Accounting Office in this issue, judge advocates at VII Corps had concluded that this incident demonstrated the need for rapid, objective judge advocate participation in such situations. The technical expertise required to investigate such an incident, however, meant that a lawyer should *not* serve as the investigating officer. Nevertheless, judge advocates had to participate in such an inquiry, since their education and training made them well suited to ask the right questions and to organize an investigation in a logical fashion.[36]

The VII Corps judge advocates practicing administrative law also wrestled with the issue of casualty classification. Although this was ordinarily a personnel (G–1) function, General Franks directed his judge advocates to assume this responsibility after it became an increasingly controversial issue. Proper classification was important for determining notification of next-of-kin, for the Army wanted to provide families with information concerning a soldier's death or injuries quickly and accurately. Additionally, accurate casualty classification was important for Purple Heart entitlement.

In one case, for example, a soldier riding in a convoy in a combat zone was carrying a piece of unexploded ordnance. It exploded, killing him and wounding others. Should the death and injuries have been classified as accidental? Or killed in action, or wounded in action? In this

VII Corps judge advocates. Left to right: *Capt. Denise R. Lind, Maj. Calvin L. Lewis, Col. Walter B. Huffman, Lt. Col. Charles E. Trant, and Maj. George B. Thomson, Jr.*

instance, the corps judge advocates concluded that the death and injuries were accidental. The scenarios in other casualty classification cases, however, were often more complex, and classification remained both difficult and problematic.[37]

The issue of war trophies was not a significant problem for the first eight months that American forces were in Southwest Asia, as the language of CENTCOM General Order no. 1 was very clear regarding this matter. However, no one had anticipated the significant amount of Iraqi equipment and weapons that would fall into American hands, and after the fighting ended corps lawyers responded to more questions concerning war trophies than almost any other five administrative law issues combined. For example, American soldiers were unhappy when they learned that Iraqi bayonets could not be kept as a reminder of their soldiering in the Persian Gulf. Their complaints, combined with those of several commanders, resulted in General Schwarzkopf's modifying General Order no. 1 to permit the taking of an enemy bayonet as a souvenir. This and other similar issues led the corps staff judge advocate to refer to war trophies as "the greatest tar baby of all time."[38]

The issue of individual war trophies was easy compared with the question of unit trophies. Some commanders wished to return a few AK–47 rifles for display in their U.S. offices or headquarters buildings, and others, notably VII Corps armored divisions, wished to return to Germany with captured Iraqi T–62 and T–72 Main Battle Tanks. After examining the Army regulation governing the establishment of post museums, corps judge advocates advised that the best course of action in dealing with unit trophies would be the establishment of a VII Corps Museum. Thus, General Franks issued a regulation creating the museum and, in accordance with Army regulatory guidance on museum operations, established a property accountability procedure for each item of materiel that a unit wished to seize as a war trophy. This injected some discipline into the war trophy issue.[39]

1st Armored Division

Based on discussions with Colonel Bozeman at XVIII Airborne Corps, Colonel Huffman at VII Corps, and Colonel Crean at U.S. Army, Europe, Colonel Altenburg determined that he should deploy to Saudi Arabia with all the judge advocate personnel authorized on the division's table of manpower and equipment. Altenburg believed that building the best legal team in Saudi Arabia required taking as many judge advocates as possible. Thus, although some lawyers stayed behind to support those elements of the 1st Armored Division remaining in Germany, the majority of the division's judge advocates flew to Saudi Arabia on 26 December.[40]

Soon after arriving with the division commander, Maj. Gen. Ronald H. Griffith, Colonel Altenburg visited every XVIII Airborne Corps unit in the theater. In a 24-hour period he observed how each staff judge advocate had configured his legal operations. Conditions varied. Some lawyers lived and worked in tents, while others were housed in trailers or warehouses. But their tasks were similar. The tempo of DESERT SHIELD and the effect of General Order no. 1 meant that there was little military justice work. However, much legal assistance was provided, and almost everyone was standing guard duty, digging bunkers, and filling sandbags. Seeing the manner in which other staff judge advocates provided legal services assisted Altenburg in organizing those of the 1st Armored Division. However, regardless of how legal operations were structured, communication problems and the long distances between units made work difficult.[41]

The 1st Armored Division's deputy staff judge advocate, Maj. Gilpin R. Fegley, supervised the Legal Assistance Branch headed by

Capt. John Dunlap; the Military and Operational Law Branch headed by Capt. Mark Cremin; and the Administrative Law, Claims, and Contract Law Branch headed by Maj. James M. Coyne.[42] In setting up their offices in the field, the division lawyers learned one lesson understood by all staff judge advocates in Saudi Arabia. Although an operational law attorney should be collocated with the tactical operations center, legal operations as a whole should be structured so that they were "outside the wire"—outside the secured area containing the operational planning cells and related personnel. This arrangement provided soldiers ease of access for their personal legal problems.[43]

Contracting Issues

When an item was unavailable through the Army supply system, a field ordering officer could purchase it from local vendors. Recognizing that units deploying to Southwest Asia likely would have difficulty procuring items through normal supply channels, the 1st Armored Division appointed one field ordering officer for each battalion and two for each brigade. Because these officers had considerable discretion in making purchases and could contract for goods and services up to $2,500 per transaction, Colonel Altenburg believed that it was likely that legal issues could arise in this area, including possible criminal misconduct. To ensure that his judge advocates understood the field ordering officer process and would be prepared to respond to questions, Altenburg arranged for more than half of the division's military lawyers to be appointed as field ordering officers.

Later, Major Coyne, using his appointment as a field ordering officer, spent thousands of dollars on bright orange cloth. Cut into strips and attached to division vehicles, the cloth aided in identifying these vehicles and preventing friendly fire incidents. Thus, although the intent behind certifying judge advocates as field ordering officers was to give them a familiarity with this type of contracting, Army lawyers like Coyne were able to use their combat contracting expertise to resolve a variety of problems.[44]

Discipline and Military Justice

General Griffith, the division commander, recognized that continuing to try courts-martial during DESERT SHIELD and DESERT STORM would enhance discipline in his unit. As a result, the practice of military criminal law intentionally reflected a "business as usual" approach. Disciplinary matters were handled as they would have been in Germany, so that soldiers would see that the operational tempo of the mission would not result in lower standards of discipline. On the advice of his staff judge advocate, Griffith continued court-martial activity throughout DESERT SHIELD and DESERT STORM.

As head of the criminal law section of the Military and Operational Law Branch, Capt. Michael E. Smith supervised five judge advocates. Capts. Scott F. Romans, Tara O. Hawk, and Michael H. Leonard were with the three maneuver brigades and traveled with those units at all times. Capts. James F. Garrett and Irvin M. Allen, at the division rear headquarters, provided trial counsel support to the other three division brigades, the separate battalions, and the reserve units attached to the 1st Armored Division. For Captain Smith, overseeing the work of these trial counsels and monitoring disciplinary actions in the geographically dispersed division was not the toughest challenge. Rather, the greatest obstacle to success was logistical—finding a tent or other suitable facility in which to set up a field courtroom; obtaining electrical generators for power (critical for court reporting equipment); locating witnesses and arranging for their appearance; and transporting the accused, counsel, court members, military judge, and court reporter to the field for trial.

Despite these difficulties, the 1st Armored Division conducted three general courts, one special court, and six summary courts in Saudi Arabia, Iraq, and Kuwait and conducted over 100 nonjudicial proceedings under Article 15 of the Uniform Code. All these military justice activities occurred between the end of December 1990 and the end of April 1991—the four months that the division participated in DESERT SHIELD and DESERT STORM. And two of the general courts-martial and the special court were held within days of the start of the ground war on 24 February.[45]

On 22 February 1991, for example, two soldiers were convicted at separate general courts-martial. The first case was that of a combat signaler in the 1st Brigade. He had shot himself in the foot with his M16 rifle, hoping that this injury would result in his early return to Germany—and his wife. But his misconduct instead resulted in a prosecution for intentionally inflicting self-injury for the purpose of avoiding hazardous duty. Court was held in a general purpose, medium tent ten kilometers from the Iraqi border. The tent, ordinarily used to store mail, had been emptied, and tables and chairs set up for the court. Additionally, the lawyers had brought their own electrical power generator, as the unit had only 110-volt generators, incompatible with the 220-volt court reporting equipment used for transcribing the record. A UH–60 Black Hawk helicopter, dedicated that day to judge advocate use, flew Captain Allen, the trial counsel; Capt. Walter S. Wallace, the defense counsel; and the military judge, Col. Frank B. Ecker, Jr., to the field location. After accepting a plea of guilty to the charged offense, Ecker sentenced the accused to two years' confinement, total forfeiture of all pay and allowances, and a bad conduct discharge.[46]

Judge Ecker and the assistant trial counsel then boarded the helicopter and flew to a second location for the court-martial of another soldier. The accused had claimed conscientious objector status while in Germany and had initially refused to deploy to Saudi Arabia. He changed his mind after being ordered to deploy, but continued to be a disruptive influence in the engineer battalion to which he belonged. Charges against him were referred to a general court-martial for trial, and on 22 February he pleaded guilty to a number of offenses, including disobedience of a lawful order. The following day, Colonel Ecker sentenced the soldier to three years' confinement, total forfeiture of all pay and allowances, and a dishonorable discharge.[47]

These trials demonstrated that courts-martial might be successfully prosecuted even in the midst of final preparations for battle. However, *United States* v. *Presbury*, although prosecuted after the shooting was over, also illustrated how military justice functioned during the campaign in Southwest Asia. Not only was *Presbury* a fully contested trial, with an officer and enlisted panel, but the court proceedings were held in three countries—Saudi Arabia, Iraq, and Kuwait.

At the start of the air offensive on 16 January 1991, S. Sgt. Maurice Presbury chambered a round in his M16 rifle, pointed it at a soldier in his platoon, and threatened to shoot him. Charges of aggravated assault and communicating a threat were referred to trial at a general court. Early proceedings, held on 11 March in Kuwait, required flying the accused, a military judge, and a court reporter there. Two days later, another hearing was held at Log Base ECHO in Saudi Arabia. When the trial finally began at 8 A.M. on 29 March, the accused and all court members were located in Iraq at a base camp about forty kilometers west of Basra. Presbury's unit set up one tent as a field courtroom. A second tent was erected so that the court members would have a waiting area and a place to sleep during the night before and the night of the court-martial. The trial ended at 2 A.M. on 30 March, with Presbury convicted of simple assault and sentenced to be reduced to private (E–1), to forfeit all pay and allowances, and to be discharged from the service with a bad conduct discharge.[48]

Did trials like those conducted at the 1st Armored Division mean that the military justice system functioned well during the Persian Gulf War? Some observers of military criminal law during the war in Vietnam had sharply disagreed over whether the Uniform Code could work in a combat zone. Thus, this was an important question for both commanders and judge advocates. The consensus in the Army was that the Uniform Code did work well during these operations. It is true that the short duration of combat activities may well have obscured problems that might have arisen in a longer war. Moreover, in contrast to Vietnam, units participating in DESERT SHIELD and DESERT STORM consisted overwhelmingly of highly trained and motivated volunteers, rather than draftees, and the Southwest Asia area of operations was free of drugs and alcohol, all of which contributed to the low number of disciplinary problems.

However, when there were violations of the Uniform Code, especially challenges to military authority like those that arose in the 1st Armored Division, the Persian Gulf experience demonstrated that the military criminal justice system not only functioned fairly, but also served an important role in enforcing discipline. In the 1st Armored Division, for example, junior enlisted soldiers "were surprised, if not shocked" upon hearing that a trial by court-martial was being conducted the night before the attack on Iraq was to begin. These young soldiers no doubt assumed that military justice matters would be postponed until combat operations had ended. Holding a trial on the eve of battle, however, demonstrated to every division soldier that the maintenance of discipline was an integral part of preparing for the upcoming attack. High standards of military discipline would remain in place.[49]

The role played by the law and lawyers in preserving good order and discipline also meant that the prosecution of Capt. Yolanda Huet-Vaughn was closely watched in the theater, even though the trial occurred at Fort Leonard Wood, Missouri. Like Wiggins, Huet-Vaughn refused to perform her duties as an Army doctor. A Reserve Medical Corps officer, Captain Huet-Vaughn was ordered to active duty in late December 1990. Refusing to deploy to Saudi Arabia, she deserted on 31 December, remaining absent until 2 February 1991, when she returned to military control. At her trial by general court-martial, Huet-Vaughn alleged that her belief that war crimes would be committed during Operation DESERT SHIELD permitted her, "under the Nuremberg principles," to refuse to participate in the mission. A jury of officers rejected this defense. Convicted of desertion with an intent to avoid hazardous duty and to shirk important service, she was sentenced to thirty months' confinement and dismissal from the service.[50]

Likewise, *United States* v. *Brown*, a court-martial arising out of mutinous behavior by a member of the Louisiana National Guard, also served to demonstrate the important role played by the Uniform Code of Military Justice in promoting discipline. After Spc. Dwayne K. Brown, activated along with the other members of the 156th Infantry, arrived for training at Fort Hood in November 1990, he and a small group of his fellow soldiers decided that the chain of command was not addressing their complaints concerning long hours, inadequate time off, pay problems, alleged racial discrimination, and perceived poor leadership. Unhappy with this situation, Brown solicited a number of his fellow soldiers to absent themselves without authority ("go AWOL") and to return to Louisiana. Taking such action, Brown believed, would force the chain of command to address their grievances, particularly since the disgruntled soldiers intended to publicize their plight with the news media after they returned home. To further this protest action, Specialist Brown arranged for the hiring of buses to transport as many as 100 soldiers who intended to go AWOL with him. As a result of this conduct and because, after being personally admonished by Col. Frank Catalano, the deputy brigade commander, to cease his disruptive behavior, Brown nonetheless persisted in soliciting his comrades to go AWOL en masse, he was prosecuted by general court-martial. Despite his pleas, a panel of officers convicted him of conspiring to organize a strike and soliciting others to participate. He was sentenced to a dishonorable discharge, confinement for one year, and forfeiture of all pay and allowances. Brown's conviction and sentence signaled to other soldiers that those who defied their commander and engaged in coordinated efforts to disrupt combat training could face the sanction of a court-martial.[51]

A matter of particular concern in the area of military justice—an issue also faced by judge advocates in Vietnam—was the lack of authority to exercise criminal jurisdiction over civilians accompanying the force. Over one thousand civilians deployed to Southwest Asia on temporary duty. Twenty-two either left their positions without permission or refused to deploy.[52] Only administrative action could be taken against those who departed Saudi Arabia without authority, as applying the criminal jurisdiction of Article 2, Uniform Code of Military Justice, to civilians is possible only when Congress has declared war. This jurisdictional void remains a matter of ongoing concern.

War Crimes Investigations

At the end of March 1991, the 1st Armored Division was still in Iraq. Some 100 of its military police were manning Checkpoint Bravo, located approximately forty-five miles northwest of Basra on the highway to Baghdad. The 300 to 500 Iraqi soldiers who daily passed through the checkpoint on their way to camps in Saudi Arabia were asked if they had knowledge of war crimes or other misconduct. Those who said they did were separated from their fellow Iraqis for further questioning. Although division intelligence (G–2) exercised overall responsibility for this war crimes evidence gathering, judge advocate Captains Allen and Garrett actually questioned the Iraqis. Having been assigned by Colonel Altenburg to investigate war crimes allegations, the two men flew by helicopter to the checkpoint. Over a two-week period, Allen and Garrett lived and worked in a tent, interviewing and videotaping the statements of Iraqi soldiers. They documented approximately eighty war crimes cases.

The interviewees, whether victims or witnesses, told horrifying tales. In one instance, it was alleged that 50 to 100 civilians had been massacred in An Nasiriyah by the Iraqi Republican Guard. Having been identified as troublemakers, the civilians were herded into a mosque and killed by well-aimed shells fired from tanks. Other cases involved the brutal torture and execution of Iraqi deserters by Republican Guard members. In each case, Captains Allen and Garrett asked questions that would establish the identities of the perpetrators and victims and establish the specific elements of the offenses, with a view toward possible prosecution. Because those Iraqis responsible for such war crimes never came under American control, the evidence gathered by Allen and Garrett has not been used. It remains available, however.[53]

2d Armored Cavalry Regiment

A number of units below division level were provided their own legal support, particularly those operating independently in the theater. At the 2d Cavalry, for example, Capt. Matthew L. Dana served as the unit's command judge advocate. Deploying with the regiment to Saudi Arabia on 9 December, Dana was the lone legal adviser to the unit commander, Col. Leonard D. Holder, Jr. He provided command advice on military justice, administrative, and operational law matters, as well as personal legal assistance and claims advice to the roughly 11,800 troops in the regiment and units under its operational control. While Captain Dana officially remained a one-lawyer organization for the duration of DESERT SHIELD and DESERT STORM, he was joined by another judge advocate, Capt. Michael T. Burmeister, on 1 January 1991. Burmeister was attached to the regiment by the Army Trial Defense Service and, until the unit's redeployment to Germany, served as a defense counsel to troops requiring advice on courts-martial, nonjudicial proceedings, or adverse administrative actions. Together, Captains Dana and Burmeister lived and worked in a tent. The conditions were austere, and contact with other judge advocates generally was limited to weekly conversations via tactical telephone with either VII Corps or Trial Defense Service, Saudi Arabia.[54]

Administrative Law, Operational Law, and Military Justice

On 24 February 1991, Captains Dana and Burmeister, traveling in a High Mobility Multipurpose Wheeled Vehicle, or "HUMVEE," crossed into Iraq with the regimental support squadron. For the next four days, they moved with the armored cavalry regiment ever deeper into enemy territory. The judge advocates did no shooting and, without radios, were not in communication with the command group. But Captain Dana followed the advice given to him by Colonel Huffman: "Wherever you go," the VII Corps staff judge advocate had said, "dig a hole."

On 28 February, with shooting at an end, Captain Dana was immediately called upon to provide legal advice. After arriving at the tactical operations center, Dana learned that an M113 armored personnel carrier belonging to an American engineer unit had been mistakenly attacked by a friendly unit. That unit had recognized it as an M113, but was advised by the squadron intelligence officer that the enemy might have obtained the vehicle after overrunning Kuwait. A high-explosive antitank shell hit the M113, killing an American soldier.

Although another officer was detailed as the investigating officer, Captain Dana, assisted by two Criminal Investigation Command agents attached to the regiment, organized the inquiry into the fratricide. Their investigation revealed that the incident had occurred at roughly the same time as an Iraqi attack on the regiment, and that vision on the battlefield was poor due to smoke coming from an oil field fire. These two factors led Captain Dana to conclude that the fratricide resulted from the "fog of war." He recommended a finding of no wrongdoing and that all those involved be absolved of any responsibility for the tragic event. Dana's recommendation was approved.

Captain Dana's most time-consuming administrative law issue was that of war trophies. The 2d Cavalry's presence deep in Iraqi territory resulted in large amounts of enemy equipment and materiel falling into American hands. Believing that CENTCOM General Order no. 1 lacked sufficient detail on what constituted a war trophy, Captain Dana, using a commonsense approach, compiled a list of those items that he deemed to be war souvenirs. Berets and hats, binoculars, license plates, maps, gas masks, uniforms, patches, flags—all were deemed legitimate war trophies for individual soldiers. High-value items, such as a captured Soviet-made military surveying kit worth $30,000, were not permissible as war trophies. Nor were any optical or telescopic sights from an individual weapon or weapons system that, after being taken to the United States, might be used on a weapons system there. Captain Dana's list was not exhaustive, however, and if any question arose concerning an item not listed, Dana was the regiment's approval authority for the item. Both Colonel Holder and VII Corps G–1 endorsed these procedures.

Upon redeployment to Germany in late April, Dana's approved list became critical when military police enforcing customs regulations objected to license plates being taken back to Germany as war trophies. According to the police, license plates were not specifically identified in General Order no. 1. These objections were swept aside when Captain Dana—with the support of Colonels Huffman and Trant at VII Corps—presented his approved war trophy list to the officer in charge of military police operations.

In addition to administrative law issues, Captain Dana busied himself in operational law matters. At the end of the fighting, the 2d Cavalry moved to the Euphrates River in Iraq, near An Nasiriyah and the temple of Ur, where it became a magnet for Iraqi refugees. Some three thousand passed by the regiment's cantonment area each day, and almost all were in need of food, water, and medical care. But what could be provided? And what did the law require? Dana telephoned Colonel Huffman and was informed that U.S. forces were not in Iraq as an occupying force. Yet, while they did not have the legal responsibility of an occupying power, Huffman advised Dana that the 2d Cavalry should do "everything" it could to alleviate suffering. As a result, the regiment provided food, water, and medical care to all Iraqis who requested it.

An issue related to the feeding and care of Iraqi refugees was that of political asylum. Refugees repeatedly requested the cavalry troops for asylum, and the latter came to Captain Dana for guidance. After conferring with the VII Corps judge advocates, Dana advised that no political asylum could be granted. A granting of asylum could be made only by U.S. authorities in an area over which the United States exercised sovereignty, and, as this was not the case in the area of Iraq under the control of the 2d Cavalry, no asylum could be granted. Dana advised, however, that temporary refuge could be granted, but only in cases involving imminent danger to the life or safety of the person concerned—and when the person seeking refuge was not fleeing duly constituted law enforcement officials.

Summing Up

Colonel Ruppert, the CENTCOM staff judge advocate, called Desert Storm "the most legalistic war we've ever fought."[55] More than ever before, lawyers and the law were an integral part of Army planning, training, and warfighting. In part, this extensive role for judge advocates resulted from the enhanced Law of War instruction and training that had been required by the Department of Defense since 1974. However, increased judge advocate involvement in operations at all levels was primarily the result of the metamorphosis that had occurred in the Judge Advocate General's Corps. That institution no longer viewed its role as one of providing only traditional legal support. Although administrative law, military justice, legal assistance, and claims remained important to the practice of Army law, the corps now recognized operational law as central to the Army mission.

In both DESERT SHIELD and DESERT STORM, commanders willingly sought legal advice at every stage of operational planning, for they realized that judge advocates substantially contributed to the successful conduct of the operation through their knowledge of fiscal law, combat contracting, intelligence law, the Law of War, and rules of engagement. With judge advocates now a presence at all operational levels, however, commanders also routinely began to turn to them to resolve important nonlegal matters as well. In two years, Operation RESTORE HOPE would demonstrate the extent to which commanders relied on the expertise of their judge advocates.

Notes

[1] Interv, author with Maj Mark Cremin, 22 Nov 96, Historians files, OTJAG.

[2] Schubert and Kraus, *The Whirlwind War*, p. 177.

[3] Ibid., p. 201.

[4] Interv, author with Ruppert, 2 Dec 96.

[5] Report to Congress, App. O, pp. 614–15; interv, author with Ruppert, 2 Dec 96.

[6] DSAT Rpt, Operational Law, pp. 12–13.

[7] Interv, author with Ruppert, 2 Dec 96.

[8] Report to Congress, App. O, pp. 615–16; interv, author with Ruppert, 2 Dec 96.

[9] Interv, author with Ruppert, 2 Dec 96.

[10] DSAT Rpt, Operational Law, p. 16.

[11] Tom Clancy with Frederick M. Franks, Jr., *Into the Storm: A Study in Command* (New York: Putnam Publishing Group, 1997), p. 485.

[12] For a brief look at judge advocates in Operation PROVIDE COMFORT, see Chapter 8.

[13] Report to Congress, App. O, p. 610.

[14] JULLS Long Rpt 62154–20717, 27 Jun 91, sub: Authority to Use DOD Resources in Liberated Territory, Historians files, OTJAG.

[15] For a discussion of environmental damage during the Persian Gulf War, see Walter G. Sharp, "The Effective Deterrence of Environmental Damage During Armed Conflict: A Case Analysis of the Persian Gulf War," *Military Law Review* 137 (1992): 40–41.

[16] Report to Congress, App. O, p. 625.

[17] Contract Law Branch, HQ, US Forces Central Command Support Command, C+180 Lessons Learned Report, 5 Feb 91, pp. 1–3, Historians files, OTJAG; interv, author with Hagan, 13 Feb 97.

[18] C+180 Lessons Learned Rpt, 5 Feb 91, p. 103; interv, author with Hagan, 13 Feb 97; Memo, Maj Gen William G. Pagonis, HQ, Army Forces Central Command Support Command, for Support Command personnel, 29 Jan 91, sub: Seizure of Property, Historians files, OTJAG.

[19] Interv, author with Hagan, 13 Feb 97; Ltr, Col William R. Hagan to author, 10 Feb 97, Historians files, OTJAG.

[20] Memo, Pagonis for Support Command personnel, 29 Jan 91.

[21] Ibid.; interv, author with Hagan, 27 Jan 97.

[22] Schubert and Kraus, *The Whirlwind War*, p. 162.

[23] Interv, author with Hagan, 27 Jan 97.

[24] DSAT Rpt, Operational Law, p. 15; JULLS Long Rpt 62155–64792, 2 Jul 91, sub: Use of EPW Labor, Historians files, OTJAG.

[25] DSAT Rpt, Operational Law, p. 21; Memo, Ch, Military Justice, Off of SJA, 22d SUPCOM, for Commanding General, 22d SUPCOM, 26 Apr 91, sub: Allegation of EPW Misconduct, Historians files, OTJAG.

[26] Report to Congress, App. O, p. 617.

[27] Ibid., p. 620.

[28] DSAT Rpt, Operational Law, p. 5.

[29] Interv, author with Huffman, 4 Nov 96.

[30] Report to Congress, App. O, p. 630; Barton Gellman, "Reaction to Tactic They Invented Baffles 1st Division Members," *Washington Post*, 13 Sep 91, p. A21; Eric Schmitt, "U.S. Army Buried Iraqi Soldiers Alive in Gulf War," *New York Times*, 15 Sep 91, p. 10.

[31] Interv, author with Huffman, 4 Nov 96.

[32] Ibid.

[33] U.S. General Accounting Office (GAO), Report to the Honorable Fred Thompson, *Operation DESERT STORM: Investigation of a U.S. Fratricide Incident*, Apr 95, pp. 2–6; Bradley Graham, "Army Revokes Citations in Friendly Fire Incident," *Washington Post*, 19 Apr 95, pp. A1, A14.

[34] GAO, *Investigation of a U.S. Army Fratricide Incident*, pp. 45–47.

[35] Ibid., p. 5.

[36] Interv, author with Huffman, 4 Nov 96.

[37] Ibid.

[38] Ibid.; DSAT Rpt, Administrative Law, p. 1.

[39] Interv, author with Huffman, 4 Nov 96.

[40] Interv, author with Brig Gen John D. Altenburg, Jr., 30 Oct 96, Historians files, OTJAG.

[41] Ibid.; interv, author with Col Gilpin R. Fegley, 12 Dec 96, Historians files, OTJAG.

[42] Interv, author with Fegley, 12 Dec 96.

[43] DSAT Rpt, Operational Law, p. 4.

[44] Interv, author with Fegley, 12 Dec 96.

[45] Memo, Ch, Criminal Law, for SJA, 1st Armd Div, 6 Sep 91, sub: Criminal Law Portion of DESERT SHIELD/DESERT STORM After Action Report, Historians files, OTJAG.

[46] Ibid.; *United States* v. *Green*, ACMR 9100972, 22 Feb 91.

[47] Memo, Ch, Criminal Law, for SJA, 1st Armd Div, 6 Sep 91, sub: Criminal Law Portion of After Action Report.

[48] *United States* v. *Presbury*, ACMR 9100658, 29–30 Mar 91.

[49] Interv, author with Maj Michael E. Smith, 19 Feb 97, Historians files, OTJAG.

[50] *United States* v. *Yolanda Huet-Vaughn*, ACMR 9101873, *Military Justice Reporter* 43: 114 (Army Court of Criminal Appeals, 1995). But some who refused to deploy were not prosecuted. See Schubert and Kraus, *The Whirlwind War*, pp. 224–25.

[51] *United States* v. *Dwayne K. Brown*, ACMR 9100880, *Military Justice Reporter* 41: 504 (Army Court of Criminal Appeals, 1994).

[52] Info paper, DAPE–Civilian Personnel Law, 14 Jun 91, sub: Civilian Departures from Southwest Asia, Historians files, OTJAG.

[53] Interv, author with Irvin M. Allen, 7 Mar 97, Historians files, OTJAG.

[54] Interv, author with Maj Matthew L. Dana, 28 Jan 97, Historians files, OTJAG. The material in this section and the following section on administrative and operational law and military justice in Iraq is drawn from the interview with Major Dana.

[55] Steven Keeva, "Lawyers in the War Room," *American Bar Association Journal*, Dec 91, p. 52.

6

Judge Advocates in Africa, 1992–1994

> Somalia was a real culture shock. No water, no electricity, no roads, raw human waste and garbage along every road. Then came the reality of children armed with hand grenades and AK–47s, mobs throwing stones, and Somalis tearing wristwatches and sunglasses from the arms and faces of American soldiers.[1]
>
> —*Capt. Karen V. Fair, Command Judge Advocate JTF Support Command and UN Logistics Support Command, Somalia*

In April 1992, after years of civil war and a decade-long drought resulted in the death by starvation of more than half a million Somalis, the United Nations (UN) Security Council established the United Nations Operation in Somalia (UNOSOM). Since its primary mission was providing humanitarian aid, an airlift of food and medical supplies began shortly thereafter. Direct U.S. participation started in July, when President George Bush authorized American air flights for emergency humanitarian relief. From 15 August to 9 December 1992, as part of UNOSOM's Operation PROVIDE RELIEF, C–141 Starlifter and C–130 Hercules aircraft flew daily relief sorties into Somalia during daylight hours. They eventually delivered some 28,000 metric tons of critically needed relief supplies.[2]

Despite the success of PROVIDE RELIEF, the security situation in Somalia grew steadily worse. In November a ship laden with relief material was fired upon in the harbor at Mogadishu, forcing its withdrawal before badly needed supplies could be brought ashore. In the

United States and elsewhere, public distress grew, and on 4 December 1992, President Bush announced the initiation of Operation RESTORE HOPE. Under the terms of UN Resolution 794 (passed the previous day), the United States would both lead and provide military forces to a multinational coalition effort to be known as the Unified Task Force (UNITAF). This temporary force, under the command of U.S. Marine Lt. Gen. Robert B. Johnston, would provide security, restore order, and assist humanitarian organizations in their relief efforts until the situation stabilized enough for the mission to be turned over to a more permanent UN peacekeeping force. The UN mandate called for two important missions: providing humanitarian assistance to the Somali people and restoring order in southern Somalia.

From 9 December 1992 through 4 May 1993, the Unified Task Force in Somalia involved more than 38,000 troops from twenty-one coalition nations, including 28,000 Americans. It succeeded in stabilizing the security situation and was particularly successful in confiscating "technicals," crew-served weapons mounted on trucks and other wheeled vehicles. With better security, more relief supplies were distributed throughout Somalia, ending the immediate threat of starvation in many areas. However, plans for terminating the Unified Task Force and transferring its functions to the permanent peacekeeping force, UNOSOM II, were repeatedly postponed.

On 4 May 1993, UNOSOM II (the "II" distinguished it from UNOSOM "I" activities between July and December 1992) officially began. With a mission significantly different from UNOSOM I, UNOSOM II was tasked with rehabilitating the political institutions and economy of Somalia and building a secure environment throughout the country. These far-reaching goals went well beyond the more limited scope of RESTORE HOPE, as well as those of any previous UN operation. To implement these objectives, a full UN peacekeeping structure was created in Somalia. It was headed by retired U.S. Navy Admiral Jonathan Howe, a special representative of UN Secretary General Boutros Boutros-Ghali. Turkish Lt. Gen. Cevik Bir was selected as commander of the UN Multinational Force, and U.S. Army Maj. Gen. Thomas M. Montgomery was "dual-hatted" as the deputy commander of the UN force and the commander of all U.S. forces.

The ambitious United Nations goals in Somalia particularly threatened the Mogadishu power base of one clan warlord, Mohammed Farrah Aideed. In June 1993, after twenty-four Pakistani soldiers were killed in an ambush by Aideed supporters, the UN Security Council called for the immediate apprehension of those responsible. This led to U.S. forces being employed in a manhunt for Aideed and his chief deputies. After a series of clashes between armed Somali clans and U.S. Army Rangers and other units, a major engagement occurred on 3 October.[3] The Somalis killed eighteen Americans and wounded seventy-five others. Shortly thereafter, President Bill Clinton announced that American participation in UNOSOM II would end on 31 March 1994.

U.S. Army participation in Operation Restore Hope was markedly different from its later involvement in UNOSOM II. Some 12,000 U.S. Army soldiers took part in Restore Hope, most from the 10th Mountain Division (Light) based at Fort Drum, New York. Army participation in UNOSOM II's multinational contingent, however, was primarily logistical in nature, with some 3,000 U.S. personnel committed to that support mission. The United States did, however, provide a Quick Reaction Force—some 1,150 soldiers from the 10th Mountain Division—that operated under the tactical control of the commander, U.S. Forces, Somalia, a position created on 4 May 1993. The participation of these 4,150 American personnel in United Nations Operation Somalia II ended when the last of these troops departed on 28 March 1994.

Organization of Legal Services

As Somalia fell within the responsibility of the commander in chief, U.S. Central Command (CENTCOM), it was his staff judge advocate, Army Col. Walter B. Huffman, who was primarily responsible for formulating legal policy for U.S. forces in Somalia. Thus, Colonel Huffman's staff issued initial guidance on rules of engagement (ROE) to the UNITAF staff judge advocate, Marine Col. F. M. Lorenz. Lorenz then implemented this CENTCOM guidance throughout the force, ensuring that all twenty-one coalition forces participating in Restore Hope operated with the same rules of engagement.

Both the CENTCOM and the UNITAF legal operations were multiservice. Colonel Huffman had 1 Air Force, 1 Marine, 1 Navy, and 1 subordinate Army judge advocate on his staff at CENTCOM headquarters at MacDill Air Force Base, Florida. Colonel Lorenz also had a joint staff. Army Lt. Col. Francis R. Moulin was the UNITAF deputy staff judge advocate, having deployed from Fort Carson, Colorado, to Mogadishu on 11 December 1992. Lorenz also had Marine judge advocate Maj. Walter G. Sharp handling international and operational law issues and Army judge advocate Maj. Sarah S. Greene serving as his claims attorney.[4]

During Operation Restore Hope, the U.S. Army Forces, Somalia, legal operation, headed by Lt. Col. John M. Smith III, provided legal

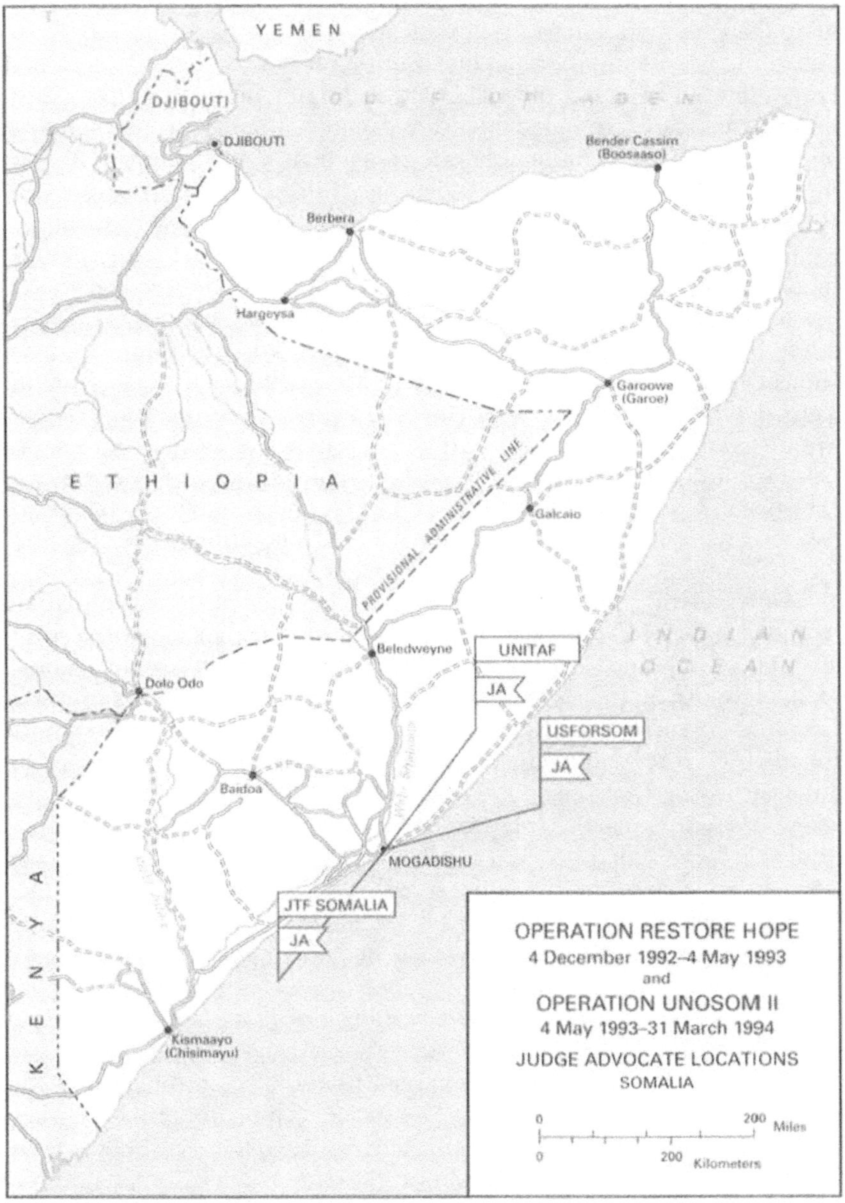
YEMEN
DJIBOUTI
DJIBOUTI
Bender Cassim
(Boosaaso)
Berbera
Hargeysa
Garoowe
(Garoe)
PROVISIONAL ADMINISTRATIVE LINE
ETHIOPIA
Galcaio
INDIAN
OCEAN
Beledweyne
UNITAF
JA
Dolo Odo
USFORSOM
JA
Baidoa
MOGADISHU
JTF SOMALIA
JA
KENYA
Kismaayo
(Chisimayu)
OPERATION RESTORE HOPE
4 December 1992–4 May 1993
and
OPERATION UNOSOM II
4 May 1993–31 March 1994
JUDGE ADVOCATE LOCATIONS
SOMALIA
0
200 Miles
0
200 Kilometers

MAP 10

support to some twelve thousand soldiers on a 24-hour-a-day basis. As Smith was the 10th Mountain Division staff judge advocate, and as more than six thousand of the American soldiers in Somalia were from that unit, it was appropriate that Smith served as the senior Army Forces lawyer in the country. The remainder of Colonel Smith's operation consisted of five judge advocate captains, three of whom handled international and operational law, legal assistance, and claims issues at Smith's headquarters. To ensure that the 10th Mountain Division's 2d Brigade and Task Force MOUNTAIN's artillery brigade received timely legal advice, a judge advocate captain was also located with each of these separate brigades.

Other members of the 10th Mountain Division legal office deploying to Somalia at various times were Maj. Richard E. Gordon, the division's deputy staff judge advocate, who first replaced Colonel Moulin as the UNITAF deputy staff judge advocate and later replaced Colonel Smith as the Army Forces staff judge advocate; Maj. Michael A. Corbin; and Capts. Joseph A. Dewoskin, John M. Bickers, and Edward J. O'Brien. Corbin and Bickers served as command judge advocates to two brigade commanders; O'Brien handled operational law matters; Dewoskin was responsible for legal assistance and claims.

Other Army judge advocates participating in RESTORE HOPE included Capt. Karen V. Fair, the command judge advocate of the Joint Task Force Support Command. This unit of some 4,500 soldiers, commanded by Army Brig. Gen. Billy K. Solomon, had deployed from III Corps at Fort Hood, Texas, to provide logistical and medical support for all U.S. forces in Somalia. Captain Fair, as the command judge advocate from December 1992 to March 1993, dealt with a broad range of legal assistance, claims, operational law, and military justice issues. Joining Fair in Somalia were Maj. John H. Belser and Capt. William C. Peters, both members of the U.S. Army Trial Defense Service. These attorneys deployed to ensure that criminal defense advice was always available; however, they also provided legal assistance for individual soldiers.

As RESTORE HOPE ended and United Nations Operation in Somalia II began, Colonel Huffman; his successor, Army Col. Alexander M. Walczak; and the CENTCOM legal staff continued formulating legal policy for all U.S. troops in the UN peacekeeping force. On 4 May 1993, as UNOSOM II and U.S. Forces, Somalia (USFORSOM), replaced the Unified Task Force and U.S. Army Forces, Somalia, judge advocates like Army Maj. Frank W. Fountain, working at CENTCOM headquarters in Florida, found themselves assisting the USFORSOM staff judge advocate, Lt. Col. Dale N. Woodling. *(Map 10)*

Pfc. Joseph Carter, an Army legal specialist, left, *Navy judge advocate Lt. Richard Zeigler,* center, *and Maj. Richard E. Gordon, UNITAF deputy staff judge advocate, in Mogadishu, November 1993*

An Army judge advocate who deployed to Somalia in March 1993, Woodling first served simultaneously as UNITAF deputy staff judge advocate and staff judge advocate, U.S. Army Forces, Somalia. With the activation of U.S. Forces, Somalia, Colonel Woodling became its staff judge advocate. U.S. Forces, Somalia, was the highest American headquarters, and Woodling's status as the senior judge advocate in theater enabled him to exercise technical supervision over all U.S. legal assets in theater except those belonging to the U.S. Special Operations Command. Consequently, while Navy and Marine judge advocates did not work for Colonel Woodling directly, they coordinated their legal work with him to ensure the provision of uniform legal advice throughout Somalia. Similarly, Major Gordon, who returned to Somalia in October 1993 to take up duties as the staff judge advocate of Joint Task Force SOMALIA, coordinated regularly with Colonel Woodling's successor, Army Lt. Col. Victor L. Horton, who replaced Colonel Woodling in September 1993. *(Chart 10)*

The U.S. Forces, Somalia, staff judge advocate directly supervised the command judge advocate at the UN Logistics Support Command. Thus, Colonel Woodling supervised Captain Fair, who remained in Somalia for the first few weeks of UNOSOM II operations, and Fair's replacement, Capt. Jody M. Hehr. Colonel Horton supervised Capt.

John A. Schill, who deployed from XVIII Airborne Corps in August 1993 to replace Captain Hehr. When Schill left in January 1994, he was replaced by Capt. Walter M. Hudson of the 24th Infantry Division.

Although Colonels Woodling and Horton were the lone judge advocates at U.S. Forces, Somalia, the staff judge advocate of Joint Task Force Somalia supervised a larger organization. Thus, Major Gordon, the task force staff judge advocate from October 1993 to January 1994, had two Army judge advocates working for him. Captain Dewoskin, who had returned to Somalia for a second tour, worked with Major Gordon in the same office. Capt. Charles N. Pede, however, as the command judge advocate of the U.S. Quick Reaction Force, was located at the U.S. embassy compound in Mogadishu. Pede reported directly to Major Gordon, as did Pede's replacement, Captain Peters. Peters, whose first deployment to Somalia had been as a defense counsel, returned for a second tour in January 1994.

Several judge advocates operated outside of these technical chain frameworks. For example, Maj. Gary L. Walsh served as legal adviser to Task Force Ranger in Mogadishu from August to November 1993. Walsh was ideally suited for this position, having deployed as a judge advocate during Operation Urgent Fury some ten years earlier and having spent most of his time as an Army lawyer with special operations units. Walsh provided the full range of legal support to the Rangers and special operations units involved in the ultimately unsuccessful search for Mohammed Aideed. Army Lt. Col. Daniel V. Wright provided legal advice to other special operations units.

While the organization of legal services during Restore Hope worked well, the same cannot be said for United Nations Operation in Somalia II. After May 1993, there were never more than six Army lawyers in theater, yet four separate organizations existed. While there may have been some theoretical merit to having separate legal organizations at U.S. Forces, Somalia, Joint Task Force Somalia, the Quick Reaction Force, and the UN Logistics Support Command, in practice this dispersal of limited legal assets did not work well. In retrospect, a better approach would probably have been the centralization of all legal assets at U.S. Forces, Somalia.

Operation Restore Hope, December 1992–May 1993

In early December 1992, as planning for the deployment to Somalia accelerated, Army judge advocates at the Office of the Judge Advocate General and U.S. Central Command examined the laws governing anticipated American military operations in Somalia. They

CHART 10—LEGAL ORGANIZATION OF ARMY JUDGE ADVOCATES IN SOMALIA, MARCH 1993

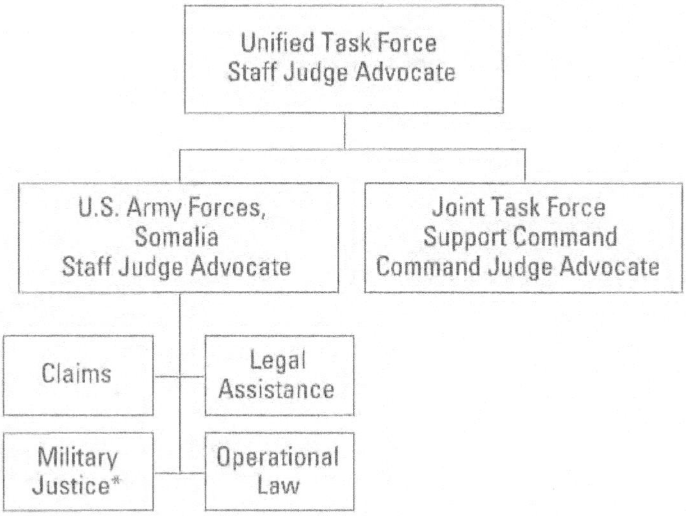

* Trial counsel were located with 2d Brigade, 10th Mountain Division, and Task Force Mountain's artillery brigade.

concluded that the legal basis for RESTORE HOPE was UN Security Council Resolution 794, adopted on 3 December. What was not clear was the impact of customary international law on the conduct and employment of forces in Somalia. As Somalia was a country in chaos, if not anarchy, there was no local law upon which to base standards of conduct for UNITAF personnel. The lack of a functioning Somali government meant that neither the UN nor any individual country could negotiate status arrangements and other forms of agreements with a host government.

Judge advocates also questioned the applicability of the 1949 Geneva Conventions to the UN forces. These forces had not been "invited" to Somalia, nor had they invaded that country. The Somalis were not considered to be an "enemy" or "hostile"; arguably, there were no "belligerents." Thus, except for the applicability of Common Article 3 of the conventions to the various Somali clans, the conventions, as a matter of policy and law, were judged not to apply. Additionally, the Law of Occupation, as established in the Fourth Geneva Convention and in chapter 6 of Field Manual 27–10, was deemed inapplicable to UN forces.

While determining that the Geneva Conventions and Law of Occupation did not expressly apply, judge advocates at the Office of the

Judge Advocate General and at Central Command determined that customary international law did govern the conduct of UNITAF operations. The policy decision to apply only customary law was significant, as it relieved the UNITAF commander of any responsibility to establish a military government in Somalia. This in turn meant that UN forces did not have to attempt to exercise complete authority over localities in southern Somalia. This benefited RESTORE HOPE's mission, as commanders could focus limited resources solely on facilitating the delivery of humanitarian relief, particularly in the areas of providing safe passage and other protections for nongovernmental and official aid organizations.

CENTCOM and UNITAF judge advocates advised, however, that while the inapplicability of the Law of Occupation meant that no significant civil affairs operation was required in Somalia, this fact did not completely relieve the UN forces of legal obligations to the Somali people. On the contrary, in those areas under the actual control of the UN Task Force, or where the activities of UN forces resulted in damage or harm, the American military lawyers found there existed customary international law requirements to provide medical care for affected Somalis and to resolve the claims of those Somalis dislocated from their homes or land, or whose property had been damaged. The UN forces, they asserted, were also responsible for the health and welfare of any Somali that they detained.[5]

While judge advocates at higher command levels worked to articulate the legal basis for RESTORE HOPE and the law applicable to the military operations being undertaken, judge advocates at 10th Mountain Division were more concerned with determining the rules of engagement (ROE) to be used for the upcoming deployment. In approaching this issue, Colonel Smith and the division G–3 conferred with Maj. Gen. Stephen Arnold, the 10th Mountain Division commander, who emphasized that he desired flexible rules that would allow his soldiers to accomplish their mission but also ensure force protection. In accordance with this guidance, Colonel Smith obtained permission from the UNITAF staff judge advocate, Colonel Lorenz, to contact U.S. Central Command directly regarding rules of engagement.

Smith spoke with Colonel Huffman who, remembering that British forces operating with VII Corps units in DESERT STORM had vastly different rules of engagement, determined that all military forces in Somalia would function under uniform rules. Additionally, the rules of engagement for RESTORE HOPE would differ from the type of rules ordinarily used by American forces, as these rules would focus on "conduct" rather than "status." Thus, while wartime rules of engagement like those used in DESERT STORM regulated the use of force based on the status of a readily identifiable enemy, the lack of an enemy or hostile foe in Somalia meant that RESTORE HOPE's rules must necessarily focus on threatening conduct in their regulation of the use of force. Consequently, the CENTCOM guidance provided to Colonel Smith stressed the development of rules of engagement dealing with the use of force based on threatening conduct, especially actions involving the so-called technicals, the vehicles mounted with crew-served weapons so popular with Somali clan members.

Since Colonel Smith and his staff had significant operational law experience, they spearheaded the drafting of rules of engagement for RESTORE HOPE. They closely coordinated their efforts with the 10th Mountain Division's operations section (G–3), with Colonel Lorenz at the Unified Task Force, and with the CENTCOM staff judge advocate. These combined efforts resulted in rules of engagement for the entire Unified Task Force. Colonel Lorenz then arranged for the base printing shop at Camp Pendleton, California, to print 35,000 cards containing these rules. The cards were flown to Somalia and distributed to the marines who went ashore on D-Day, 9 December.[6]

The manner in which rules of engagement were developed for RESTORE HOPE illustrated the practical approach taken toward the development of legal policy on the one hand and the issuance of specific legal guidance on the other. That is, while CENTCOM lawyers were responsible for formulating legal policy based on directives from the Joint Chiefs of Staff, no legal guidance was issued by Central Command or implemented at the Unified Task Force or Army force level absent close coordination between Colonels Huffman, Lorenz, and Smith. Constant coordination among judge advocates ensured that legal policy and guidance were not created in a vacuum, and this led to the smooth implementation of the resulting policy and guidance.

With the approval of rules of engagement for RESTORE HOPE, the 10th Mountain Division judge advocates quickly implemented a training program that ensured that every U.S. soldier deploying to Somalia understood the rules and that they had trained with them. Division lawyers wrote thirteen situation training exercises, each of which emphasized different aspects of the rules. Each vignette was based on a tactical situation that division intelligence officers believed might occur in Somalia. These training exercises were conducted in training lanes using an opposing force that simulated hostile Somali citizens; judge advocate Captains Bickers and O'Brien assisted in this lane training. As a result, every unit of squad size and larger could practice not only the rules of engagement, but also the tactical maneuvers necessary to respond to threats to the force. This process was most successful, both in providing realistic training on the rules of engagement and in identifying problem areas for small unit leaders.[7]

Commanders deploying to Somalia also placed a great deal of importance on establishing guidelines essential to the enforcement of good order and discipline throughout Somalia. At the direction of the UNITAF commander, General Johnston, the UNITAF lawyers drafted and General Johnston approved UNITAF General Order no. 1. Published in electronic message format and distributed through command channels prior to deployment, UNITAF General Order no. 1 was based on the General Order no. 1 issued by U.S. Central Command during DESERT SHIELD. The UNITAF order prohibited the use of alcohol, nonprescription drugs (including "khat," an amphetamine widely used in Somalia), and the taking of captured weapons as "war trophies." With a view toward preventing black marketeering, General Order no. 1 also restricted the sale and barter of some food items. Some exceptions to the order were requested and granted. For example, the restriction on alcohol was lifted for UNITAF personnel in Kenya, as that country was not predominantly Muslim and no local prohibition on alcohol existed. Infractions of General Order no. 1 generally were punished at proceedings held under Article 15, Uniform Code of Military Justice.

With the 10th Mountain Division providing the majority of U.S. troops participating in RESTORE HOPE, a judge advocate from that unit was the first attorney to deploy. This was Colonel Smith, who arrived with the division advance party at the Mogadishu airport on 7 December. A former military police officer with more than fifteen years of experience as a judge advocate, Smith had been the division staff judge advocate only since July. Shortly after joining the division, however, Colonel Smith had deployed with it to Florida, where he participated in relief efforts provided in the aftermath of Hurricane Andrew. Having seen his staff judge advocate in action, General Arnold decided that Smith, despite his brief tenure in the division, should serve as the chief of staff for the 25-man advance party deploying to Somalia a few days before the main body of troops. Thus, in addition to providing legal advice to Brig. Gen. Lawson Magruder, the advance party commander, Colonel Smith supervised and coordinated the work of Magruder's principal staff (G–1 through G–5), as well as the activities of the engineers, communications cell, and security force.

As the rest of the 10th Mountain Division deployed to Somalia over the next few weeks and Unified Task Force and U.S. Army Forces, Somalia, operations expanded in size and scope, judge advocates handled a variety of legal issues. It was challenging work, made even more difficult by the adverse living conditions. Located near the equator, Mogadishu was dry and dusty, with daytime summer temperatures well over 100 degrees. There was no air conditioning. The judge advocates lived in their offices in Mogadishu—sleeping in sleeping bags on cots. The use of mosquito netting was imperative in order to prevent malaria. No one could leave the American compound without his bulletproof vest, Kevlar helmet, and loaded weapon. Additionally, vehicles could leave the compound only in pairs, with guards. This resulted in an exceptionally austere living and working environment, especially as the violence against American and other foreign personnel escalated after June 1993.[8]

During RESTORE HOPE, Army lawyers handled claims, fiscal law, and criminal and administrative law matters. The U.S. Air Force had single-service claims responsibility for Somalia, and an Air Force headquarters in South Carolina exercised this responsibility. Colonel Smith quickly realized, however, that processing claims at such long distances would not work smoothly. Consequently, Smith requested that the Air Force appoint a number of deployed Army judge advocates as foreign claims commissions. The Air Force concurred and appointed both Army and Marine lawyers. While initially granting each commission a $2,500 settlement authority, the Air Force later increased this authority to $12,500. This greatly increased the flexibility of claims operations in Somalia, as UNITAF judge advocates could more quickly investigate and adjudicate a significant number of claims.[9]

Initially, a major claims policy issue existed in the form of whether Somalis who suffered injury or damage from hostile action taken by U.S. forces were proper claimants under the Foreign Claims Act. Because Operation RESTORE HOPE was a humanitarian rather than a combat operation, some felt that any claim for injury or damage was cognizable. Eventually, however, the UNITAF and Army Forces judge advocates concluded that the Foreign Claims Act excluded payment for damages arising out of force protection and other similar security or combat-related activities. Thus, despite the noncombat character of Operation RESTORE HOPE, no claims were paid for damages arising out of security missions conducted by U.S. troops against hostile Somali clans, bandits, and other armed individuals. The type of activity constituting a security mission, however, was narrowly defined, and this ensured the resolution of meritorious claims. Consequently, when Somalis were killed or injured in accidents involving U.S.-operated equipment and vehicles on routine missions, claims were paid. Several claims also were paid for Somalis killed or wounded by the accidental discharge of weapons.

As judge advocates had learned earlier in Vietnam, even when the Foreign Claims Act did not permit the payment of a claim, the payment of solatia was an option. After U.S. Ambassador Robert K. Oakley, the American special envoy to Somalia, approved the concept of such payments to Somalis, the authority to make solatia payments was delegated to the UNITAF commander and chief of staff. This provided U.S. forces with the flexibility to pay solatia in those situations in which a claim otherwise could not be paid, while demonstrating to town elders and the families of victims that the United States took responsibility for the actions of its troops. For example, a solatia payment was made in one combat situation in which compensation clearly was appropriate. In this case, a squad of U.S. soldiers in four HMMWVs came upon a truck while on a night tactical move. Unbeknownst to the Americans, the truck had just been hijacked, and the truck's owner and his family were in the truck. One of the hijackers panicked and fired at the squad. The soldiers returned fire, killing the hijackers and the owner's family. As this was a combat situation, a claim was not payable. The death of the innocent Somali family, however, made a solatia payment appropriate, and one was made. Additionally, a claim was paid for damage done to the truck after the firefight. This occurred when a U.S. tow truck sent to recover the truck damaged it as a result of improper towing. The Somali owner thus received both solatia for combat-related losses and compensation for the noncombat damage that he suffered.[10]

Judge advocates also learned that determining the value of property damaged in Somalia often presented unique problems. Political and economic anarchy in Somalia meant that there was no market to buy or sell many items. In one case, an Army HMMWV collided with a Somali tractor, one of the few still running in the country. Ordinarily, a few estimates from local repair shops would be used as the basis for paying a claim for damage to the tractor. In Somalia, however, repair services did not exist. Thus, it was difficult to determine reasonable compensation for the loss of the tractor—and the services that it could provide. The solution was for a nongovernmental organization to purchase a used tractor and to deliver it to Mogadishu. The Unified Task Force then provided follow-on transport of the tractor, taking it 200 miles south to Kismayo. In this way, the claim was settled.

Other valuation problems arose in "wrongful death" claims. Somali culture placed the value of a working man's life at 100 camels, while a woman's life was valued at only half this amount. Children were valued at even less. Before the central government collapsed, a camel sold for less than $100. At the time that Restore Hope was conducted, however, a camel sold on the black market for $800. As the annual per capita income in Somalia was approximately $500, payment of a claim based on the black-market price of a camel would result in the relatives of an accident victim becoming extraordinarily wealthy. As the Unified Task Force wished to avoid this result, it established an upper payment limit of $10,000, after coordination with the Air Force claims headquarters on this matter. This upper limit was accepted by the Somalis and the village elders who "negotiated" such claims and who, under Somali tradition, received a portion of the money to be used for the benefit of the village.[11]

RESTORE HOPE judge advocates also wrestled with funding issues. While all agreed that Operation and Maintenance (O&M) funds could be used to accomplish the mission, it was not always clear as to what activities were mission related. A question arose, for example, concerning providing fuel to clan factions to assist them in departing Mogadishu and moving back to outlying areas. CENTCOM judge advocates concluded that, as such support directly enhanced the mission to ensure safe and secure passage of humanitarian relief, O&M monies could be used for this purpose. Similarly, the judge advocates advised that O&M funds could be used to train and equip a Somali security force. Although U.S. forces generally are statutorily prohibited from training, equipping, or funding foreign police forces,[12] the prohibition in issue was not viewed as applying to the unique situation existing in Somalia, where no government police force existed. Absent the organization of some form of security force, humanitarian relief operations—the principal mission of the Unified Task Force—could not be effected. Later, UN funding became available to support this project.

While operational law, claims, and fiscal law issues were of critical importance to Army lawyers, their talents in military justice matters also contributed to mission success. On 8 December 1992, acting upon a request from the CENTCOM staff judge advocate, the secretary of defense empowered General Johnston to convene general courts-martial. Thus, from the moment that American troops arrived in Somalia, a criminal jurisdictional framework was in place.[13]

The first significant criminal charges did not arise, however, until 2 February, when Marine Gunnery Sgt. Harry Conde shot a Somali teenager in an incident that captured worldwide media attention. Conde, a 33-year-old radar technician, was a passenger in a convoy traversing central Mogadishu. A seventeen-year-old Somali youth reached through the open window of Conde's HMMWV and snatched his sunglasses from his face. While such an incident was a daily occurrence on the crowded streets of Mogadishu, what happened next was not: Conde fired his buckshot-filled M79 grenade launcher at the youth. The boy was wounded in the abdomen, as was another youth standing nearby. At issue was whether Sergeant Conde had followed the established rules of engagement—firing because he feared for his safety—or whether he had used excessive force in shooting the Somali teenager. This was an important point for, while the shooting itself merited prosecution, the circumstances surrounding it also indicated a violation of the rules of engagement. While it was critical that Americans not hesitate to use deadly force when necessary, it also was clear that undisciplined fire would undermine the success of the humanitarian relief mission if the Somalis—and international public opinion—came to view the Americans as "trigger-happy." Consequently, Sergeant Conde's court-martial was significant, as it signaled to all—UNITAF personnel, the Somalis, and the world—that the Americans were committed to a strict application of rules on the use of force.

Major Gordon, the UNITAF deputy staff judge advocate, assisted in the court-martial of Sergeant Conde. While he did not participate in the courtroom proceedings, Gordon worked with the Marine Corps trial counsel in developing a case strategy, preparing effective direct and cross-examination questions, and organizing evidence. At his trial, Sergeant Conde stated that he had acted in self-defense. However, he was found guilty of assault with a dangerous weapon, sentenced to forfeit $1,706 of his pay, and reduced one grade.[14]

While Major Gordon provided only limited assistance in *United States* v. *Conde*, Army judge advocates participated fully in *United States* v. *Mowris*, another undisciplined fire case. On the morning of 14 February 1993, Spc. James Mowris and his platoon—all military policemen—were conducting a sweep of a Somali village in order to seize weapons and munitions that observers had sighted there. If necessary, the platoon was also to disarm members of one of the Somali bands that had been interfering with international famine relief efforts in the country. After initially sweeping the village and finding a few small arms and live mortar rounds but no armed Somalis, the platoon paused while an interpreter questioned a villager. The platoon leader then noticed two Somali men running between the buildings of a nearby abandoned military compound and ordered the platoon to chase them. In the ensuing chase, as one of the men ran from members of the platoon, the platoon leader and a sergeant fired shots into the air in an attempt to get the Somalis to halt. Specialist Mowris pursued one of the men into a busy area away from the buildings and, after shouting "There he is," fired what he later said was "a warning shot in the dirt" to convince the Somali to stop running. The man was discovered with a bullet hole in his back. He suffered severe blood loss and died while en route to a UN hospital.[15]

At an investigation conducted pursuant to Article 32, Uniform Code of Military Justice, Captains Fair and Bickers served as government representatives, while Captain Peters served as Mowris' defense counsel. After examining the ballistics and medical evidence and hearing testimony, the prosecution team recommended trial by court-martial. As a result, Captain Fair, the JTF Support Command judge advocate, spent more than an hour briefing the JTF Support Command commander, General Solomon, on an appropriate disposition of the case. While recognizing that there was some evidence that Mowris and the members of his platoon had not received adequate training on the rules of engagement, and that this weighed against any finding that Mowris had killed the Somali with criminal intent, she nonetheless expressed the opinion that Mowris had used excessive force in firing his weapon. Under the circumstances, the evidence appeared to support a prosecution for negligent homicide. General Solomon concurred with this advice and recommended to General Johnston, the UNITAF commander, that a trial by court-martial was appropriate. No proceedings took place in Somalia, as Mowris redeployed to Fort Carson. However, after General Johnston's legal staff transmitted the court-martial packet to the commanding general of the 4th Infantry Division, that convening authority referred the charges against Mowris to trial. He was convicted of negligent homicide; however, the convening authority later set aside the conviction.[16]

Finally, in the area of administrative law, Army judge advocates focused on a number of significant issues involving real estate use, weapons seizure and destruction, environmental law, war trophies, detainees, and the disposition of bodies. With regard to taking and using Somali realty, the lawyers advised that it was best to avoid taking and using any private real estate. Rather, it was preferable for the Unified Task Force, Army Forces, and other subordinate units to occupy the former U.S. embassy and to use public property such as the airport, seaport, and university campus in Mogadishu. This avoided claims that would have arisen from the use of private realty and also meant that no Somali civilians were displaced, except for those occupying abandoned government property.

Another administrative law issue involved the seizure and destruction of Somali weapons. As Ambassador Oakley remarked, "There are three things that are most important to a Somali male—his wife, his camel, and his weapon." The Somali people possessed hundreds of thousands of weapons, from handguns and semiautomatic and automatic rifles to crew-served weapons, mortars, and artillery pieces. Because these weapons posed a threat to all those involved in the humanitarian relief efforts, the UNITAF commander, at the urging of Ambassador Oakley, began an "arms reduction" program.

As Security Council Resolution 794 had tasked UN forces with creating a "secure environment for the delivery of relief supplies," Army judge advocates advised that a "guns for cash" trade was within UNITAF's force protection and security mission. This meant that O&M funds could be used in purchasing weapons. Army units first began implementing a "weapons for cash" program in the southern town of Kismayo. In the third week of January, some Marine Corps forces initiated a "guns for food" program, providing receipts to Somalis who turned in weapons. These were exchanged for bags of wheat provided by humanitarian relief organizations. In the end, however, General Johnston decided against implementing a large-scale cash for weapons incentive program. The sheer numbers of weapons involved and the Somalis' reluctance to part with them were two reasons for this decision. Additionally, it was thought that a guns for cash program might actually encourage crime if bandits began stealing weapons to turn in. Such a program also might simply encourage arms dealers to import more weapons into Somalia.[17]

Having determined that a nationwide weapons incentive program was not prudent, General Johnston made the decision to confiscate weapons. Judge advocates were thus asked for advice on the lawfulness of an aggressive weapons confiscation program. They counseled that, under the Law of War, captured military property becomes the property of the "seizing" country. While there was no "war" or "enemy" in Somalia, they nevertheless deemed seizing weapons as necessary for the accomplishment of the UN mission and, consequently, lawful. By February 1993 the Unified Task Force had instituted a comprehensive confiscation program. Military checkpoints inspected all civilian vehicles and, generally, all crew-served weapons were immediately seized. But, as almost every humanitarian relief organization in Somalia used private security guards, a number of these organizations complained that weapons carried by their guards were being confiscated. As a result, the Unified Task Force distributed pink identification cards to these private guards. Such a card permitted the bearer to possess a weapon that otherwise would be confiscated. Later, a blue weapons policy card, in both English and Somali, was issued to UNITAF and relief organization personnel. The card used both words and pictures to explain who could possess a weapon ("employees" of relief organizations), what weapons were prohibited ("crew served, anti-aircraft, anti-armor, and other similar weapons"), how weapons could be carried, and what acts would result in the confiscation of a weapon. The card was a great success as all parties came to understand the applicable rules.[18]

While it might seem surprising that judge advocates faced environmental law issues in Somalia, a number of significant questions did arise. For example, confiscated weapons and ammunition could not be dumped at sea, as this would violate the London Anti-Dumping Convention and the U.S. Ocean Dumping Ban Act. Additionally, Executive Order 12114, Environmental Effects Abroad of Major Federal Actions, requires that major military overseas deployments adhere to the more stringent of local foreign law or U.S. law. Consequently, judge advocates advised UNITAF logistics (J–4) that it had to follow U.S. standards for environmental actions when those procedures did not interfere with mission accomplishment. As an example, engineers attempted using environmentally sound dust suppressants at airfields instead of waste oil.

Judge advocates in every contingency operation in the 1980s and 1990s have advised commanders on the legality of taking souvenirs, or "war trophies," from the theater of operations. Somalia was no exception. As all public property captured by American forces became U.S. property as a matter of law, no individual "right" to a war trophy existed. Thus, when UNITAF General Order no. 1 was issued on 6 December 1992, it provided that "no weapon, munitions, military document, or equipment captured or acquired by any means . . . may be retained for personal use." There were, however, two "limited exceptions": First, "souvenirs, other than weapons or munitions," were allowed to be taken if they had been "legitimately purchased." Second, "abandoned" property of "minimal intrinsic value" could be retained as a souvenir with the approval of the "unit commander." A favorite item in this last category turned out to be a piece of the marble floor of the U.S. embassy in Mogadishu. The floor had been broken by squatters and anarchists during the two years since the embassy had been evacuated, and American troops took pieces of it as mementos.[19]

For judge advocates, the more difficult war trophy issues arose with regard to "unit historical property." Some commanders wished to return captured weapons to the United States, especially the Somali mobile crew-served weapon systems popularly known as technicals. Army judge advocates advised that Army Regulation 840–20 permitted the seizure of such unit war trophies and their return to a unit or post museum for display as historical artifacts. This determination, however, did not resolve the issue—at least in the view of the U.S. Customs Service. Its officials insisted that, as RESTORE HOPE was not a "declared war," no captured weapons carried "war trophy" status.

Instead, weapons seized from Somalis by U.S. troops were said to have the status of "surplus weapons," and these could not enter the United States under the provisions of the Gun Control Act of 1968. In the view of the Customs Service, only the approval of a war trophy program by the Department of Defense would allow for the return of captured weapons as historical artifacts. Questions concerning the correctness of this Customs Service interpretation of the law were rendered moot, however, when General Johnston, the UNITAF commander, decided as a matter of policy that no weapons seized in Somalia would be retained as unit war trophies.[20]

Another issue in which judge advocates figured prominently was in the area of law enforcement, particularly in the formulation of a UNITAF detainee policy. For Restore Hope's humanitarian mission to succeed, the establishment of some degree of law and order was required. Yet Somalia had no police and no prisons, much less a judicial system. This meant that the United Nations Task Force had to create and operate a law enforcement infrastructure. Effective policing would require the detention of Somalis who had committed serious criminal offenses or had attacked UNITAF personnel and whose continued freedom likely would endanger UNITAF forces or innocent third parties.

Judge advocates determined that the legal basis for the UNITAF law enforcement operations—and the authority to implement a detainee policy—flowed from general principles of international law and Security Council Resolution 794. The absence of a government in Somalia made the United Nations Task Force a de facto sovereign, and the inherent powers of a sovereign included the power to arrest and detain wrongdoers in order to protect the population. Additionally, Resolution 794 directed that the Unified Task Force use "all necessary means" to ensure the passage of relief supplies. This logically implied the power to arrest and detain those interfering with or otherwise jeopardizing the delivery of humanitarian goods. Still another legal basis for UNITAF law enforcement operations was the commander's inherent authority to take those actions necessary to protect the personnel under his command and those civilians present in areas under his control.

Using these legal bases, and recognizing that all detainee procedures must adhere to certain due process standards, judge advocates advised that Somalis who committed serious offenses, such as murder and rape, could be detained. Additionally, those who attacked UNITAF forces or posed a future threat to UN troops or innocent Somali citizens also could be detained. UNITAF's Colonel Moulin, working closely with Marine judge advocate Maj. Walter G. Sharp, formulated rules for detainee treatment. These included, among other provisions, that all detentions that were to be in excess of twenty-four hours would require a probable cause determination reviewed and approved by the UNITAF chief of staff. Detainees were to be housed in a facility under the control of the Joint Task Force Support Command, and Captain Fair, as the command judge advocate, was to monitor activities at the detention center in order to ensure compliance with minimum humanitarian standards, particularly the provision of adequate food and medical care for all detainees. When those detained in the UNITAF facility could be transferred to the newly formed Somali law enforcement forces, UNITAF and Army Forces judge advocates were to inspect the Somali facilities, ensure that adequate food was available, and require the Somalis to adhere to minimum standards of humane treatment. Some of these judge advocates later accompanied representatives of the International Committee of the Red Cross during their inspections of these Somali prisons.[21]

United Nations Operation in Somalia II, May 1993–March 1994

When Restore Hope came to an end, the legal issues that had challenged judge advocates in that operation did not. As UNOSOM II's peacekeeping mission got under way, Army lawyers in Somalia faced most of the same questions that had confronted their predecessors in Restore Hope. The staff judge advocate of U.S. Forces, Somalia, provided advice on all legal matters—from military justice and operational law to fiscal, contract, and administrative law. Colonel Woodling, and later Colonel Horton, also coordinated all USFORSOM legal actions with the United Nations, the U.S. Department of Justice, and UNOSOM II headquarters. Initially, the absence of a UNOSOM II staff judge advocate resulted in Colonel Woodling's advising General Bir directly. By the time Colonel Horton replaced Colonel Woodling, however, a Pakistani Army lawyer had been appointed as UNOSOM II staff judge advocate. Nevertheless, Horton continued to work closely with General Bir's staff. In some instances, the UNOSOM II and USFORSOM staff judge advocates worked jointly in responding to legal issues confronting UN operations.[22] *(Chart 11)*

The first important legal issue requiring a decision by UNOSOM II was the drafting of rules of engagement for the new peacekeeping mission. As there was no UNOSOM II staff judge advocate, and as the United States still contributed the largest contingent of UN peacekeep-

CHART 11—LEGAL ORGANIZATION OF ARMY JUDGE ADVOCATES IN UNOSOM II, NOVEMBER 1993

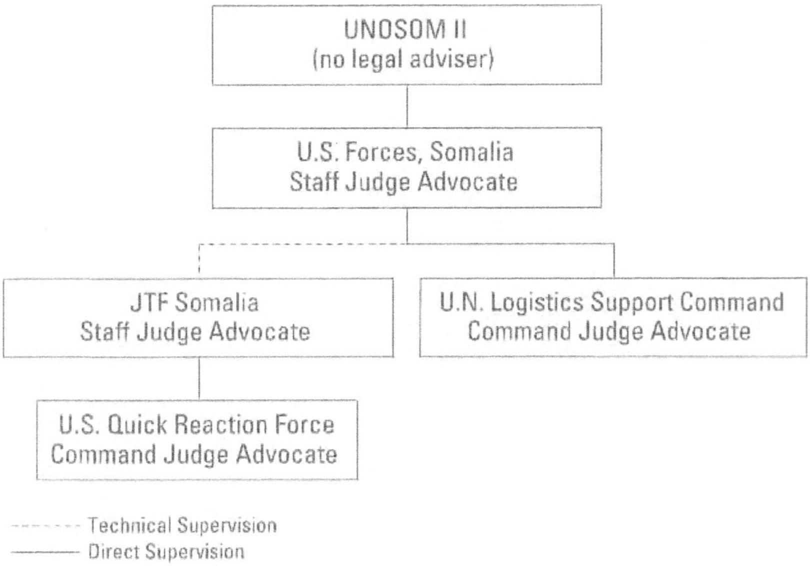

Note: SJA, Task Force Ranger, in theater during UNOSOM II, but not under technical supervision of U.S. Forces Somalia SJA

ers in Somalia, Colonel Woodling coordinated the development of these new rules. He worked closely with Major Fountain at U.S. Central Command and with Maj. Bradley P. Stai, the chief of operations law at XVIII Airborne Corps. The new rules were published by General Bir on 2 May 1993 in a two-page legal appendix to the UNOSOM II Operations Plan. The rules of engagement, like those used in RESTORE HOPE, focused on "conduct" rather than "status." The American lawyers counseled speedy dissemination of the new rules to all UNOSOM personnel. But they recognized that simply distributing copies of the two-page appendix was an impractical approach. Consequently, Major Stai designed a small "ROE card" that reduced the rules of engagement to a practical, easy-to-use format. This card, approved by USFORSOM operations (G–3) on 30 July 1993, was distributed to all soldiers deploying to Somalia.

The card emphasized that nothing in the rules limited the right "to take all necessary and appropriate action to defend yourself and your unit." U.S. Forces, Somalia, were to use the minimum force necessary "under the circumstances." In furtherance of the peacekeeping mission, however, "all necessary force, including deadly force," was permitted "to confiscate . . . crew served weapons," and "to disarm . . . individuals in areas under UNOSOM control." That said, if the tactical situation permitted, USFORSOM personnel were required to shout a warning in either English or Somali or to fire warning shots prior to using deadly force. The ROE card prohibited the use of "unattended weapons," such as booby traps, mines, and trip guns. It also prohibited the seizure of civilian property without command authorization. Finally, while authorizing the detention of individuals interfering with mission accomplishment or committing "criminal acts in areas under UNOSOM control," the rules card reminded all UN forces to treat all persons with dignity and respect.[23]

Although RESTORE HOPE's 28,000 American soldiers were replaced by a smaller force of 4,150 American soldiers for the UN mission, the UNOSOM II rules of engagement initially were not significantly different from those used during RESTORE HOPE. Escalating violence, however, resulted in Fragmentary Order 39, issued by General Bir after coordination with U.S. Central Command. This order stated that "organized, armed militias, technicals, and other crew served weapons are considered a threat to UNOSOM Forces and *may be engaged without provocation*" (emphasis added). This order was a significant departure from the threat-based rules of engagement that had been adopted during RESTORE HOPE. Given the increasingly deadly attacks on UN forces, however, General Bir viewed this modification in the rules on the use of force as necessary. Fragmentary Order 39 was still in effect when the last U.S. soldier left Somalia in March 1994.[24]

Ensuring that all USFORSOM personnel were familiar with the rules of engagement and that they possessed ROE cards for quick reference was only part of the judge advocate mission. The other part—of no less importance—was realistic ROE training. Captains Pede and Fair, command judge advocates at the Quick Reaction Force and Logistics Support Command, respectively, developed realistic scenarios that provided practical guidance to soldiers and marines patrolling the streets of Mogadishu. Scenarios on the use of force, for example, stressed that deadly force was authorized against technicals and snipers. On the other hand, no more than "the minimum force necessary to repel the threat" was authorized against "unarmed mobs." However, if an individual with an AK–47 rifle, in a crowd of civilians, demonstrated "hostile intent" by pointing his weapon at the Americans, deadly force was authorized.

All American soldiers received briefings on the UNOSOM II rules of engagement. At the UN Logistics Command, Captains Fair, Schill, and Hudson continued the practice of intensive training, using "current" fact-driven scenarios. This meant basing ROE training on the "real life incidents experienced by soldiers within the area of operations." Thus, Fair, Schill, and Hudson challenged the soldiers that they were training to imagine a hostile scenario—whether an attack by rock-throwing Somalis or an assault to steal sunglasses or wristwatches—and then apply the rules of engagement to that situation. At the end of each training session, judge advocates emphasized that envisioning such hostile scenarios and methodically applying the rules to them would "ensure no unnecessary taking of human life," as well as guaranteeing that "every soldier came home alive."[25]

Similarly, at the Quick Reaction Force, every newly assigned soldier received an initial briefing on the rules immediately upon arriving at the airfield in Mogadishu. More important, however, refresher training in the rules was provided after each engagement involving Quick Reaction Force troops. In September 1993, for example, after a UH–60 Black Hawk helicopter was shot down by Somalis using a rocket-propelled grenade, a company from the Quick Reaction Force was dispatched to rescue the crew and secure the aircraft. Arriving on the site of the crash, the Americans were ambushed, and they returned fire. When the men returned to the embassy compound, they were highly agitated, not only because they had been attacked, but also because an American riding in the helicopter had been killed when it was shot down. To counteract this stress, Captain Pede immediately briefed the soldiers on the ROE, emphasizing the humanitarian aspects of the mission and the threat-based nature of the rules. This refresher training helped soldiers maintain a proper perspective in a very stressful environment.[26]

In addition to their rules of engagement work, Army lawyers provided other operational law advice. Thus, at the Quick Reaction Force, Capt. Roger C. Cartwright and his successor, Captain Pede, were always present in the tactical operations center during a previously scheduled mission or an unanticipated "event," such as an ambush. Their duty was to maintain the operations log, and this meant summarizing radio messages and orders and recording impact rounds. If a judge advocate was in the center during a mission or event, he could provide immediate legal advice to the commander and "battle captains" who required it. Moreover, Captains Cartwright and Pede were able to record very precise information in the operations log and to prepare a command summary after each operation or event.

As the violence in Somalia escalated, judge advocates were confronted with a variety of complex issues. In several firefights, Somali warlords used women and children as human shields to protect the gunmen engaging UN forces with small-arms fire. Another Somali tactic involved shelling U.S. positions using mortars fired from populated areas. After firing several rounds, these crew-served 60-mm. and 81-mm. mortars, mounted on the rear of Toyota trucks, would simply retreat into garages in populated areas. Finally, upon learning that medical facilities were protected places under the Law of War, the Somalis chose the Benadir Hospital in Mogadishu as a site from which to fire rocket-propelled grenades at American helicopters. In response, judge advocates prepared leaflets advising hospital personnel that they would lose their protected status if this improper use of the hospital continued, and these were airdropped on the facility.

In each of these situations, judge advocates advised commanders on an appropriate response, based on both the established rules of engagement and the principles of military necessity and proportionality. Thus, for example, when the Somalis began illuminating U.S. helicopters at night with searchlights, Army lawyers were asked whether this illumination constituted a "hostile act," permitting deadly force in response. Judge advocates, after considering the UNOSOM II rules of engagement, determined that using deadly force would be lawful and appropriate.

The UNOSOM II judge advocates also continued the work done by their predecessors with Somali detainees, whose numbers grew significantly during the UN mission. Not only did they ensure that all Somalis in custody received appropriate humanitarian treatment, but Colonel Woodling, and later Colonel Horton, personally escorted representatives of the International Committee of the Red Cross during their regular inspections of the detention facility.

In addition to their operational law work involving rules of engagement and detainees, Army judge advocates used their talents to solve operational problems not ordinarily handled by lawyers. Thus, after U.S. military police traveling in a convoy were killed by a land mine at a traffic circle in Mogadishu, Colonel Woodling and his civil affairs counterpart negotiated a right-of-way for a new road that would permit future convoys to avoid this hazardous traffic circle. In exchange for some flour, molasses, and sugar, the Somali owner of the land agreed to the construction of a bypass across his land. When completed, this bypass significantly improved force protection, as UNOSOM II forces controlled access and use of the road. Later, Colonel Horton and Captain Schill conducted similar force protection negotiations. In exchange for food, Somali farmers agreed to move from the perimeters of compounds housing the Americans. In addition to enhancing base security, the removal of the local Somali populace had the added benefit of reducing the theft of U.S. property.[27]

Lt. Col. Victor L. Horton at University Compound, Mogadishu

Army lawyers also crafted an efficient jurisdictional scheme for U.S. Forces, Somalia. On 14 June 1993, the USFORSOM staff judge advocate prepared a memorandum recommending that General Montgomery, the USFORSOM commander and general court-martial convening authority, authorize three special court-martial convening authorities and nine summary court-martial convening authorities in U.S. Forces, Somalia. General Montgomery approved the scheme that same day, and from that time on military justice ran along command lines.[28]

During the United Nations peacekeeping operation, most misconduct was handled administratively or with nonjudicial proceedings under Article 15, Uniform Code of Military Justice. A summarized or company grade Article 15 was typically used for offenses involving disrespect to a superior noncommissioned officer, dereliction of duty, accidental discharge of a firearm, and loss of government property. Drinking alcohol in violation of General Order no. 1, being absent without leave, and using marijuana, however, ordinarily were handled as field grade–level Article 15s.

Only a limited number of courts-martial were conducted, and these involved serious misconduct. In *United States* v. *Brewer*, for example, a soldier whose day-to-day work in Somalia required him to guard vital convoys and supply routes decided to "get out of guard duties" forever. In his own words, he was dissatisfied with the "stupid people" and "didn't want to be an infantryman." As a result, Private Brewer determined that the best way for him to escape service in Somalia was to be injured. He solicited another soldier to shoot him, but the man declined to help. Deciding that he would have to act himself, Brewer "intentionally and purposely" shot himself in the leg with his 9-mm. pistol. This injury did result in Private Brewer's immediate return to Fort Drum, New York; however, it also earned him a general court-martial. Captain Pede prosecuted the case, with Captain Bickers serving as Brewer's defense counsel. At the June 1994 trial, Private Brewer elected to plead guilty to the offense of "malingering in a hostile fire pay zone" and was sentenced by the military judge to a dishonorable discharge and five years' confinement.[29]

Finally, as in RESTORE HOPE, all judge advocates dealt with claims issues. At the Quick Reaction Force, Captain Pede managed all claims arising out of 10th Mountain Division activities, while Captain Fair, and later Captains Hehr, Schill, and Hudson, processed claims at the Logistics Support Command. Pede accepted claims at the front gate of the U.S. embassy, while Fair, Hehr, Schill, and Hudson processed claims at the front gate of the University of Mogadishu.

The construction of "Victory Base" by 24th Infantry Division personnel resulted in the largest number of claims. In late October and early November 1993, following the major engagement in which eighteen Americans were killed, UN forces in Somalia were reinforced by troops and M1 Abrams tanks, M2 Bradley fighting vehicles, and 155-mm. M198 howitzers from the 24th Infantry Division. Building a base camp for these large American armored vehicles and widening the main road and other avenues of approach between Mogadishu and the camp meant seizing land belonging to Somali farmers. Recognizing that these farmers deserved compensation for the destruction of existing crops and for the loss of the use of their land, judge advocates took the lead in assisting local clan leaders in filing claims. Major Gordon and Captains Dewoskin, Pede, and Schill, working closely with civil affairs officers, Somali translators, and clan leaders, adjudicated some fifty claims. In some instances, claims were "paid" through a barter system, whereby Somali farmers received flour, sugar, molasses, lumber, and other similar material instead of U.S. currency. Fair, prompt, and equitable claims adjudication not only demonstrated to the Somalis that the United States assumed responsibility for the actions of its soldiers, it also promoted force protection, as Somalis living near Victory Base were friendly toward the Americans in their midst.[30]

Left to right: *Capts. Charles N. Pede and Joseph A. Dewoskin, Col. Edward Frothingham III, and Maj. Richard E. Gordon aboard the aircraft carrier USS* America *off the Somali coast, November 1993*

In adjudicating claims, Somali claimants received the benefit of the doubt. A significant problem existed, however, when a claim for damage to a single piece of property was filed by more than one person, as there existed no method for verifying who actually owned the property. In these situations, judge advocates used a common-sense approach in arriving at a fair solution. In firefights in which innocent Somalis were killed, Americans continued the practice of providing a payment of between $3,000 and $5,000 per individual—applying the earlier noted standard based on the value of a camel.[31]

By the time U.S. forces redeployed from Somalia, the claims program administered by the judge advocates was so successful that the UNOSOM II headquarters in Mogadishu requested that Colonel Horton draft a claims policy for use by UN forces. The American policy of paying Somalis for noncombat damage had promoted such goodwill that the UN forces remaining in Somalia feared that the lack of some mechanism on their part to pay claims would have a negative impact. Using Army Regulation 27–20, *Claims*, as his guide, Colonel Horton authored a claims policy for use by the United Nations, and this was forwarded to New York for approval.[32]

Conclusion

Judge advocate involvement in Operation Restore Hope was unlike any other deployment. The expedition itself was unprecedented, as there existed no Somali government to request intervention—or object to it. Additionally, despite the open warfare present throughout Somalia, the UN Security Council Resolution made clear that Restore Hope was a humanitarian operation. Thus, there existed no "armed conflict" in Somalia as a matter of international law, and American soldiers deployed to Somalia for the principal purpose of assisting a multitude of private organizations.

The humanitarian purpose underlying Restore Hope and UNOSOM II meant that judge advocates applied the law in support of military operations designed to accomplish humanitarian goals. Yet the anarchy in Somalia also meant that Army lawyers operated in an environment in which the rule of law had been replaced by the law of the gun. Thus, judge advocates also applied the law in ways that assisted commanders in identifying the threat, protecting the force, and minimizing collateral damage.

In Restore Hope and UNOSOM II, judge advocates once again had deployed thousands of miles from home in support of Army operations. Those deploying with 10th Mountain Division would learn that the experience gained in Somalia would be needed sooner, rather than later. In a matter of months, Army units from Fort Drum were in Haiti spearheading Operation Uphold Democracy.

Notes

[1] Interv, author with Maj Karen V. Fair, 16 Dec 96, Historians files, OTJAG.

[2] Kenneth Allard, *Somalia Operations: Lessons Learned* (Washington, D.C.: National Defense University Press, 1995), p. 13.

[3] On Sunday, 3 October 1993, about 100 members of Task Force RANGER were sent on a mission to capture two top aides of warlord Mohammed Farrah Aideed. The mission, which was supposed to last about an hour, went terribly awry: the Americans were ambushed by Somali gunmen, and two Black Hawk helicopters were shot down. Eighteen soldiers were killed and dozens more badly wounded. The U.S. Quick Reaction Force mounted a successful rescue of the survivors. An infantry platoon leader participating in that rescue mission, 2d Lt. Timothy J. Ryan, would later serve in the Judge Advocate General's Corps. For the story of the events of 3–4 October, see Mark Bowden, *Black Hawk Down: A Study of Modern War* (New York: Atlantic Monthly Press, 1999).

[4] F. M. Lorenz, "Law and Anarchy in Somalia," *Parameters* (Winter 1993–94): 28.

[5] Memo, Ofc of General Counsel, Dept of Defense, for Secretary of Defense, 5 Dec 92, sub: Legal Authority for Somalia Relief Operations; After Action Report, SJA, 10th Mtn Div, n.d. [1993], sub: US Army Legal Operations, Operation RESTORE HOPE, 5 December 1992–5 May 1993, p. 3, both in Historians files, OTJAG.

[6] Interv, author with Col John M. Smith III, 23 May 96, Historians files, OTJAG; Lorenz, "Law and Anarchy in Somalia," p. 29. For the text of the JTF Somalia rules of engagement, see Appendix B.

[7] After Action Rpt, SJA, 10th Mtn Div, p. 31; interv, author with Smith, 23 May 96.

[8] Interv, author with Maj Karen V. Fair, 3 Mar 98, Historians files, OTJAG.

[9] Interv, author with Smith, 23 May 96.

[10] After Action Rpt, SJA, 10th Mtn Div, p. 17.

[11] Ibid., p. 16.

[12] Title 22, U.S. Code, sec. 2420.

[13] Memo, SJA, CENTCOM, for CJCS Legal Counsel, 7 Dec 92, sub: General Courts-Martial (GCM) Authority for the Commander, JTF Somalia, Historians files, OTJAG.

[14] Interv, author with Lt Col Richard E. Gordon, 3 Feb 98, Historians files, OTJAG; Mark S. Martins, "Rules of Engagement for Land Forces: A Matter of Training, Not Lawyering," *Military Law Review* 143 (1994): 64; Keith B. Richburg, "Marine Is Convicted in 'Sunglasses' Case," *Washington Post*, 7 Apr 93, p. A20.

[15] *United States* v. *Mowris*, GCM no. 68 (Fort Carson and 4th Inf Div [Mech], 1 Jul 93).

[16] See Martins, "Rules of Engagement for Land Forces: A Matter of Training, Not Lawyering," p. 18; Peter Sleeth, "A Guilty Verdict Thrown Out in Somali Death," *Denver Post*, 5 Oct 93, p. A5.

[17] After Action Rpt, SJA, 10th Mtn Div, pp. 21–24.

[18] Ibid., p. 22; Memo, UNITAF Commander, n.d., sub: Identification and Weapons Policies, Historians files, OTJAG; Lorenz, "Anarchy and Law in Somalia," pp. 31–32.

[19] After Action Rpt, SJA, 10th Mtn Div, p. 24; Msg, Commander, Joint Task Force Somalia, 6 Dec 92, sub: JTF Somalia General Order no. 1, Historians files, OTJAG.

[20] After Action Rpt, SJA, 10th Mtn Div, pp. 24–25; Memo, J. F. Rhinebeck, U.S. Customs Advisor Somalia, for Lt Col Spataro, Provost Marshal CJTF Somalia, 25 Feb 93, sub: Captured Firearms (Surplus Military Weapons), Historians files, OTJAG.

[21] After Action Rpt, SJA, 10th Mtn Div, p. 26.

[22] Memo, SJA, USFORSOM, for Ch, Operations, UNOSOM II, 15 Dec 93, sub: UNOSOM II After Action Report, p. 2, Historians files, OTJAG.

[23] Memos, Ch, Operations Law, USFORSOM, for ACofS, G–3, USFORSOM, 30 Jul 93, sub: Rules of Engagement (ROE) Card for Operations in Support of UNOSOM II; 2 May 93, sub: Rules of Engagement (ROE) Card Based on Appendix 6 to Annex C to UNOSOM II OPLAN 1; 2 May 93, sub: Appendix 6 to Annex C to UNOSOM II OPLAN 1. All in Historians files, OTJAG. For the text of the approved rules of engagement, see Appendix B.

[24] Allard, *Somalia Operations: Lessons Learned*, p. 37.

[25] Interv, author with Fair, 16 Dec 96; Karen V. Fair, "The Rules of Engagement in Somalia—A Judge Advocate's Primer," *Small Wars and Insurgencies* 8 (Spring 1997): 107–26.

[26] Interv, author with Maj Charles N. Pede, 24 Feb 98, Historians files, OTJAG.

[27] Ltr, Col Victor L. Horton to author, 20 Jan 99, Historians files, OTJAG.

[28] Memo, SJA, HQ, USFORSOM, for Cdr, USFORSOM, 14 Jun 93, sub: UCMJ Jurisdictional Scheme for U.S. Forces, Somalia, Historians files, OTJAG.

[29] *United States* v. *PV2 Cody M. Brewer*, Army 9401039, unpublished opinion, 23 May 95.

[30] Interv, author with Lt Col Richard E. Gordon, 9 Apr 98, Historians files, OTJAG.

[31] Interv, author with Lt Col Dale N. Woodling, 25 Jun 96, Historians files, OTJAG.

[32] Ltr, Horton to author, 20 Jan 99, Historians files, OTJAG.

7

Judge Advocates in Haiti, 1994–1996

My experiences in Haiti were absolutely incredible ... the most memorable event was when, while providing security in front of the Presidential Palace, I shook the hands of both Presidents Clinton and Aristide . . . this was a moment in history ... seeing the leader of the wealthiest nation of the hemisphere alongside the leader of the poorest nation in the hemisphere.[1]

—Maj. Catherine M. With
Judge Advocate, 25th Infantry Division
and UN Mission in Haiti (1995)

On 16 December 1990, Reverend Jean-Bertrand Aristide captured an overwhelming majority of votes to become the president of Haiti. After assuming office on 7 February 1991, the new president instituted a major reorganization of the army. Unhappy with this change, the army staged a coup d'etat on 30 September 1991. Aristide was forced into exile, and Lt. Gen. Raoul Cedras and a military junta instituted a dictatorship.

In June 1993, after the Cedras regime had rebuffed a series of diplomatic efforts to restore Aristide to power, the United Nations Security Council declared an oil and arms embargo on Haiti. This prompted a change of heart in Haiti, and in July General Cedras traveled to Governors Island, New York, where he signed an agreement with Aristide. Among the provisions of this Governors Island Agreement were pledges by Cedras to exercise "his right to early retirement" and to allow Aristide's return to Haiti on 30 October 1993.[2]

Within weeks, however, it was clear that the military dictatorship was not going to honor the Governors Island Agreement. On 11 October, as part of the plan for Aristide's return to Haiti, some 200 U.S. troops arrived in Haiti aboard the USS *Harlan County*. After a confrontation with a small group of gunmen, the troops and the ship hastily left Haiti and returned to the United States. In response to this episode and to two days of violence instigated by this same group of gunmen, the United Nations Security Council adopted Resolution 873, which renewed sanctions against the Cedras dictatorship. All member states, acting either nationally or through regional organizations, were to halt all maritime shipping inbound to Haiti for inspection for arms, military and police supplies, and petroleum and, if necessary, to order the diversion of vessels transporting embargoed goods. By 19 October U.S. and Canadian naval vessels and aircraft were enforcing the embargo. President Clinton also froze assets and revoked the visas of officials in the military regime. These efforts, however, did not bring quick results; at the end of 1993 Aristide was still unable to return to Haiti.[3]

U.S. and UN efforts to restore democracy in Haiti were soon overshadowed by a more serious threat: a mass exodus of Haitians. Haitians had been fleeing their homeland by boat since 1992. U.S. policy at that time had been for U.S. Coast Guard vessels to intercept these migrants on the high seas and return them to Haiti. Despite a legal challenge to this policy brought by human rights activists, this remained U.S. policy until the first half of 1994. However, when members of the military junta orchestrated an increase in politically motivated intimidation and repression against Aristide supporters on 8 May 1994, President Clinton announced that the United States would hear claims for asylum from Haitian boat people. This provoked a tremendous flood of Haitian migrants. Once again, the United States changed its policy: Haitian migrants would be returned to Haiti or taken to "safe havens" in Guantanamo Bay Naval Base in Cuba, in Panama, and elsewhere.

By July 1994 it was evident that international sanctions alone would not restore democracy in Haiti. It also was apparent that the exodus of Haitians by boat would continue to grow as long as the Cedras regime remained in power. Consequently, on 31 July 1994, the United Nations Security Council approved Resolution 940, which mandated a multinational force, led by the United States, to use "all necessary means" to remove the military junta, return Aristide to power, and establish a secure and stable environment in Haiti. On 15 September, President Clinton announced on television that the United States would use military force to oust Cedras from power. On 17 September, in a final attempt to persuade the junta to step down, President Clinton dispatched former President Jimmy Carter, retired General Colin L. Powell, and Senator Sam Nunn to Haiti. The next day, in the very hour that paratroopers from the 82d Airborne Division were flying toward drop zones in Haiti, Cedras and his colleagues agreed to relinquish power. But while an invasion of Haiti was no longer necessary, the United States did not trust the military dictatorship to keep its promise to restore Aristide as Haiti's president. Consequently, U.S. forces entered Haiti in large numbers on 19 September 1994. This was D-Day of Operation UPHOLD DEMOCRACY.[4]

Given the fluid political situation in Haiti, U.S. war planners had not known whether American forces would conduct a "forced" or "semi-permissive" entry into Haiti. Consequently, they drafted two alternative plans. The first plan, Operation UPHOLD DEMOCRACY, provided for forced entry and would be executed by Combined Joint Task Force 180. This force, under the command of Lt. Gen. Henry H. Shelton, Commanding General, XVIII Airborne Corps, consisted primarily of Army paratroopers who would airdrop into Haiti while some 1,800 marines conducted an amphibious landing. The second plan, Operation MAINTAIN DEMOCRACY, provided for semi-permissive entry and would be executed by Combined Joint Task Force 190. Maj. Gen. David C. Meade, Commanding General, 10th Mountain Division (Light), headed this force of one aviation and two infantry brigades.[5]

In the end, Operation UPHOLD DEMOCRACY was a blend of both plans, as the forced entry operation was already under way when former President Carter announced that General Cedras would step down. Admiral Paul D. Miller, the commander in chief of the U.S. Atlantic Command, quickly halted the forced entry, organized Combined Joint Task Force 190 as a subordinate command to Force 180, and ordered a semi-permissive entry. General Shelton promptly recalled the 82d Airborne Division to Fort Bragg and directed Joint Task Force 190 to land at Port-au-Prince airport. The two brigades of the 10th Mountain Division executed this modified Operation UPHOLD DEMOCRACY on 19 September; by nightfall some 2,000 soldiers had been ferried to Haiti by helicopter from the USS *Eisenhower*.

The next day another 3,000 soldiers from the 10th Mountain Division deployed in Port-au-Prince, while about 1,800 marines launched an amphibious landing onto Cap Haitien from the USS *Wasp*. Over 10,000 American troops were ashore by D+2, 21 September, and this number increased to nearly 21,000 by 4 October, when the first soldiers from other coalition countries arrived. Eventually, about 3,600

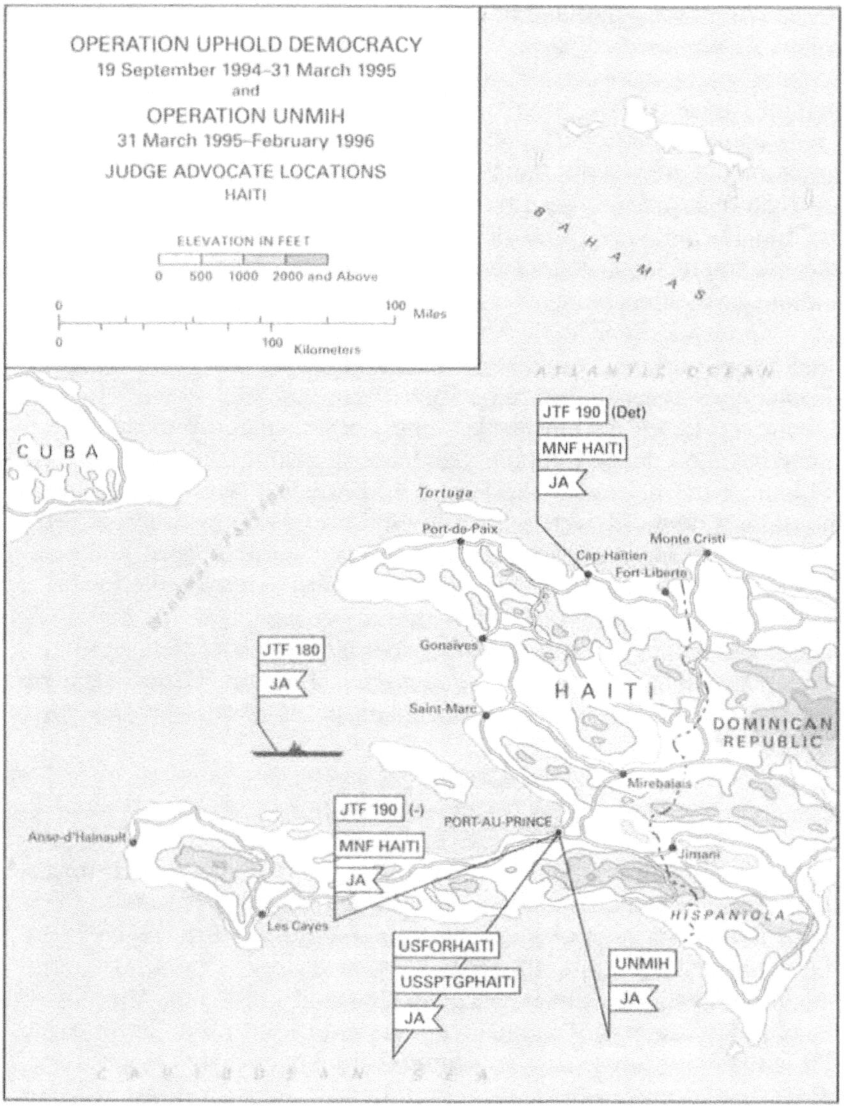

MAP 11

personnel from Argentina, Bangladesh, Belgium, The Netherlands, Bolivia, and other nations participated as part of the UN-mandated Multinational Force (MNF) Haiti.

Cedras soon resigned and left Haiti, and President Aristide returned on 15 October. Some ten days later Task Force 180 ceased its operations, handing over command of Multinational Force Haiti to the commander of Combined Joint Task Force 190. In mid-January 1995, the 25th Infantry Division deployed from Hawaii to replace the 10th Mountain Division in the MNF.

A few months later, on 31 March, UPHOLD DEMOCRACY ended and Multinational Force Haiti transitioned to a peacekeeping force, the United Nations Mission in Haiti Force (UNMIH). Reflecting the broad international support for the deployment, some thirty-three nations would eventually contribute men and women to this military force, including the United States, which contributed approximately 2,400. Army Maj. Gen. Joseph W. Kinzer became the commander of the UN military force and was tasked with maintaining the stable and secure environment established during UPHOLD DEMOCRACY.

From April 1995 to June 1995 some 1,300 soldiers from the 25th Infantry Division donned the blue berets of the United Nations Force. In June these troops returned to Schofield Barracks, Hawaii, and were replaced by soldiers from the 2d Armored Cavalry Regiment from Fort Polk, Louisiana. Then, in September 1995, a smaller number of soldiers from the 101st Airborne Division (Air Assault) replaced these Fort Polk–based troopers. Although U.S. military participation in the UN peacekeeping force officially ended in February 1996, a smaller group of American soldiers comprising the U.S. Support Group, Haiti (USSPTGRP-Haiti), continued working on public works projects in Haiti through 1997.[6] *(Map 11)*

Organization of Legal Services

As General Shelton commanded Combined Joint Task Force 180, it made sense for Colonel Altenburg, now the staff judge advocate of the XVIII Airborne Corps, to assume duties as Combined Task Force 180 staff judge advocate. Joining Altenburg's staff were Comdr. Joseph Callahan, a Navy judge advocate who deployed from Norfolk, Virginia, and served as deputy staff judge advocate, and Capt. Joseph P. Bialke, an Air Force judge advocate who deployed from Minot Air Force Base, North Dakota. Callahan brought with him Law of the Sea expertise; Bialke was an expert on Air Force–related legal issues. Other judge advocates in Joint Task Force 180 included Majs. Bradley P. Stai and Kyle D. Smith and Capts. Peter G. Becker, Margaret Baines, Allan D. Berger, James A. Martin, and Kerry L. Erisman. Although the Joint Task Force 180 judge advocates necessarily engaged in some legal work in Port-au-Prince, Colonel Altenburg's judge advocates operated primarily from offices aboard the USS *Mount Whitney*—probably the first time in Army history that an Army staff judge advocate had operated from a naval vessel.

The Combined Joint Task Force 190 commander, General Meade, naturally chose Lt. Col. Karl K. Warner, the staff judge advocate of the 10th Mountain Division, as his staff judge advocate. Warner, calculating that he would need one judge advocate for every 4,000 to 5,000 soldiers, decided that he would deploy with the headquarters, along with an operational law attorney, claims judge advocate, and legal assistance officer. Since Task Force MOUNTAIN, a separate command and the largest Army component of Joint Task Force 190, would also require legal support, Warner assigned his deputy staff judge advocate, Major Gordon, as its command judge advocate. He also decided to deploy another judge advocate, Capt. Cheryl Bullard, with Task Force MOUNTAIN. This would provide Gordon with an assistant who could handle any military justice matters that arose. In addition to these six judge advocates, Army lawyers deployed as legal advisers with the 10th Mountain's 1st and 2d Brigades. Consequently, the original "legal package" for Task Force 190 consisted of Warner, Gordon, and six captains, two of whom were attached to brigades.

By the time Joint Task Force 180 dissolved in late October and Joint Task Force 190 and the 10th Mountain Division had grown to become the U.S. element of Multinational Force (MNF) Haiti, Colonel Warner's legal operation had expanded significantly. By the first week of December 1994, nine subordinate commands had judge advocates. *(Chart 12)* At Task Force MOUNTAIN, Major Gordon continued to serve as the command judge advocate. Capt. Christopher B. Valentino was the legal adviser for the 1st Brigade Combat Team, located in Port-au-Prince. Capt. Edward J. O'Brien was at 2d Brigade in Cap Haitien. Capt. Thomas J. Barrett was the lone judge advocate at the 10th Aviation Brigade, while Capt. James M. Patterson was with the Special Operations Task Force. Captain Erisman, who had initially deployed as part of Task Force 180, now provided legal advice to the 16th Military Police Brigade and the Joint Interrogation Facility, while Lt. Col. Arthur L. Passar, a contract law specialist, served as the staff judge advocate at the Joint Logistics Support Command. The Army lawyer at the 20th Engineer Brigade was Captain Martin, who, like Captain Becker at the 625th Military Intelligence Brigade, had earlier arrived in Haiti as part of Task Force 180. Capt. Darryl R. Wishard, who had deployed with Warner as part of Task Force 190's advance party on 19 September, continued to handle operational and criminal law matters at the task force headquarters. Capt. Joseph A. Dewoskin dealt with claims, and Capts. Krista B. Edgette and Sean K. Howe focused on

Chart 12—Organization of Legal Services in Uphold Democracy, December 1994

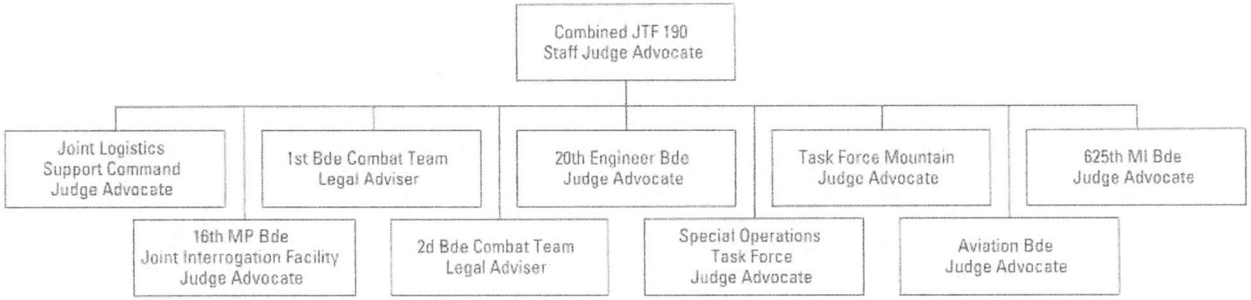

legal assistance, claims, and administrative law. Finally, Capt. Nicholas J. Lorusso provided legal assistance, as well as legal support to the Joint Detention Facility.

The 25th Infantry Division (Light) deployed from Hawaii to Haiti in early January 1995. On 20 January 1995, Maj. Gen. George A. Fisher, the division commander, replaced General Meade as the Multinational Force Haiti commander, and Fisher's top lawyer, Col. Brian X. Bush, replaced Colonel Warner as the MNF staff judge advocate. Also deploying from Hawaii were Maj. Mark P. Sposato and Capts. Kenneth E. Patton, John P. Coakley, Fred K. Ford, and Catherine M. With. Sposato assumed duties as the deputy staff judge advocate, and Patton served as the legal adviser for the 2d Brigade, headquartered in Port-au-Prince. Captain Coakley was the legal adviser for the 3d Brigade at Cap Haitien, while Captain Ford worked legal assistance and claims issues at the MNF headquarters, along with Captain With, who handled operational and administrative law matters.

The U.S. Army Trial Defense Service also sent counsel to Haiti during Operation UPHOLD DEMOCRACY. While 10th Mountain Division troops were deployed, Capt. Norman F. Allen, normally located at the 82d Airborne Division, provided defense counsel support in Haiti. Later, he was replaced by Capt. John M. Bickers, the senior defense counsel at 10th Mountain Division. Capt. Judith L. Camarella, also from the Fort Drum Trial Defense Service office, traveled to Port-au-Prince for a short period while representing one soldier as an individually requested defense counsel. With the deployment of the 25th Infantry Division, Capt. Steven E. Engle deployed from Fort Lewis, Washington, where he was serving as a trial defense counsel. Engle provided defense support until he returned to the United States upon the redeployment of the division.

In late February 1995, slightly more than a month before the transition of UPHOLD DEMOCRACY to the United Nations Mission in Haiti, Army Maj. William A. Hudson, Jr., deployed from Fort Bragg to assume duties as the UN Mission in Haiti Force legal adviser. For the next six months, until August 1995, Hudson headed a two-man legal operation composed of himself and a Canadian judge advocate. As the UN Force legal adviser, Major Hudson was General Kinzer's principal legal adviser on United Nations issues. This included advising the UN Mission staff on UN rules and procedures, orders and directives promulgated by the UN Force commander, and any matters of a legal or political-legal nature arising in the context of the force's mission. Major Hudson was not, however, the lone Army judge advocate in Haiti; Colonel Bush had selected Captain With to remain behind when the rest of the 25th Infantry Division's judge advocates returned to Hawaii. With had donned the distinctive UN blue beret and now served as command judge advocate at U.S. Forces Haiti (USFORHAITI), the American component of the UN Force. As its sole judge advocate, Captain With was the legal adviser to General Kinzer on U.S. contingent issues. But, as With would also serve as command judge advocate, U.S. Support Group, Haiti, she also advised its commander, a U.S. Army brigadier general, on legal issues revolving around United States–Government of Haiti bilateral assistance projects.[7]

Joining Hudson and With was Capt. Gregory G. Woods, a defense counsel deployed to Haiti by the U.S. Army Trial Defense Service. The presence of Woods, who replaced Captain Engle, ensured that defense services would remain available to all American troops wearing blue berets or otherwise participating in peacekeeping and humanitarian operations in Haiti. Unlike Major Hudson and Captain With, however, Captain Woods' defense counsel mission meant that he provided legal advice to individual soldiers, rather than to commanders. Woods did, however, assist his two Army judge advocate colleagues in providing legal assistance.

Major Hudson remained as the UN Force legal adviser until August 1995. Before Hudson redeployed, however, Captain With returned to Hawaii with the last remaining elements from the 25th Infantry Division. She was replaced at U.S. Forces Haiti by Capt. David Dahle, who deployed from Fort Polk, and at U.S. Support Group, Haiti, by Capt. Devin A. Walker. In August, Army Maj. Mark S. Ackerman, deploying from the Office of the Judge Advocate General, Washington, D.C., replaced Hudson as a "blue hatter" at UN headquarters. Like Hudson, Ackerman also had a Canadian Army lawyer as his deputy force legal adviser. And, like Major Hudson, Ackerman provided a full range of legal advice to the force commander, General Kinzer. Ackerman's return to the United States on 2 March 1996 ended the presence of Army lawyers in humanitarian and peacekeeping operations in Haiti.

Judge Advocate Operations in Uphold Democracy

The first Army lawyers in Haiti were Colonel Altenburg from Task Force 180 and Colonel Warner and Captains Valentino and Wishard from Task Force 190, all of whom arrived in Port-au-Prince on 19 September. These judge advocates, and those who followed them over the next weeks, faced austere conditions. The climate was harsh—100-degree temperatures and near 100 percent humidity—and the physical demands daunting. Those landing at the airport at Port-au-Prince carried a sixty-pound rucksack and also had a sixty-pound "A" bag, weapon, protective mask, Kevlar vest, and Kevlar helmet. It was a difficult mile and a half walk with this gear to an industrial park, where the judge advocates set up their living and working areas in stifling hot buildings. Battle dress uniforms quickly turned white from the salt stains of perspiration. There was no fresh water, and food consisted of Meals, Ready-to-Eat. Bottled water and hot meals were not available until mid-October. Of course, every soldier in Task Force 190 faced these same physical challenges.[8]

Army lawyers provided a full range of legal support in UPHOLD DEMOCRACY, with the most important legal issues falling into four categories: operational law, law and order, contract and fiscal law, and claims. Without exception, their legal work reflected the new role of judge advocates in the Army: working closely with operators to achieve mission success by providing practical, lawful solutions to both legal and nonlegal problems. Like Operations JUST CAUSE, DESERT SHIELD, DESERT STORM, and RESTORE HOPE, legal activities in Operation UPHOLD DEMOCRACY again demonstrated that the practice of operational law within the Army was firmly in place.

Operational Law

Army attorneys began work on the Haiti operation some seven months prior to the deployment. At that time, Major Stai, the chief of operational law at XVIII Airborne Corps, began drafting rules of engagement and a legal appendix to an operations plan. Stai's work, coordinated with Lt. Col. Carl Woods, a Marine Corps judge advocate at U.S. Atlantic Command, became the basis for UPHOLD DEMOCRACY'S rules on the use of force. In the forced entry plan, the rules declared the Haitian armed forces and national police a "hostile force." Consequently, the Americans were to treat these forces as hostile and "attack [them] . . . until neutralized, destroyed or captured." Moreover, soldiers could "presume that civilians in public, armed with crew-served weapons, automatic weapons, or rifles" were members of the Haitian military or "paramilitary groups" and therefore could treat them as hostile.[9]

While Stai and the staff officers working on UPHOLD DEMOCRACY planned for forced entry, judge advocates at the 10th Mountain Division drafted alternative rules of engagement for Operation MAINTAIN DEMOCRACY, the code name for the permissive entry operation. Starting in July 1994, Captain Wishard spearheaded the effort to draft these rules of engagement. In contrast to Stai's forced entry rules,

Wishard's rules declared no forces hostile. Rather, his rules permitted the use of force only in response to hostile acts or indications of hostile intent. Thus, the rules expressly stated that "no forces have been declared hostile" and that no offensive military operations, such as raids or assaults, could be conducted without Combined Joint Task Force 190 approval. Consequently, using force against "members of the Haitian military, police, or other armed Haitians" was permitted only if they "commit[ted] hostile acts or show[ed] hostile intent." After the rules were approved, a card reflecting these rules was distributed to each 10th Mountain Division soldier in early September. Just as at Task Force 180, ROE training was provided for Task Force 190 soldiers. At Fort Drum, an Army lawyer assisted each company commander and another company grade officer in conducting ROE training for each company. Veterans of the Somalia deployment learned these new rules, as well as those 10th Mountain troops who had been in uniform for only a short time. It was a massive training effort, and it continued even when the combat teams were aboard Navy vessels en route to Haiti.[10]

With the transformation of the forced entry mission into a permissive entry operation, only one set of rules of engagement survived—those approved for Task Force 190's operations plan. Unfortunately for the soldiers who had trained on these rules—and who were now arriving in Haiti on D-Day believing that they understood the command guidance on the use of force—the rules of engagement were modified at the last minute to deal with a practical and political dilemma that confronted American soldiers immediately upon arriving in Haiti—the lack of express guidance in the rules for resolving Haitian-on-Haitian violence.

On 18 September, D minus 1, judge advocates and planners at U.S. Atlantic Command and the 10th Mountain Division realized that establishing a secure and stable environment in Haiti would be difficult unless American personnel could use force against persons committing serious criminal acts. While there was no wish that soldiers and marines police the streets of Port-au-Prince, the Joint Chiefs of Staff did approve a change to the rules of engagement, allowing the use of force against Haitians responsible for civil unrest. While the U.S. Atlantic Command quickly transmitted this new guidance to Task Force 180, and the latter relayed the information to Task Force 190, new cards containing these modified rules of engagement were not distributed until 21 September. The revised rules were welcomed by all, however, especially after several episodes in which Haitian police and militia brutally beat demonstrating Aristide supporters, followed by news reports that American troops had not intervened to stop the bloodshed. Judge advocates ensured that troops received training in the new rules of engagement and understood that they were to detain Haitians committing serious criminal acts such as homicide, aggravated assault, rape, arson, and robbery.[11]

While the rules of engagement applicable to U.S. troops in Haiti did not change again during UPHOLD DEMOCRACY, soldiers continued to receive briefings and training on these rules throughout the deployment. The emphasis was on situational training exercises, as judge advocates counseled that requiring soldiers to apply rules of engagement in realistic scenarios was the best way to translate abstract concepts into practical lessons. At the 10th Mountain Division, for example, soldiers were confronted with a vignette in which a speeding vehicle crashed through a traffic checkpoint barrier. A judge advocate assisting with the training evaluated the soldiers' response and discussed alternative responses available within the limits set by the rules of engagement. Similarly, 25th Infantry Division judge advocates like Major Sposato and Captains Ford and With briefed deploying soldiers on the rules of engagement, arranged for the printing and distribution of ROE cards, and drafted training vignettes for use by "Tropic Lightning" personnel.[12]

Significantly, scenario training in the rules of engagement was not confined to U.S. troops. Army judge advocate Capts. Thomas N. Auble and Michael A. Newton, for example, developed a human rights training package that they used to instruct non-U.S. troops in the Law of War generally, and the rules of engagement in particular. This instruction, conducted at Camp Santiago, Puerto Rico, was provided to Combined Caribbean, Bangladeshi, and Guatemalan soldiers, as well as to a group of international civilian police officers—all of whom subsequently deployed to Haiti as part of MNF Haiti. Auble and Newton began their training with a twenty- to thirty-minute lecture, and then followed up with carefully planned lane training. Training scenarios included a "riot," in which a hostile group of civilians threw rocks, bottles, and sticks at the soldiers, and an "incident," in which three armed "Haitians" threatened relief workers at a Red Cross food distribution site. In both vignettes, the soldiers being trained were tasked with applying the rules of engagement in controlling the rioters or protecting the humanitarian relief workers.

Captains Auble and Newton also created and distributed a pamphlet entitled "Ten Commandments of Human Rights for Soldiers," used during the training provided the Combined Caribbean Force (CARICOM). This publication discussed, in clear and easily understood language, "ten commandments" that soldiers must obey. These ranged from "thou shall respect individual integrity and human dignity" to "thou shall not commit, nor tolerate, murder, rape, torture, or the excessive use of force." The last page of the pamphlet contained two pocket-size cards on which the ten commandments were printed, thus providing a ready reference for future use.[13]

As a result of this training in Puerto Rico, soldiers and police from Jamaica, Grenada, Barbados, Guatemala, St. Vincent, and other countries not only became familiar with the rules of engagement, but practiced them in realistic settings. Consequently, when these personnel arrived in Haiti, they not only understood the rules, but had some experience in achieving the balance between initiative and restraint so important to success in operations other than war.[14]

While most Army judge advocates practicing operational law dealt with rules of engagement and related issues, commanders also looked to judge advocates for advice and counsel on operational matters that were more political than legal. Thus, very late on Sunday evening, 18 September, after Task Force 180's airborne assault into Haiti had been canceled and last-minute planning for a peaceful entry the next morning was under way, General Shelton spoke with Colonel Altenburg. He advised Altenburg that the national command authorities had directed him to meet with General Cedras and his colleagues. With this meeting scheduled to occur at 9:00 the next morning, Shelton told Altenburg to accompany him to the meeting. Additionally, General Shelton tasked Colonel Altenburg to prepare an outline for the meeting that would include statements that Shelton should make to Cedras.[15]

This meeting was extremely important. While General Cedras and his colleagues had agreed—once again—to relinquish power peacefully, no one could be sure that this promise would be honored. Consequently, the next morning's soldier-to-soldier meeting between General Shelton and General Cedras could well determine whether American troops would enter Haiti unopposed or face armed resistance. In any event, as the United States was committed to deploying large numbers of forces to Haiti the next morning, General Shelton sought a meeting that, while demonstrating U.S. resolve, encouraged the Cedras regime to accept a peaceful entry.

Working most of the night, Colonel Altenburg and his deputy, Commander Callahan, produced a talking paper. The document consisted of three parts. The first provided General Shelton with an outline of U.S. authority for deploying troops to Haiti, including a mission statement based on UN Security Council Resolution 940. The second section of the document was in the form of a statement that Shelton would make to Cedras, explaining what U.S. troops intended to do in Haiti. The third and final part of the talking paper contained taskings for Cedras. In drafting this document for the meeting, Colonel Altenburg ensured that its tone communicated the fact that General Shelton was not requesting permission from the junta to enter Haiti and conduct operations. In describing the event some weeks later, Altenburg recollected: "We weren't asking for anything; we were just telling [Cedras], This is what we are going to do, 'A' through whatever. . . . And this is what you're going to do. . . . it wasn't [General Shelton] asking General Cedras to please do something; it was telling him this is what you're going to do."[16]

When General Shelton met with General Cedras in Port-au-Prince at 10:30 A.M. on 19 September, he had Altenburg's talking paper on the table in front of him. Colonel Altenburg was also present and, anticipating that Cedras might invoke some provision of the Haitian constitution or Haitian law during the meeting, Altenburg was prepared to advise Shelton on a response to any legalistic arguments. There was no such discussion, however. On the contrary, while treating Cedras with dignity and respect, General Shelton quickly took charge of the situation.

The D-Day meeting at Haitian Army headquarters was a complete success. Colonel Altenburg, while never called upon to give legal advice at the conference, did contribute something else. During the meeting, General Cedras pointed his finger at Altenburg and the cloth scuba badge that he wore on his battle dress uniform. Cedras, who also was scuba-qualified, wanted a badge like Colonel Altenburg's. Consequently, when Altenburg returned to the USS *Whitney* some hours later, he talked a young sailor into giving up his metal scuba badge—a badge that General Shelton formally presented to an appreciative General Cedras a few days later.[17]

Perhaps because of his success in this nontraditional role, Colonel Altenburg also was called upon to explain UPHOLD DEMOCRACY's rules of engagement to the news media. Thus, on D+1, at a Port-au-Prince press conference arranged by the Joint Task Force 180 information bureau, Altenburg addressed an international group of reporters for some forty-five minutes. His clarification of the rules governing the use of force and his answers to a variety of questions resulted in favorable newspaper, television, and radio reports over the next several days. More important, Colonel Altenburg's performance demonstrated the appropriateness of using a judge advocate as a subject matter expert in the public affairs arena.[18]

Law and Order

While U.S. troops entered Haiti without firing a shot and captured no prisoners of war, it was clear within a few days that certain Haitians posed a threat to the Americans, as well as to their fellow citizens. At Task Forces 180 and 190, Colonels Altenburg and Warner both advised that Security Council Resolution 940, combined with the inherent authority of Generals Shelton and Meade to protect their forces, constituted more than enough legal authority to detain Haitian troublemakers. Additionally, they advised that, as existing Haitian jails and prisons were neither humane nor reliable locations to house these detainees, the Americans could lawfully detain them in a U.S.-operated facility.

Some American officers proposed detaining Haitians in the brig aboard the USS *America*. But this idea was rejected, as it would have tied the naval vessel to Haitian waters and would have used space required to confine sailors facing courts-martial. Additionally, this arrangement would make visits by representatives of the International Committee of the Red Cross (ICRC) difficult. Finally, while detainees would not be considered prisoners of war, and were not legally entitled to such status, a policy decision had already been made that U.S. forces would treat Haitian detainees as having a status "equivalent to" prisoners of war. The fact that the 1949 Geneva Convention Relative to the Treatment of Prisoners of War (GPW) required internment "only in premises located on land" provided yet another reason to avoid using the *America*'s brig.[19]

The Americans, instead, made a decision to locate a detention facility at the Light Industrial Complex in Port-au-Prince. Operations at the Joint Detention Facility began on 30 September. A military police company commander was placed in charge of the facility, and his unit provided the manpower necessary to run its daily operations. A small military intelligence cell operated in the facility as well, and its several interrogation teams gathered intelligence from the detainees.

As Task Force 190 was responsible for the detention facility, its judge advocates assumed the lead in ensuring that humane treatment and due process were afforded to all detainees. Colonel Warner and Major Gordon, aware of the lessons learned by 10th Mountain troops in detaining Somalis during RESTORE HOPE, designed a procedure for detaining Haitians. As a matter of policy, they determined that detainees should be afforded the same treatment accorded detained persons under the 1949 GPW. Such treatment included decent clothing, an examination by a medical doctor and a dentist, and adequate food in the form of the Meals, Ready-to-Eat, consumed by U.S. soldiers. As a practical matter, these standards of treatment resulted in some problems: poor living conditions in Port-au-Prince meant that some Haitians preferred detention to freedom, and a number of Haitians admitted that they committed minor criminal acts hoping to be caught and detained.[20]

A list containing the name of each detainee was maintained and updated daily. A judge advocate conferred with each detainee on the list and, while not acting as a defense counsel, provided the detainee with an opportunity to offer any reason why he should be released from the facility. This information was relayed to Colonel Warner. At the same time, a second Army lawyer, who acted as legal adviser to the facility, obtained information from law enforcement and intelligence personnel as to why a particular Haitian should be detained. Thus, for example, Captain Lorusso, after discussions with a detained Haitian, might advocate his immediate release. On the other hand, Captain Becker, representing the command, would present the views of police and intelligence personnel concerning a particular Haitian detainee and often argued that continued detention was necessary. After hearing from both Lorusso and Becker, Colonel Warner would brief General Meade. After Meade had heard the substance of the arguments presented by both sides and Colonel Warner's recommended course of action, he would determine who merited release and who should continue to be detained. Between forty and fifty individuals were held in the detention facility, and, while Meade and Warner made no written record of their discussions, General Meade made daily decisions regarding detainees.[21]

The Joint Detention Facility was an unqualified success. Its operations protected the force and, because detaining those who would harm their fellow Haitians also enhanced law and order generally, it also aided Aristide's return to power. Additionally, the procedural safeguards implemented by Army judge advocates stood in stark contrast to Haiti's legacy of arbitrary and sometimes brutal detention. This demonstrated to the Haitian people that the law could be a source of good, rather than a tool of oppression. The ICRC stated publicly that the Joint Detention Facility adhered to the highest standards of humane treatment. Later, when some members of the media joined relatives of detainees in criticizing detention facility operations, Red Cross personnel spoke out in the facility's defense.[22]

A "cash for guns" program was another law and order success achieved through significant judge advocate involvement. Judge advocates advised that the Law of War permitted the confiscation of weapons belonging to the Haitian armed forces, police, and paramilitary organizations. But most recognized that seizing publicly owned firearms and explosives would prove to be inadequate in terms of creating a stable and secure environment; weapons would also have to be collected from those Haitians who simply had no legitimate need for them. Based on their knowledge of similar programs implemented in JUST CAUSE and RESTORE HOPE, the Task Force 180 and 190 staffs established a weapons buy-back program. In September and October, psychological operations soldiers informed Haitians that the Americans would pay from $100 per handgun to $600 for large-caliber machine guns. Fixed collection points were later supplemented with "roving weapons collection teams," as this facilitated the collection of weapons from Haitians who were unable or unwilling to turn in weapons at a fixed location. By early January confiscation and purchase activities had resulted in the collection of more than 15,000 weapons and explosive devices. By 31 March, when Uphold Democracy ended and the United Nations Mission in Haiti began, more than 33,000 rifles, handguns, shotguns, heavy weapons, and explosives had been recovered.[23]

The Joint Detention Facility and weapons buy-back and control program were two methods of promoting a safe and secure environment in Haiti. A third was the American effort to assist the Aristide government in restructuring the Haitian criminal justice system—to build a competent police force, mentor judges and other court officials, and improve conditions in prisons. Again, as with the detention facility and weapons control program, Army lawyers played a significant role.

As U.S. law generally prohibits American military forces from training and equipping foreign police forces, the International Criminal Investigation and Training Assistance Program of the U.S. Department of Justice spearheaded efforts to build a professional, corruption-free, Haitian police force. Judge advocates, however, were instrumental in advising that firearms obtained from Haitians—through confiscation, purchase, or other means—could be transferred to the new Haitian police force. Army lawyers also assisted commanders and military police in planning and executing operations to quell the vigilante violence that resulted when the new Haitian police force initially began to assume its duties while still inadequately prepared to fight crime. Finally, judge advocates advised military police engaged in investigating and tracking Haitian criminals during the train-up period for the Haitian police force.

Army lawyers also served as "judicial mentors," assisting Haitian judges and court officials in revitalizing the Haitian criminal justice system. By the second week of Uphold Democracy, rudimentary efforts were under way, as Colonel Warner and his staff made their first contacts with the Haitian courts in Port-au-Prince. In one instance, the American lawyers introduced themselves to a Haitian magistrate who advised them that she had wanted to leave her position, but had been informed by the Ministry of Justice that she would be imprisoned if she did so. The Haitian official agreed to stay on after judge advocates "worked with her, told her about the [American] system of justice, brought her in to view

[a] court-martial, and discussed [possible improvements to] the Haitian system." These same judge advocates also made periodic visits to the Palace of Justice in Port-au-Prince, where they observed appellate proceedings. Later, Colonel Warner remarked that "many Haitian lawyers credited our simple presence [there] and discussions with judges with giving their system legitimacy with the Haitian people, [and also] showing U.S. respect for Haitian institutions."[24]

Based on recommendations from Colonel Warner and others, a more formal judicial mentoring program was established in February 1995. Active component judge advocates, such as Army Lt. Col. Philip A. Savoie, and Army Reserve judge advocates, like Maj. Michele H. Altieri, spent many hours working as members of the Team of Ministerial Advisors to Haiti. They assessed the Haitian justice system by making on-site evaluations of more than 175 justices of the peace, 15 prosecutors, 15 investigating judges, and over 100 civil registrars. These judge advocates also audited the skills of court personnel, examined court records, inventoried supplies, surveyed the caseload distribution, and evaluated the scheme of compensation for judicial officers. This assessment led to a number of recommendations. Chief among them was a recommendation that the Ministry of Justice establish a court security program and renovate a number of dilapidated courthouses. The lawyers also advocated the creation of a supervision program that would audit judicial processes, investigate corruption complaints, monitor training, and develop a code of judicial ethics.

Virtually all of these recommendations were implemented to some degree. More than 5,000 copies of the Haitian Constitution were printed and distributed, along with 200 sets of legal codes containing Haitian laws. More than 25,000 legal forms were created, reproduced, and distributed to justices of the peace and others, as were some 200 manually operated typewriters. Based on an earlier recommendation, the Team of Ministerial Advisors in Haiti also planned and coordinated the transformation of the military academy into the proposed national judicial training center. Judge advocate mentoring of Haitian justices of the peace, officials at the Ministry of Justice, and other criminal justice system personnel undoubtedly contributed to the initial success of UPHOLD DEMOCRACY. First, it increased Haitian awareness of the role of law in a democracy. Second, it provided a positive image of U.S. intent in Haiti. And third, it enhanced popular support for the Haitian government. Finally, the mentoring encouraged the hope that Haiti would develop a lasting democracy.[25]

Of course, law and order issues were not restricted to Haitians. Good order and discipline also required the handling of misconduct by

U.S. members of the Multinational Force. General Order no. 1, modeled after similar orders used during earlier deployments, was promulgated at both Task Forces 180 and 190. Prohibited activities included possessing, using, or selling privately owned firearms or ammunition; using alcohol; gambling; taking war trophies; and "removing, possessing, selling, defacing, or destroying archeological artifacts or national treasures." For the first time, both general orders also contained provisions that, while not punitive, placed "further restrictions" on the activities of American troops in Haiti. In Task Force 180, for example, there existed a "restriction" on adopting any animal as a pet or mascot, and at Task Force 190 personnel were restricted, for safety reasons, from providing food directly to civilians. For practical reasons, commanders did not want to make certain conduct criminal, but did want to discourage it.[26]

Minor violations of General Order no. 1, as well as minor offenses committed in violation of the Uniform Code of Military Justice, ordinarily were handled by nonjudicial punishment. Courts-martial were reserved for serious offenses. Ironically, the most high profile court-martial to evolve from UPHOLD DEMOCRACY began as a nonjudicial proceeding. However, after the accused, Capt. Lawrence P. Rockwood, refused a nonjudicial determination of his guilt or innocence, and instead demanded trial by court-martial, charges against him were referred to a general court.

An Army counterintelligence officer assigned to the 10th Mountain Division, Rockwood had deployed to Haiti and was serving at Task Force 190 headquarters in Port-au-Prince. On the evening of 30 September 1994, Captain Rockwood was scheduled for duty as the senior officer in charge of a counterintelligence cell. A perimeter wall surrounded the secure compound that included the headquarters, and security guards prevented anyone from leaving the area without an escort. Rockwood, armed with a loaded M16 rifle, avoided the security guards by jumping over the perimeter wall. He then hitchhiked about six kilometers to the National Penitentiary, where Haitian authorities remained responsible for the prisoners, and demanded entry.[27]

After learning that Captain Rockwood was making an unscheduled appearance at the prison, the military attaché at the U.S. embassy, an Army major, went to the prison in order to prevent an altercation. Rockwood was insubordinate to the attaché. He also alleged that President Clinton's televised speech of 15 September had provided him with the authority to prevent human rights abuses. Approximately two hours later, the attaché succeeded in calming Captain Rockwood, convinced him to unchamber the round in his weapon, and persuaded him to leave the prison.

Rockwood was charged with failure to go to his place of duty at Task Force 190 headquarters on 30 September, violation of an order not to leave the compound without a proper convoy, dereliction in the performance of his duty while at the hospital ward to which he was taken after leaving the prison, disrespect to a lieutenant colonel whom he confronted after leaving the hospital, disobedience to this same individual when he repeatedly had ordered Rockwood to "stop talking" and to "lower his voice" during the confrontation, and conduct unbecoming an officer and gentleman for the entire course of events leading up to his departure from the prison.

At his general court-martial, held at Fort Drum, New York, Rockwood defended his conduct on the grounds of justification and duress. He argued that, as President Clinton had announced at the beginning of Uphold Democracy that the primary objective of the operation was "to prevent brutal atrocities against Haitians," he was justified in leaving his place of duty in order to inspect conditions in the Port-au-Prince National Penitentiary. Captain Rockwood also asserted that his conduct resulted from duress, in that he "had no choice" but "to act," as "lack of such action would have meant an acquiescence on [his] part to the imminent and ongoing human rights violations, hypocrisy in the face of duty."

On 14 May 1995, the court members rejected Captain Rockwood's affirmative defenses and found him guilty of all charges but the two pertaining to convoy procedures. It sentenced him to dismissal and total forfeiture of all pay and allowances. The case was widely reported in the news media and was the subject of much newspaper, radio, and television commentary. Although some groups, including the American Civil Liberties Union, trumpeted Captain Rockwood as a humanitarian, the prevailing opinion was that Captain Rockwood, regardless of his intentions, had been wrong to disobey orders and to substitute his own moral judgment for that of the command.

Only one court-martial was held in Haiti. This was *United States* v. *Pacheco*. It involved a 10th Mountain Division soldier who, while guarding a warehouse full of weapons, stole a pistol from that warehouse. Military judge Col. Keith H. Hodges was deployed to Haiti and presided over a general court-martial that found Spc. Eric B. Pacheco guilty of larceny of a .357-caliber Desert Eagle pistol and of dereliction of duty. Pacheco was sentenced to be reduced to the grade of private (E–1), to forfeit all pay and allowances, to be confined for six months, and to receive a bad conduct discharge. Capt. E. J. O'Brien, the trial counsel in the case, commented that Pacheco's trial in Port-au-Prince had a visible and favorable impact on discipline in the command.[28]

Contract and Fiscal Law

Unlike troops engaged in combat operations, in which the Law of War generally governs the battlefield acquisition of property required for the mission, American forces procuring supplies for use in an operation other than war generally must comply with statutes and regulations governing military acquisition. Fortunately, when urgent operational needs warrant immediate action, these statutes and regulations provide for expedited procurement without "full and open" competition. Moreover, just prior to the initiation of UPHOLD DEMOCRACY, procurement laws were amended to allow for the use of simplified acquisition procedures for purchases up to $200,000 in contingency operations. As a result, judge advocates in Haiti found that existing laws and regulations presented no significant impediments to expedited procurements.[29]

From the beginning of UPHOLD DEMOCRACY, lawyers with special skills in contract and fiscal law provided expert advice to both Task Forces 180 and 190. During the first few days, contract and fiscal law matters were handled by Army Reserve Maj. Michael L. Larson. Larson, an Army civilian employee at Fort Bragg, possessed invaluable real-world experience because of his work as a contract attorney in Saudi Arabia during Operation DESERT STORM. Consequently, he was activated and deployed to Port-au-Prince. Larson did not remain in Haiti for long, however, as a joint logistics support element had gone ashore with the first units on D-Day. Accompanying the unit was William Harbour, a civilian attorney from Pine Bluff Arsenal, Arkansas, who served as the element's legal adviser until 23 October, when all logistics support functions were assumed by the Joint Logistics Support Command. Harbour then became the command's legal adviser, remaining in that position until he was replaced on 1 December by Lt. Col. Arthur L. Passar. Passar, designated as the staff judge advocate, advised Brig. Gen. Julian A. Sullivan, Jr., on a wide range of legal matters. However, given the primary mission of the support command, the vast majority of Colonel Passar's legal advice pertained to contract and fiscal matters. Capt. Marilyn L. Fiore, who replaced Passar in February 1995, provided the same type of advice until she returned to the United States in March. Finally, Captain With provided legal advice on contract and fiscal law matters to the command element throughout and following the transition to United Nations Mission in Haiti.

The primary mission of the lawyers at the Joint Logistics Support Command was to be the provision of legal advice to support units providing contingency contract support for UPHOLD DEMOCRACY. As a practical matter, however, the existence of the Logistics Civil Augmentation Program (LOGCAP) meant that there was a greatly reduced need for contingency contracting in Haiti. This resulted from the fact that, under the auspices of LOGCAP, the Corps of Engineers had earlier awarded a large contract to Brown and Root Services Corporation. This company was required to provide basic life support—sanitation, shelter, food, and laundry services—to troops deployed in contingency operations. Under the terms of the generic contract applicable during 1994 and 1995, Brown and Root agreed to receive and support 1,300 troops per day within fifteen days of notification of a deployment, and 20,000 troops within thirty days. Consequently, within two weeks of D-Day civilian employees of Brown and Root were in Haiti providing food and potable water, building showers, electrifying buildings, and delivering other logistical support.

While the support command was not responsible for supervising Brown and Root's performance or otherwise administering the contract, its judge advocates monitored contractor activities to ensure that U.S. forces, as the "customer" or beneficiary of the contract, were receiving high quality, responsive services. Harbour, Colonel Passar, and Captain Fiore also explained what could—or could not—be accomplished under the program and often assisted in remedying problems that arose during its administration. Colonel Passar was instrumental in suggesting, developing, and implementing a plan through which the support command could evaluate Brown and Root's performance under the contract. Until this plan was in place, there existed no method by which the Corps of Engineers could obtain General Sullivan's assessment of Brown and Root's performance of its responsibilities in Haiti, despite the fact that its performance was critical to determining the "award fee" that the firm would receive for its overall contract performance. Although the contractor's involvement in planning for and executing missions in other geographic areas meant that its award fee would not be based solely on its performance in Haiti, this operation was far and away the most significant. Consequently, Colonel Passar's plan for formal input into the fee-determination process was an essential tool for ensuring high quality performance.[30]

Related contract law issues concerned the scope of the LOGCAP contract. For example, Colonel Passar advised that the contract included transporting Haitian workers hired by the Joint Logistics Support Command. Passar's contract law expertise also assisted in ensuring that the United States met certain international law obligations. When a local Brown and Root representative declined to provide rations to

Haitian detainees in the Joint Detention Facility on the grounds that this was outside the scope of the contract, Colonel Passar intervened immediately. He insisted that, given the legal requirement that all detainees receive humane treatment, this meant that U.S. resources must be used to meet this obligation. But Passar did not simply issue a legal opinion stating that providing Meals, Ready-to-Eat, to the Haitian detainees was within the scope of the contract; he ordered the issuance of the rations and advised the recipient of his order that he would assume responsibility for any arguable contract scope infractions. The contracting officer administering the contract subsequently agreed with Colonel Passar's actions, but Passar's timely, on-the-spot legal advice illustrated the need for the involvement of experienced contract lawyers in operations such as UPHOLD DEMOCRACY.[31]

The largest contract negotiated in Haiti—and one that involved intensive judge advocate participation—concerned the services of linguists and interpreters. As there were too few Creole-speaking soldiers, civilian interpreters were hired. Again, Colonel Passar provided invaluable assistance. Initially, he assisted in drafting the scope of work. The scope and estimated cost of the contract led the contracting officer to conclude that it could be awarded using simplified acquisition procedures. It became apparent after the receipt of proposals, however, that the contract cost would far exceed the $200,000 limitation on the use of the simplified procedures. Consequently, Colonel Passar advised that full and open competition had in fact been obtained in soliciting the proposals, allowing the action to be converted into a regular negotiated procurement. Passar also recommended that certain clauses be included in a solicitation amendment, and in the contract itself, to ensure that the procurement of interpreter services could be completed expeditiously and properly.

Just as the nature of the deployment to Haiti did not change the way in which contracts were negotiated for goods and services, UPHOLD DEMOCRACY did not result in Congress' enacting special legislation providing the president—or the heads of military departments—with the express authority to spend appropriated funds as they deemed necessary for the operation. This resulted in existing fiscal constraints remaining applicable to the expenditure of such monies during UPHOLD DEMOCRACY. As with procurement issues, the challenge for judge advocates was to ensure that the legal advice provided on fiscal law matters facilitated mission accomplishment to the greatest possible degree.

Fiscal law issues centered around the expenditure of either operation and maintenance or military construction appropriations. As Congress intended for these monies to fund the military's daily operations, maintenance, and construction needs—and not to assist a foreign country in rebuilding or improving its infrastructure—Army lawyers ensured that no Multinational Force contract deviated from this congressional intent. Thus, judge advocates advised that no appropriated monies could be spent to build a basketball court for non-U.S. military personnel. Nor could Defense Department funds be used to provide supplies for the Department of Justice training of the new Haitian police force. Additionally, absent some military purpose, monies could not be used to improve roads or other similar Haitian infrastructure. Consequently, while Colonel Passar concurred in a road improvement project involving a main supply route critical to the overall Army mission, Army appropriated funds could not be spent on other types of road work. Looking for alternatives that would support the humanitarian goals of UPHOLD DEMOCRACY, judge advocates advised that other appropriated monies, such as the Economic Support Fund administered by the U.S. Agency for International Development, could be used to build bridges, pave roads, and accomplish similar projects.[32]

Claims

Claims operations in Haiti were unlike those of previous deployments in a number of respects. First, the short duration of UPHOLD DEMOCRACY meant that, in contrast to longer and larger deployments such as DESERT SHIELD and DESERT STORM, substantially fewer claims for losses of soldiers' property were filed. Similarly, the rarity of hostile encounters in UPHOLD DEMOCRACY meant that judge advocates in Haiti never had to resolve the issue faced by their colleagues in Vietnam, Grenada, and Panama: compensating claimants for damage resulting from American combat operations when U.S. law generally prohibits the payment of such claims.

But while judge advocates processed fewer claims from soldiers and did not wrestle with the problem of combat-related claims, UPHOLD DEMOCRACY still demonstrated that efficient claims operations contribute to the success of a mission. As usual, anticipating claims requirements was the key to success. Thus, prior to deploying from Fort Drum, Colonel Warner arranged for the appointment of four 10th Mountain Division judge advocates as one-member foreign claims commissions. At his request, the U.S. Army Claims Service also appointed three additional division judge advocates as a three-member commission. As a result, Haitian claims could be adjudicated at Task Force 190 from D-Day on. Later, as the 25th Infantry Division replaced the 10th Mountain Division, Colonel Bush arranged for an identical number of his judge advocates to be appointed as one-member and three-member foreign claims commissions. At both divisions, Colonels Warner and Bush determined that they and their deputy staff judge advocates, Majors Gordon and Sposato, should be appointed to the three-member commissions. This ensured that experienced judge advocates were involved in high-dollar claims adjudication.

Additionally, after 22 September 1994, when the Department of Defense designated the Army as the single-service claims authority for Haiti, the foreign claims commissions at the 10th Mountain Division—and later at the 25th Infantry Division—functioned as the exclusive mechanisms for providing U.S. compensation to Haitian claimants. As a practical matter, this meant that most Haitians seeking compensation for personal injury, death, or property loss came to the front gate of "Camp Democracy" at the Light Industrial Complex in Port-au-Prince. However, while most claims were received in this way, brigade legal advisers in Port-au-Prince and Cap Haitien also received and adjudicated claims stemming from incidents for which their units were responsible. By August 1995 Haitians had filed some 295 claims and had received more than $175,000 in compensation.[33]

Three claims issues were of particular importance in UPHOLD DEMOCRACY: claims for compensation for wrongful death, exaggerated claims, and misdirected claims. As judge advocates had discovered in Somalia during Operation RESTORE HOPE, local law provided little or no guidance concerning the amount to be paid for loss of life. In Haiti, judge advocates learned that the mix of civil law and tradition in the Haitian legal system made this task particularly difficult. As a result, Army lawyers selected $5,000 to $14,000 as the amount that should be paid in a wrongful death claim, depending on the facts and circumstances. These monies were paid in addition to any solatia payments made to the victim's family, as such payments were made without regard to liability.[34]

Exaggerated claims were also an issue for judge advocates. The extreme poverty of many Haitians, combined with their belief that the U.S. forces were a source of great wealth, made it tempting for claimants to inflate damage claims or to "create" losses. Thus, for example, Americans conducting operations on an athletic field under a lease from the purported owner also received a claim from another individual who alleged that he was the rightful owner of the field. Additionally, this claimant sought damages for the disappearance of an elaborate and expensive multisport athletic complex said to have been on the property prior to the Americans' locating there. Creative claims also surfaced in the form of exaggerations of the extent of damage allegedly suffered by Haitians involved in traffic accidents with U.S. vehicles.

Capt. Catherine M. With with children at Nos Petits Freres et Soeurs, a Haitian orphanage, February 1995

While the operational tempo of UPHOLD DEMOCRACY meant that the policing of claims fraud necessarily took a low priority, certain actions could be taken that would assist in ensuring the best possible claims adjudication. Consequently, judge advocates advised that photographs be taken to record the condition of rented real estate facilities prior to setting up operations. They also counseled that written proof of property ownership should be obtained prior to signing leases.

Finally, claims judge advocates had to deal with misdirected claims. These occurred in the form of claims filed against the United States, when the actual damage or loss in issue had been caused by non-U.S. members of the Multinational Force. Some Haitians viewed the Americans—and the Foreign Claims Act—as tempting targets for claims. Additionally, as the Americans conducted "claims days" twice a week in Port-au-Prince and quickly acquired a reputation for promptness and efficiency, it was common for Haitians to file a claim with the U.S. Forces even when they knew that their claims had resulted from the activities of non-Americans. As the 25th Infantry Division's Captain With explained to Haitian claimants, however, even when Americans had caused the damage or loss in issue, the U.S. military would pay compensation only for the activities of military personnel. With could not process a claim for damage caused by members of the police training program administered by the Department of Justice.[35]

Judge Advocates in the United Nations Mission in Haiti

On the last day of March 1995, as UPHOLD DEMOCRACY ended, Captain With remained in Port-au-Prince—the only judge advocate with the Multinational Force to exchange her battle dress uniform cap for a blue beret. With was not, however, the only Army lawyer in Haiti, as Maj.

Bill Hudson had arrived in Port-au-Prince in February and had assumed his duties as the United Nations Force legal adviser. While a third Army judge advocate was also present in Port-au-Prince to provide defense counsel support, the Army command judge advocate presence was now considerably smaller. Yet, as UN operations began in earnest, there was no reduction in the variety and intensity of the legal work confronting Major Hudson and Captain With. The same was true for their successors, Major Ackerman and Captains Dahle and Walker. *(Chart 13)*

For the Army lawyers at United Nations headquarters, living conditions were fairly good. Major Ackerman, for example, lived in an air-conditioned hotel in Port-au-Prince and, although he had deployed with a flak jacket and protective mask, he never wore them. However, in contrast to other UN Force officers, Ackerman did carry a weapon, but he was required to wear this 9-mm. pistol concealed under his battle dress uniform shirt. Working conditions were also good. Again, Major Ackerman worked in an air-conditioned office and had the use of a four-wheel-drive utility vehicle—with radio and air conditioning—in order to better perform his duties as UN Force legal adviser. He worked long hours, however, seven days a week. And, although Sunday usually meant a shorter workday, every day brought numerous questions from UN peacekeepers of many nationalities.[36] Legal issues generally fell into three categories: administrative law, law and order, and operational and international law.

Administrative Law

An early issue of significance was the legal ramification of General Kinzer's appointment as commander of the UN Mission in Haiti. Because this command was a creation of the United Nations, the secretary general and the under-secretary general for peacekeeping operations expected General Kinzer, as force commander, to keep them fully informed concerning organizational, deployment, and operational matters. Thus, with a view toward formalizing a chain of command with General Kinzer, UN officials suggested that he sign an employment contract, accept a letter of appointment, and take an oath of loyalty to the United Nations.

After being requested by the Legal Advisor's Office, Office of the Chairman, Joint Chiefs of Staff, to examine the question, Army Col. David E. Graham, head of the International and Operational Law Division in the Office of the Judge Advocate General, advised that both law and policy precluded General Kinzer from signing an employment contract or letter of appointment with the United Nations. Moreover, in view of the oath of allegiance to the U.S. Constitution required of all

CHART 13—ORGANIZATION OF LEGAL SERVICES FOR UNITED NATIONS MISSION IN HAITI, APRIL 1995

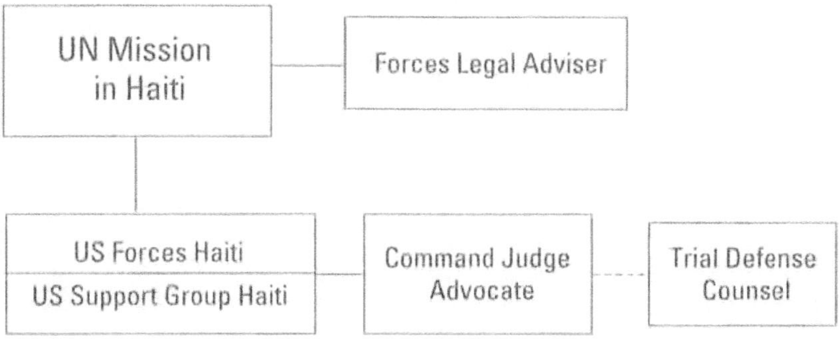

U.S. soldiers, General Kinzer could not take an oath of allegiance to the United Nations. In the end, a high-level exchange of communications between the United States and the United Nations satisfied all parties, and Kinzer assumed command of the UN Force.[37]

Another issue of importance was the fact that domestic U.S. law affected American participation in the United Nations Mission. Under the United Nations Participation Act, Congress limited to 1,000 the number of U.S. military personnel that could be assigned to UN peacekeeping operations worldwide. As more than 800 U.S. troops were serving in other United Nations operations prior to the creation of the UN Mission in Haiti, this cap imposed a severe constraint on Americans donning blue berets. The 1,000-person cap, however, did not apply to personnel detailed, under the applicable provisions of the Foreign Assistance Act, to international organizations in order "to render any technical, scientific, or professional advice." Consequently, judge advocates on the chairman's legal staff in the Pentagon—and those in the International and Operational Law Division of the Army Judge Advocate General's Office—advised that the majority of the 2,400-person American contribution to UN operations in Haiti would have to fall within the ambit of the Foreign Assistance Act.[38]

Other legal matters faced by judge advocates during UN operations included issues concerning the "retrograde" of captured weapons, the repatriation of Haitians, and the acceptance of gifts. General Kinzer, for example, was repeatedly presented high-value gifts. Given his status as U.S. Forces commander, these items could be accepted only on behalf of the United States. Thus, when General Kinzer wished to donate several valuable paintings that he had been given to a local church in Port-au-Prince, Major Ackerman advised that he could not do so; the paintings were U.S. government property.

Army lawyers also acted as legal advisers in formal investigations, reports of survey, and line of duty determinations. Major Ackerman, for example, assisted a Canadian officer in his administrative inquiry into allegations that some UN troops were trading food and money for sex. After conducting extensive interviews with the help of three civilian police investigators, Ackerman and the Canadian investigator concluded that there had been no wrongdoing.

Finally, legal issues arose in the area of UN and U.S. government personnel use of U.S. military medical care, post exchange facilities, and military aircraft. Judge advocates determined that, generally, all UN Mission in Haiti personnel were authorized U.S. medical care. Other UN employees and non–Department of Defense personnel, however, were not. With regard to post exchange eligibility, all U.S. government employees, Department of Defense contractor employees, and Red Cross personnel accompanying U.S. forces could lawfully use these facilities. Foreign nationals and non–Department of Defense personnel were not authorized travel on military aircraft unless Captain With, the command judge advocate for U.S. Forces Haiti, concurred that such travel was mission essential.[39]

Law and Order

To maintain good order and discipline among United Nations troops, General Kinzer published "Force Commander Directive Number 1." Modeled after the General Order no. 1 that had listed prohibited activities for U.S. troops during UPHOLD DEMOCRACY, Directive Number 1 proscribed the following: possessing, using, or consuming alcohol without approval of a "Contingent Commander"; selling or exchanging currency other than at the official exchange rate; purchasing or selling privately owned firearms; selling or reselling relief supplies outside official relief channels; and "throwing any food or beverage to local civilians from a vehicle, either stationary or moving." Given that General Kinzer lacked the authority to take criminal action against any UN Force personnel unless that individual was also a member of U.S. Forces Haiti, Directive Number 1 provided that General Kinzer could "recommend to the Under Secretary-General for Peacekeeping Operations the repatriation of any member of [the] military component in violation of this Force Directive." On a number of occasions, repatriation was recommended and did occur.[40]

Kinzer imposed more stringent rules on American personnel at U.S. Forces Haiti. In promulgating USFORHAITI General Order no. 1, he prohibited certain activities that were permissible for UN personnel. For example, USFORHAITI troops were forbidden to eat in local restaurants or to enter Haitian churches. Due to the high risk of contracting HIV in Haiti, the men and women assigned to U.S. Forces Haiti were prohibited from engaging in sexual relations with Haitians. Some USFORHAITI troops complained of these more restrictive rules when they learned that the general order governing their conduct did not apply to U.S. forces assigned to the UN headquarters staff. However, while it would have been preferable to have only one set of rules for all U.S. forces in Haiti, this was not possible given the different composition of the UN and U.S. forces.

Judge advocate involvement in law and order matters went beyond maintaining good order and discipline in the force, however. During Major Ackerman's tenure as force legal adviser, for example, he was an integral part of a security operation implemented to protect UN personnel from a gang of thieves. About twenty to thirty young Haitian men, operating in concert, were boarding UN vehicles at a busy intersection and stealing property—such as wristwatches and sunglasses—from drivers and passengers. Moreover, the young Haitians also periodically threw rocks at the UN vehicles, injuring UN troops. A contingent of Bangladeshi soldiers was responsible for patrolling the intersection, but the thieves were quick and persistent, and they consistently eluded capture.

As the newly constituted Haitian National Police had neither the assets nor the ability to capture these criminals or to provide safe passage for the UN Mission vehicles, the force chief of current operations, a Canadian Army major, teamed with Major Ackerman to develop a plan that would bring a halt to the activities in issue. As the Bangladeshi soldiers had been able to photograph some of the thieves and thus could identify these individuals, it was agreed that the Bangladeshis would cordon off the intersection at a time certain and detain the wrongdoers. Major Ackerman and his Canadian colleague, however, realized that the plan would be effective only if the Haitian thieves were prosecuted for their crimes. Thus, Ackerman visited the Haitian justice of the peace with jurisdiction over the area in which the road intersection was situated. He agreed to exercise jurisdiction over the thieves. Additionally, as the Haitian Constitution required that any detained citizen appear before a judge within forty-eight hours of detention, the magistrate also agreed to clear his docket so that he could receive and hear the cases of the accused within the first two days of their detention.

Major Ackerman also coordinated with the responsible Haitian National Police commander, and that official agreed to ensure that all

From left, Maj. Mark S. Ackerman, UN Mission in Haiti Force Legal Adviser; Colonel Khatak, Commander, Pakistan Contingent, UNMIH Military Force; and Canadian Maj. Marc B. Philippe, Deputy Force Legal Adviser, Port-au-Prince, 1995

detainees would appear in court within the required time period. Ackerman next visited the Bangladeshi commander and gained his assurance that the soldiers who had witnessed the stealing and rock throwing would be available to testify before the Haitian judge within forty-eight hours. Finally, he arranged for translators who would translate Bengali into English, and English into Creole French.

As a result, when Operation T-JUNCTION was executed on 9 January 1996, Ackerman's coordination paid off. Sixteen troublemakers were detained and turned over to the Haitian police for incarceration. The justice of the peace found eight of them guilty and sentenced them to forty-five days in jail. Operation T-JUNCTION also illustrated the value of having a judge advocate involved in the planning of security operations. His legal skills made him the ideal liaison between the UN Force, local law enforcement, and the local judiciary.[41]

Major Ackerman also assisted in a number of homicide investigations, not only by providing legal advice but also by suggesting questions and methods of interrogation. In January 1996, for example, after a Guyana Army soldier murdered a comrade by shooting him with a submachine gun, General Kinzer tasked U.S. Army Criminal Investigation Command agents with the investigation of the homicide. When these agents wished to interview the suspect, Major Ackerman consulted with Guyanese officers and examined an old manual on British military law before determining that a rights advisement was necessary. Although the shooter subsequently declined to answer questions and was returned to Guyana to face trial for murder, Major Ackerman had ensured that his rights were protected and that any statement obtained could have been admitted into evidence.[42]

Finally, Majors Hudson and Ackerman, together with the deputy force legal adviser, Canadian Maj. Marc B. Philippe, took a number of actions to improve the administration of criminal justice in Haiti. The three lawyers recognized that crowded conditions in the National Penitentiary might result in rioting and adversely impact the UN mission of maintaining a stable and secure environment. They also understood that, despite the constitutional requirement that a justice of the peace determine the lawfulness of an arrest and detention within forty-eight hours, prisoners were not being taken to the courthouse to appear before judges. Consequently, on their own initiative, Majors Hudson and Philippe began transporting Haitian justices of the peace to the penitentiary so that hearings could be conducted there. They also encouraged the creation of an ad hoc public defender organization, thus ensuring that prisoners requiring legal counsel received assistance. Finally, Majors Hudson and Philippe initiated—and Majors Ackerman and Philippe continued—the practice of visiting the National Penitentiary weekly. They monitored conditions in the prison, including sanitation, food, and medical treatment. On one occasion, concerned that the prisoners were not being supplied with fresh water, they arranged for a private water company to deliver water to the prison.[43]

Operational and International Law

In early 1995, during the transition from the Multinational Force to the United Nations peacekeeping operation, Army judge advocates took an active role in drafting rules of engagement (ROE) that would apply to UN forces. The greatest challenge facing these lawyers was the fact that while UN officials wanted to have rules of engagement in place that resembled those used in UN peacekeeping operations in other geographic areas, the U.S. component wished to have UN Force rules consistent with a peace enforcement mission. This was an important distinction, as troops in a UN peacekeeping operation generally are more lightly armed than soldiers in "peace enforcement" units. Similarly, rules of engagement for peacekeeping operations are more restrictive than those for peace enforcement missions. For example, other countries deploying units to Haiti resisted the idea that deadly force should be permitted to protect mission-essential property. Still another contentious issue was that of whether troops should be permitted to intervene in order to prevent harm to civilians.[44]

As it had been decided at the highest policy levels that U.S. interests would best be served by having United Nations rules of engagement that permitted the use of force—even deadly force—to protect certain military property and to halt Haitian-on-Haitian violence, Army judge advocates recognized that having draft rules of engagement in hand was essential to the American point of view receiving full consideration. Thus, in January and February Captain With reviewed several versions of proposed UN rules of engagement and assisted in drafting the comments on these rules forwarded to the U.S. delegation to the United Nations in New York City. As a result of her efforts—and those of operational law experts in the Pentagon—the rules of engagement approved by the United Nations for the UN Mission in Haiti adopted the U.S. perspective on the use of force. Deadly force was authorized "in defense of key installations" essential to the success of the UN mission. Additionally, force was authorized to halt civilian violence. Over time, even those countries who initially had been the most opposed to using force to halt Haitian-on-Haitian violence altered their views after their forces experienced firsthand the terror of such behavior on the streets of Port-au-Prince.[45]

Captain With personally typed an ROE card setting forth the new United Nations rules of engagement and arranged for its printing. She also reviewed the UN Mission in Haiti operations plan, worked on its legal annex, and reviewed draft standard operating procedures for the UN Force Legal Advisor's Office prior to the arrival of Major Hudson on 24 February.

When Hudson arrived in Port-au-Prince and assumed his duties as UN Force legal adviser, he immediately took charge of developing situational training exercises that would effectively reinforce these rules. Later, as the June 1995 parliamentary elections neared, Hudson and With also developed vignettes emphasizing potential problems that might occur at ballot sites. These scenarios addressed handling a "noisy demonstration" outside the election bureau, "voters complaining of intimidation," "shots fired at registration and voting bureaus," and "violent demonstrations" at a polling site. As a result of their efforts, UN troops stationed at polling places were well prepared to handle any situation.[46]

In addition to her work on rules of engagement, Captain With also played a significant role in the negotiation of two bilateral agreements.

Given her position as the command judge advocate at U.S. Support Group, Haiti, Captain With assisted a political officer at the U.S. embassy in structuring a support agreement under the provisions of the U.S. Foreign Assistance Act. Under the terms of this agreement, the Department of Defense—through U.S. Support Group, Haiti—provided support on a reimbursable basis to the government of Haiti. These negotiations, conducted in French, English, and Spanish between Captain With, her State Department colleague, and an official from the Haitian Ministry of Foreign Affairs, were critical to ensuring that the Haitian government received support essential to the restoration of democracy and economic stability.

Later, Captain With also assisted in drafting and negotiating an agreement "representing a bilateral Status of Forces Agreement" between Haiti and the United States. This agreement, concluded through an exchange of diplomatic notes in May 1995, provided U.S. Support Group, Haiti, personnel with the "same status as that provided to the administrative and technical staff" of the U.S. embassy. As a result, U.S. personnel enjoyed immunity from both criminal and civil process for any act arising out of their official duties. The agreement also permitted U.S. servicemembers participating in humanitarian and nation-building activities in Haiti to "enter, leave, and freely circulate in Haiti" and to "wear uniforms in the exercise of their official duties and to carry weapons." U.S. troops were also granted exemption from Haitian taxes for any activities related to the military mission in Haiti. Once again, having a judge advocate available to assist in negotiating an arrangement protected both U.S. personnel and U.S. interests.

Conclusion

Judge advocates in Haiti faced many of the same challenges that confronted their colleagues in JUST CAUSE, DESERT SHIELD, DESERT STORM, and RESTORE HOPE. Army lawyers deploying as part of Task Forces 180 and 190 and the Multinational Force, Haiti, faced tough conditions and great uncertainty, given the last-minute decision of the Cedras regime not to resist the entry of U.S. combat troops. But these judge advocates, as well as those who served as legal advisers in the UN Mission, Haiti, peace enforcement operation and the U.S. Support Group in Haiti, demonstrated that attorneys were now fully prepared to play essential roles in planning for and conducting military operations.

Whether drafting rules of engagement, advising on contingency contracting issues, planning security operations, or assisting in negotiating bilateral agreements, judge advocates reaffirmed the fact that their role now went far beyond that of the narrow, traditional combat service support and military justice function. Clearly, Operation UPHOLD DEMOCRACY demonstrated that the commitment of the Judge Advocate General's Corps to the concept of operational law was paying tremendous dividends. Having built on their experiences in deployments starting with URGENT FURY in 1983, judge advocates in Haiti provided superb legal and nonlegal support that significantly enhanced mission success. UPHOLD DEMOCRACY offered proof that the basic operational law skills of judge advocates were now in place—and that future deployments would benefit from the ongoing refinement of the expanded role of judge advocates in military operations.

Notes

[1] Interv, author with Maj Catherine M. With, 14 Apr 98, Historians files, OTJAG.

[2] Agreement of Governors Island, 3 Jul 93, reprinted in Center for Law and Military Operations, *Law and Military Operations in Haiti, 1994–95: Lessons Learned for Judge Advocates* (Charlottesville, Va.: The Judge Advocate General's School, U.S. Army, 1995), app. A, p. 174.

[3] *Law and Military Operations in Haiti*, p. 10.

[4] Ibid., pp. 12–13.

[5] Ibid., pp. 15–16.

[6] Ibid., pp. 23–24; Bradley Graham, "Clinton Decides to Maintain U.S. Military Presence in Haiti," *Washington Post*, 6 Dec 97, p. A26.

[7] USFORHAITI and USSPTGRPHAITI were two distinct organizations with different missions. To reduce confusion, USFORHAITI was commonly referred to as "the blue-hatted U.S. forces," while USSPTGRPHAITI was called the "green-hatted U.S. forces."

[8] Interv, author with Lt Col Karl K. Warner, 25 Sep 96, Historians files, OTJAG.

[9] Memo, Maj Bradley P. Stai, Chief, Civil Law, Ofc of SJA, XVIII Abn Corps, for SJA, XVIII Abn Corps, 2 Feb 95, sub: After Action Report—Operation UPHOLD DEMOCRACY, p. 20, Historians files, OTJAG.For the text of the Combined Joint Task Force rules of engagement, see Appendix B.

[10] Interv, Col Dennis P. Mroczkowski with Lt Col Karl K. Warner, 8 Oct 94, Historians files, OTJAG. See, generally, *Law and Military Operations in Haiti*, app. I, p. 213.

[11] *Law and Military Operations in Haiti*, pp. 38–39 and app. J, pp. 216–17.

[12] Memo, Maj Catherine M. With for author, 15 Apr 98, sub: My Experiences While Deployed to the Republic of Haiti During Operation UPHOLD DEMOCRACY, Historians files, OTJAG.

[13] Interv, author with Capt Thomas N. Auble, 1 Apr 98; CARICOM Forces Combined Operations Guide, "Ten Commandments of Human Rights for Soldiers," Sep 94, both in Historians files, OTJAG. For the text of the "ten commandments," see Appendix B.

[14] Interv, author with Auble, 1 Apr 98.

[15] Interv, author with Maj Gen John D. Altenburg, 13 Jun 97, Historians files, OTJAG; interv, Lt Col Steve E. Dietrich with Col John D. Altenburg, 10 Oct 94, in Cynthia L. Hayden, *JTF 180 UPHOLD DEMOCRACY Oral History Interviews* (Fort Bragg, N.C.: XVIII Airborne Corps Training Support Center, 1995), pp. 327–28.

[16] Interv, Dietrich with Altenburg, 10 Oct 94.

[17] Interv, author with Altenburg, 13 Jun 97.

[18] Interv, author with Col Barry E. Willey, 18 Jun 98, Historians files, OTJAG. Willey was the director of the Joint Information Bureau, CJTF 180, and the de facto spokesman for that joint task force during UPHOLD DEMOCRACY.

[19] *Law and Military Operations in Haiti*, pp. 63–64.

[20] Interv, author with Warner, 25 Sep 96.

[21] Ibid.

[22] *Law and Military Operations in Haiti*, p. 64.

[23] Ibid., pp. 72–74.

[24] Ibid., pp. 106–07; interv, author with Warner, 17 Apr 98, Historians files, OTJAG.

[25] *Law and Military Operations in Haiti*, pp. 106–07.

[26] The General Orders no. 1 promulgated at Task Forces 180 and 190 are reprinted in *Law and Military Operations in Haiti*, app. U.

[27] *United States* v. *Rockwood*, CM 9500872, Record of Trial, 10th Mtn Div, 22 Apr and 8–14 May 1995; *United States* v. *Rockwood*, Opinion of Army Court of Criminal Appeals, 9 Feb 98; *Law and Military Operations in Haiti*, pp. 113–15; see also Edward J. O'Brien, "The Nuremberg Principles: Command Responsibility, and the Defense of Captain Rockwood," *Military Law Review* 149 (1995): 275.

[28] *United States* v. *Pacheco*, Army 9500002, affirmed by the Army Criminal Court of Appeals, 11 Sep 98 (unpub.); *Law and Military Operations in Haiti*, p. 116.

[29] Memo, Lt Col Arthur L. Passar for author, 24 Apr 98, Historians files, OTJAG.

[30] Ibid.; see also memo, SJA, Joint Logistics Support Command, for SJA, US Army Materiel Command, 11 May 95, sub: After Action Report, Legal Support to Joint Logistics Support Command, Joint Task Force 190, Haiti, Operation UPHOLD DEMOCRACY, Sep 94–Mar 95 (hereafter cited as JLSC AAR).

[31] JLSC AAR, pp. 13–14.

[32] Ibid.

[33] Memo, SJA, 10th Mountain Division, May 95, sub: Operation UPHOLD DEMOCRACY, Multinational Force Haiti After-Action Report, 29 Jul 94–13 Jan 95, p. 15; *Law and Military Operations in Haiti*, pp. 146–47.

[34] *Law and Military Operations in Haiti*, pp. 148–49.

[35] Interv, author with With, 14 Apr 98; *Law and Military Operations in Haiti*, pp. 150–52.

[36] Interv, author with Maj Mark S. Ackerman, 19 Jun 96, Historians files, OTJAG.

[37] *Law and Military Operations in Haiti*, pp. 96–97.

[38] Ibid., pp. 49–50.

[39] Memo, With for author, 15 Apr 98.

[40] For a reprint of Force Commander Directive Number 1, see *Law and Military Operations in Haiti*, app. U.

[41] Interv, author with Ackerman, 20 Jun 96.

[42] Interv, author with Ackerman, 19 Jun 96.

[43] Ibid.

[44] *Law and Military Operations in Haiti*, pp. 44–45.

[45] Memo, With for author, 15 Apr 98; memo, Lt Col William A. Hudson, Jr., for author, 3 Mar 97, Historians files, OTJAG; UNMIH Rules of Engagement, 3 Mar 95, reprinted in *Law and Military Operations in Haiti*, app. M, pp. 237–50.

[46] Memo, Hudson for author, 3 May 97.

8

Judge Advocates in Operations Other Than War, 1965–1994

These missions will test a judge advocate's legal knowledge, interpersonal and leadership skills, and physical and mental conditioning. . . . Every day is an unexpected adventure.[1]

—*Lt. Col. Manuel E. F. Supervielle*
Staff Judge Advocate,
JTF Western Samoa and JTF Sea Angel

The sustained, large-scale combat operations in Vietnam and Southwest Asia are classic examples of American soldiers' deployments for war. With increasing frequency since 1965, however, U.S. troops also have participated in smaller, lower-intensity deployments. From peacekeeping duties in the Sinai to disaster relief in Bangladesh, these so-called operations other than war are military activities that do not involve traditional armed conflict. On the contrary, operations other than war "focus on deterring war, resolving conflict, promoting peace, and supporting civil authorities."[2] Such operations include combating terrorism, supporting counterdrug operations, enforcing sanctions and exclusion zones, humanitarian assistance, military support to civil authorities, nation building, noncombatant evacuation operations, peace operations, strikes, and raids.

While a large number of operations other than war have occurred during this period, this chapter examines only an illustrative number of them. In doing so, the experiences of some Army lawyers are discussed briefly, while others receive a more thorough examination.

Operation POWER PACK, *1965–1966*

In late April 1965, an attempted coup d'etat in the Dominican Republic quickly turned into a civil war in the streets of the capital, Santo Domingo. President Lyndon B. Johnson, believing that the rebel forces responsible for the uprising contained radical elements, ordered U.S. Army and Marine Corps units into the country to protect American lives, restore order, and, most important, prevent a Communist seizure of power. The marines arrived first, followed by combat units from the 82d Airborne Division.

At the height of the U.S. intervention, called Operation POWER PACK, nearly 24,000 U.S. troops were committed to the joint, and ultimately combined, operation. Little actual combat occurred during POWER PACK, for the presence of marines and paratroopers precluded a rebel victory. Consequently, U.S. forces under the command of Lt. Gen. Bruce Palmer, Jr., engaged in a variety of civic action, psychological warfare, civil affairs, and other noncombatant activities designed to restore stability, "win hearts and minds," and provide the foundation for a political solution.

Judge advocates participated in POWER PACK from the outset. When a "ready" force of two airborne battalions from the 82d Airborne Division deployed from Pope Air Force Base, North Carolina, on 29 April 1965, Capt. Paul H. Ray was with them. He, along with the rest of the paratroopers aboard thirty-three C–130 Hercules aircraft, wore a parachute for a possible airdrop into the Dominican Republic. But this combat jump never occurred—the Americans landed at San Isidro Airfield, nine miles east of Santo Domingo.[3] Over the next twelve months a number of Army lawyers followed Captain Ray into the Dominican Republic, including the 82d Airborne Division staff judge advocate, Lt. Col. Guy A. Hamlin, and the deputy staff judge advocate, Capt. Raymond D. Cole. Also deploying were Capts. Robert R. Aldinger, Gerald C. Coleman, and Burnett H. Radosh. All lived and worked with the rest of the headquarters staff at a military academy near San Isidro Airfield. *(Map 12)*

The judge advocates provided the full range of legal support, from military justice and international law to claims and legal assistance. But their approach to lawyering was a traditional one: providing the same legal support in the Dominican Republic as they had provided in the garrison setting at Fort Bragg. The emergence of operational law and the integration of judge advocates into military operations at all levels was still over twenty years in the future. For example, Captain Radosh, who replaced Captain Cole as the division deputy staff judge advocate

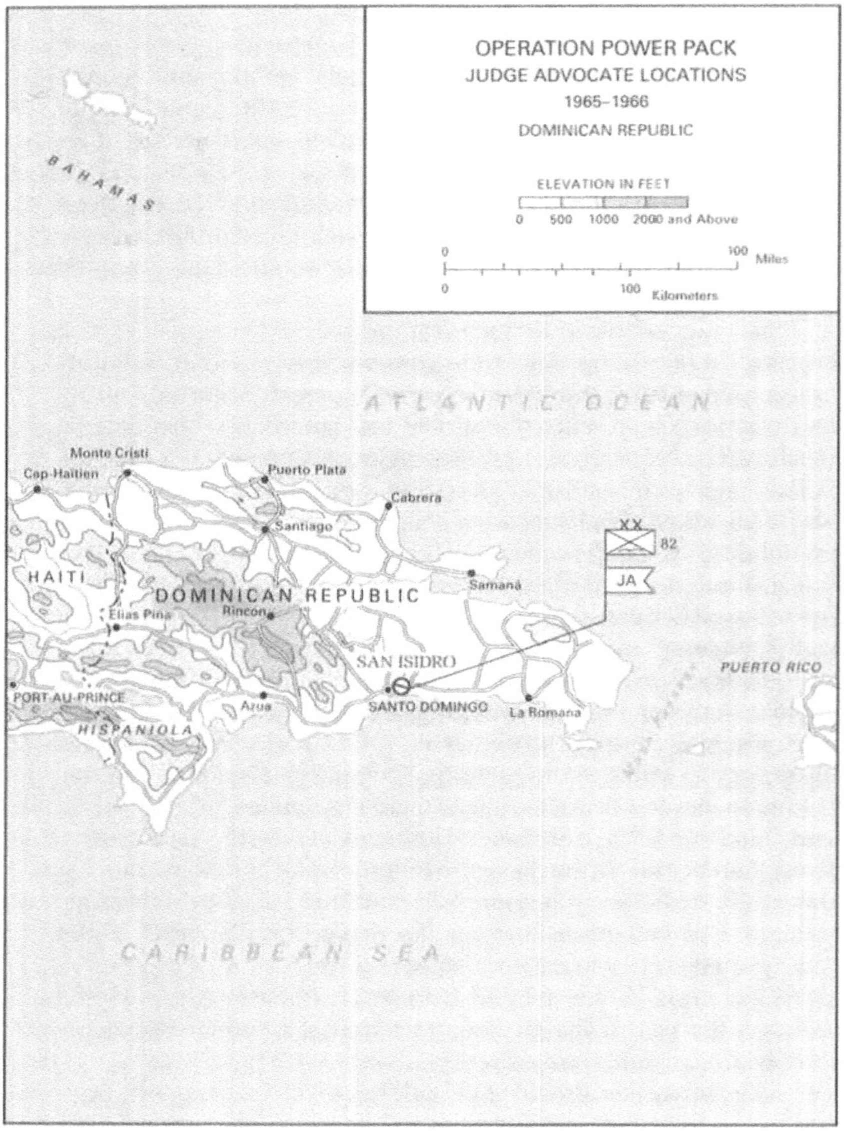

Map 12

in June 1965, spent almost all of his time working on foreign claims filed by Dominican citizens against the United States. There was a "huge claims backlog," and Captain Radosh spent hours in a jeep traveling throughout Santo Domingo investigating these claims.[4]

Courts-martial, however, took the lion's share of effort. Although there was no judge advocate participation in special courts—and would not be until 1969—those commanders with the authority to convene special courts-martial still required advice from Colonel Hamlin and his staff, as did line officers serving as trial and defense counsel in special courts. Additionally, Maj. Gen. Robert H. York, Commanding General, 82d Airborne Division, decided that general courts would be convened in the Dominican Republic. Consequently, division judge advocates prosecuted and defended paratroopers in trials held at San Isidro Air Base.

One case, occurring in June 1965, involved Pfc. Dexter M. Moore, who was court-martialed for the unpremeditated murder of a Dominican citizen named Perez. Moore, along with two other paratroopers, had left their company area without authority and gone to a nearby bar. There, the Americans met Perez. As the evening progressed, they decided that Perez might be a rebel and, along with other rebels, might be planning to ambush them. These suspicions led to an argument, and when Perez attempted to escape from the soldiers, Private Moore shot and killed him with his automatic rifle. Moore was convicted of unpremeditated murder and sentenced to a dishonorable discharge, total forfeiture of all pay and allowances, and twenty years' confinement.[5]

Another significant case was that of Pfc. Bernis J. Darling, prosecuted in October 1965 for assaulting a man with a knife and intentionally inflicting grievous bodily harm. Darling received a dishonorable discharge and two years' confinement.[6] But not all serious crimes committed in the Dominican Republic were prosecuted there. The general court-martial of Pvt. Charlie E. Monday, for example, was held at Fort Bragg rather than at San Isidro. His trial was of great import to good order and discipline in Operation POWER PACK, as Private Monday was charged with misbehavior before the enemy. On 21 June 1965, while part of a platoon occupying a defensive position in downtown Santo Domingo, Monday had walked to the front of the fortified position and intentionally cast aside his web gear and ammunition. He then "proceeded, in a rapid manner directly toward the Rebel positions," where he was captured. Monday was held briefly by the rebels and then released. Less than two weeks later, on 2 July 1965, Private Monday was court-martialed for misbehavior before the enemy and failing to obey a lawful order. Convicted by a Fort Bragg panel on 6 July, he was sentenced to a dishonorable discharge, total forfeiture of all pay and allowances, and three and a half years' confinement.[7]

During the first week of Operation POWER PACK there were claims that U.S. troops had engaged in "unauthorized trespassing, ransacking and looting of private dwellings." Although subsequent investigations did not support these allegations, General York, after consulting with his staff judge advocate, issued a letter to all subordinate commanders on 6 May 1965. It warned that looting and similar crimes were offenses under the Uniform Code of Military Justice, declared that "no such conduct will be tolerated in this division," and expressed York's "desire that all commanders prohibit such activities." General York further directed that personnel were not to use "private dwellings . . . except when required by the tactical situation." When a captain later kept radio equipment looted from a commercial radio station in Santo Domingo, he was court-martialed, and York's directive was part of the evidence used against him.[8]

Lt. Col. Guy A. Hamlin, Staff Judge Advocate, 82d Airborne Division, 1963–1966

For U.S. troops participating in Operation Power Pack, rebel snipers were a constant problem. Even judge advocates working legal issues were not immune. Colonel Hamlin, Captain Cole, and their senior legal noncommissioned officer came under fire from a sniper while near the U.S. embassy in Santo Domingo. After hearing the crack of the round go over their heads, the three men quickly took cover behind a tree. The two judge advocates chambered their .45-caliber pistols, preparing to return fire, but since no one could see the sniper they did not do so. Remembers Captain Cole: "I got shot at more in the Dominican Republic than I did in two tours in Vietnam and in Operation Desert Storm."[9]

The problems caused by snipers were greatly aggravated by the rules of engagement governing Operation Power Pack. Initial restrictions on the use of force prohibited U.S. troops from firing their weapons unless fired upon. This was a questionable rule, but most accepted it as necessary, given the humanitarian and political goals of U.S. intervention. Similarly, most understood the ban on using artillery, tanks, and mortars in urban Santo Domingo, especially as it seemed unlikely that U.S. troops would face much more than rebels with rifles and pistols. After organized resistance to the Americans ceased, however, military requirements were increasingly subordinated to diplomatic concerns, particularly when President Johnson and his advisers decided to seek a negotiated settlement to the crisis in the Dominican Republic. As a result, even more restrictive rules of engagement were issued, with the result that, by the end of Power Pack, a soldier could not fire his weapon unless his position was in imminent danger of being overrun. Once the rebels realized this new situation, they took full advantage of it. Thus, a rebel sniper, rifle in hand, would swagger down the middle of the street toward an American position, casually walk into a nearby building, choose his firing position, expend his ammunition, leave the building, and offer an obscene gesture as he departed the area.

No judge advocates had been involved in the drafting of Power Pack's rules of engagement, nor had they reviewed them. The rules were lawful; they complied with the Law of War. Yet their unduly restrictive character illustrated how lawful restrictions on the use of force might needlessly endanger lives and interfere with the military mission. Not surprisingly, soldiers complained bitterly about Power Pack's rules of engagement for many years.[10]

The last troops of the 82d Airborne Division left the Dominican Republic in September 1966. Though the lessons learned from roughly sixteen months of Power Pack and judge advocate involvement in it were largely forgotten in an Army now learning how to fight a guerrilla war in Southeast Asia, there were exceptions: When Col. John R. Bozeman, the staff judge advocate of the XVIII Airborne Corps, drafted the rules of engagement for Operation Just Cause in 1989, he was aware of the unreasonable restrictions placed on the use of force during Operation Power Pack. Additionally, Bozeman recognized that the upcoming hostilities in Panama might be similar in nature to those faced by soldiers on the streets of Santo Domingo, and that political considerations could well influence rules on the use of force. Consequently, in drafting a rules of engagement annex and in coordinating with the operations planners at the division, corps, and joint command level, Colonel Bozeman ensured that rules of engagement applicable to U.S. troops during Just Cause would not put them at undue risk.[11]

Cuban Refugee Resettlement Operation, 1980–1982

In the spring of 1980, some 125,000 Cuban refugees sailed for the United States in what became known as the Mariel boatlift. Initially, the refugees were housed at several locations: Fort Chaffee, Arkansas; Fort Indiantown Gap, Pennsylvania; Fort McCoy, Wisconsin; and Eglin Air Force Base, Florida. Approximately 30,000 refugees were present at each site.

In early June, refugees at Fort Chaffee rioted over rumors that they would not be permitted to leave Fort Chaffee in the near future. The rioters overturned motor vehicles, lit fires, and departed the post through the main gate. When some of the rioters made their way to Fort Smith, a city some ten miles distant, local and state authorities feared for the safety of that community. The federal response was quick. Fifteen hundred soldiers from Fort Sill, Oklahoma, deployed to Fort Chaffee to reinforce Arkansas National Guard troops already there. The Fort Chaffee garrison commander, an Army colonel, also received legal support when Maj. Paul Luedkte, an Army judge advocate from Fort Stewart, arrived on the scene. Luedkte also assumed duties as the legal adviser to Brig. Gen. James Drummond, commander of the task force charged with quelling the rioting and restoring order at Fort Chaffee.

On 1 July 1980, Capt. Daniel J. Dell'Orto arrived at Fort Chaffee for a thirty-day tour of duty. Dell'Orto, who replaced Major Luedkte as task force judge advocate, found that conditions at Fort Chaffee were less than ideal. While the refugees were restricted to a cantonment area, this area was defined literally by a cordon of soldiers placed at fifty-foot intervals. The soldiers sat in folding chairs in conditions in which the daytime high temperature consistently exceeded 100 degrees. Portable latrines—for use by both refugees and soldiers—were positioned every few hundred feet. This security arrangement remained in place during Captain Dell'Orto's stay at Fort Chaffee. While there, he operated out of a second-story office adjacent to the cantonment area, providing legal assistance, administrative law, and military justice advice to the command. He also acted, however, as a military defense counsel in those criminal matters in which he had not advised the command, counseling soldiers offered Article 15 nonjudicial punishment. The most frequent misconduct involved violations of the task force order prohibiting fraternization between the Americans and female refugees. Given the nature of the cantonment area, inappropriate behavior between the guards and the guarded was difficult to control.

On 3 February 1981, Dell'Orto returned to Fort Chaffee for a sixty-day tour of duty. By this time, the three other refugee centers had closed. All Cubans remaining in detention—more than 10,000, most of whom were men—were now housed at Fort Chaffee. The human cordon had been replaced by a ten- to twelve-foot-high chain link fence topped with triple concertina wire. While Dell'Orto immediately assumed duties as the task force judge advocate and provided legal support to the Army colonel who had replaced General Drummond as the force commander, his efforts were focused primarily on prosecuting Cubans for crimes committed in the camps. For the next two months, Captain Dell'Orto prosecuted misdemeanors in a U.S. magistrate's court as a special assistant U.S. attorney for the Western District of Arkansas. While Cuban refugees who had committed serious felonies, such as murder and rape, were charged in the U.S. district court in Fort Smith, virtually all other crime was handled by the U.S. magistrate. He held court two or more days every week at Fort Chaffee in a courtroom built specifically to handle refugee misconduct cases, and it was not unusual for Captain Dell'Orto to arraign thirty to forty Cubans in one day for offenses committed in the compound. With law enforcement in the refugee camp the responsibility of the National Park Police, Federal Protective Service, and Federal Bureau of Investigation, Captain Dell'Orto worked closely with these civilian law enforcement agencies in perfecting his cases for trial. Dell'Orto also served as liaison between the task force and other federal civilian agencies at Fort Chaffee, including the Department of State, the Immigration and Naturalization Service, the U.S. Customs Service, the Bureau of Prisons, and the Federal Emergency Management Agency.[12] Army judge advocates continued to support Cuban resettlement efforts until operations at Fort Chaffee ended in 1982.[13]

Multinational Force and Observers, 1981–

Although peacekeeping and similar military operations since 1945 had almost always occurred within the structure of the United Nations, the creation of the Multinational Force and Observers (MFO) in 1981 was a clear exception to this rule. A combined peacekeeping force located in the Sinai Peninsula, its mission was to supervise and administer the implementation of the 1979 Egyptian-Israeli Treaty of Peace. Created in August 1981 after the UN Security Council had failed to authorize a peacekeeping force to undertake this mission, the MFO was still operating at the end of the century.

U.S. Army judge advocates have participated in the MFO from its inception. Thus, in October 1981 Maj. David E. Graham was detailed by the Department of Defense to serve as the force legal adviser. During five months of intense activity in Washington, D.C., Graham worked to identify and resolve all legal issues associated with organizing and structuring the force. Then, at the personal request of the MFO commander, Norwegian Lt. Gen. Fredrik Bull-Hansen, Major Graham deployed to the Sinai in early 1982. He served not only as the force legal officer, but also as its political officer, as General Bull-Hansen made the decision to authorize Graham to negotiate essentially all MFO implementing arrangements with the Egyptian and Israeli governments. The legal issues were both numerous and unique—ranging from the drafting of the MFO rules of engagement to negotiating the release of Israeli contractor employees from an Egyptian jail. And, as the MFO went through an intensive start-up period, the work required a twelve- to fourteen-hour workday, seven days a week. There also was frequent travel from MFO headquarters in the northern Sinai to Tel Aviv, Jerusalem, and Cairo.[14]

As the MFO legal adviser, Major Graham was present when Israeli representatives met with Egyptian officials on 25 April 1982 to officially return to Egypt the remaining portion of the Sinai under Israeli control. At the meeting at which the transfer was to occur, Egyptian representatives balked at assuming control of the Sinai territory absent a "formal turnover document." None existed. After several anxious moments, General Bull-Hansen turned to Major Graham and said, "We'll take a 15 minute recess. Major Graham will have the necessary document ready for signature when we re-convene." And he did.[15]

Major Graham returned from the Sinai in the summer of 1982, the only military attorney to have served as the MFO legal adviser. It was a "once-in-a-lifetime experience, a baptism by fire in the peacekeeping business."[16]

In the years since, a multitude of judge advocates have deployed to the Sinai in support of U.S. participation in MFO operations. Each attorney deploys for six months with the U.S. infantry battalion task force headquartered at Sharm El Sheik, Egypt, although these judge advocates also serve the 1st Support Battalion, located several hundred miles away in El Gora, Egypt. The command judge advocate provides legal assistance, claims, administrative, operational law, and military justice advice. Additionally, the U.S. Army Trial Defense Service provides a judge advocate for defense counseling. Thus, two judge advocates always are assigned to the U.S. component of the MFO.

The experiences of Capts. Catherine M. With and Paul Fiorino illustrate the role played by judge advocates in the MFO. Captain With, who served in the Sinai from February to August 1991, deployed as part of a battalion from the 7th Infantry Division (Light). Her experiences as the MFO judge advocate were different from those of the Army lawyers who preceded her, as With's deployment occurred while Operation DESERT STORM was under way. Thus, in addition to her responsibilities as a judge advocate to force personnel, Captain With had a secondary mission of supporting other U.S. troops in the region. This meant traveling to Tel Aviv to provide legal assistance to soldiers manning Patriot missile units in Israel and to Cairo to assist Air Force transportation personnel stationed at the Cairo International Airport. Although With's legal advice was comprehensive, her toughest legal questions arose in the area of rules of engagement. The proximity of the U.S. MFO contingent to Desert Storm meant that the U.S. MFO soldiers had to be prepared to move from peacekeeping rules of engagement to rules permitting the use of force in self-defense situations. Advising the command on two markedly different sets of rules of engagement was challenging in itself; developing and then supervising realistic training in both sets of rules made Captain With's task doubly difficult.[17]

Some five years later, Captain Fiorino deployed to the Sinai. He left in January 1996, flying on a chartered commercial airliner from Pope Air Force Base to Sharm El Sheik at the southern trip of the Sinai. For the next six months, Fiorino served as the defense counsel for the roughly 1,500 soldiers of an airborne infantry battalion and a support brigade. He counseled thirty-five soldiers facing punishment under the provisions of Article 15, Uniform Code of Military Justice, and worked on a handful of administrative elimination actions. Captain Fiorino also performed general staff officer work.[18]

Other judge advocates with Sinai peacekeeping service include Capts. Ida F. Agamy, Michael R. Black, Denise A. Council-Ross, Ray P. Cox, David T. Crawford, Fred K. Ford, Mark J. Gingris, Andrew D. Hultgren, Scott L. Kilgore, David H. Lande, Roger E. Nell, James M. Patterson, Misti E. Rawles, Peter B. Ries, John M. Smith III, and Michael R. Snipes. While each served in the Sinai for six months, at least one judge advocate, Capt. Carlton L. Jackson, served two separate six-month tours with the MFO.

Operation Hawkeye, 1989

After Hurricane Hugo devastated parts of the U.S. Virgin Islands in September 1989, there was widespread looting and chaos on St. Croix. The situation was beyond local police control, and the Virgin Islands National Guard was incapable of dealing with the situation. Although the government of the islands initially resisted calling for federal troops, President George Bush ordered the XVIII Airborne Corps and other units to St. Croix to quell civil disturbances. The situation was uncertain because of the lack of communication with St. Croix. Additionally, the hurricane had virtually destroyed the federal prison, allowing 500 convicted felons to roam free and add to the chaos on the island.

At 5 P.M. on 20 September 1989, the corps was told to create a task force with a mission focus on civil disturbance operations. "Although we had virtually no planning time," remembered Lt. Col. Patrick Finnegan, the corps deputy staff judge advocate and task force legal adviser, "we were involved in legal issues from the beginning." The first question concerned rules of engagement. Some reports indicated that U.S. troops might be met by armed dissidents when they landed at the airport, and, although this threat never materialized, soldiers required guidance on the degree of force to be used when confronting looters. Using civil disturbance Operations Plan GARDEN PLOT as a guide, Colonel Finnegan drafted the rules of engagement before departing for St. Croix. After Col. John R. Bozeman, the corps staff judge advocate, reviewed Finnegan's work and the task force commander, Brig. Gen. Bruce Moore, approved the rules, all deploying task force members were briefed on them. Within seventy-two hours, however, the rules were modified further as, with so little left to loot, there was no civil unrest. Colonel Finnegan did, however, draft separate rules of engagement for those soldiers guarding the federal prison. This resulted from the fact that the hurricane had destroyed the fences and most guard towers at the prison, and, after the Federal Bureau of Investigation had captured and returned the roughly 500 missing felons to the prison grounds, U.S. soldiers and the few remaining guards had to be used as a security perimeter until engineer units could rebuild the fences.[19]

The first C–141 aircraft touched down at the devastated airport in St. Croix before 7 A.M. on 21 September, with the XVIII Airborne Corps chief of staff as the task force commander and Colonel Finnegan as its staff judge advocate. Conditions were harsh in the aftermath of Hurricane Hugo. For the first several days and nights, Colonel Finnegan and the rest of the corps staff lived in partially destroyed buildings at the airport. They slept on the concrete floor and worked without offices, desks, telephones, or electricity. There was no drinking water other than what had been brought. As a result, the focus of Operation HAWKEYE quickly shifted from civil disturbance to disaster relief operations, and as supplies began arriving from the United States living and working conditions improved markedly. The main focus of the task force now became protecting the few businesses that were still operating, guarding the condominiums and resort hotels, and assisting the territory's government to organize and control disaster relief operations and distribute food and water.

Although Colonel Finnegan continued advising on issues such as rules of engagement, his principal mission as the task force staff judge advocate was that of assisting with interagency coordination. Within an hour of arriving, the task force commander, accompanied by Colonel Finnegan, met with local government and police officials. Their initial reaction was hostile, as they were offended that federal troops had been sent into a situation that they believed they could have controlled themselves. The task force commander explained that the president had ordered the mission and offered to help. The St. Croix police, realizing that the situation was beyond their capabilities, welcomed military police assistance. As a result, joint civil-military police patrols were organized, and order was restored. Similarly, Colonel Finnegan cooperated with the Federal Bureau of Investigation and, as disaster relief became the focus of Operation HAWKEYE, with Federal Emergency Management Administration (FEMA) and local government officials to distribute food and water, to clear rubble, and to inform the populace of what assistance was available and where. Twice daily the staff judge advocate and the task force civil affairs officer (G–5) met with the lieutenant governor, FEMA representatives, and local government functionaries in order to monitor progress and to determine what additional relief actions were required.

For the next two months, Colonel Finnegan, assisted by other judge advocates deployed from Fort Bragg, such as Capts. Kevin H. Govern and Kevin A. Ohlson, confronted a variety of legal issues, including an unusual question involving intelligence law. Because the initial situation in the Virgin Islands was so unclear, two Army counterintelligence agents had arrived on the second day of the operation. It quickly became evident that there was no threat that would directly involve them. Thus, they asked if they could assist the Federal Bureau of Investigation in gathering information about and rounding up escaped prisoners. After consulting with the U.S. Army Intelligence Command, judge advocates determined that such activities would constitute an improper "intelligence activity" under Executive Order 12333. Consequently, the two agents were restricted to performing administrative functions for the bureau.[20]

Operation HAWKEYE was the first time in nearly ten years that U.S. troops had been involved in civil disturbance operations. Army judge advocates involved in the planning and execution of the mission now had practical experience in solving relevant legal and nonlegal problems and had established a close working relationship with other corps staff principals. That experience would pay dividends when XVIII Airborne Corps personnel assumed a major role in Operation JUST CAUSE just a few months later.

Disaster Relief in Western Samoa, 1990

In February 1990, Typhoon Ofa swept over the two tiny islands of Western Samoa, located in the middle of the Pacific Ocean some 2,000 miles east of Australia. The storm and an accompanying twenty-foot tidal wave destroyed all crops and badly damaged the coastal road around the main island. Miraculously, few people were killed, but some villages were washed out to sea and there was no power on either island.

Believing that disaster relief assistance was critical, and with nine American Peace Corps volunteers missing and perhaps injured or dead, the chargé d'affaires at the U.S. embassy in Western Samoa radioed the commander in chief, U.S. Pacific Command (CINCPAC), for help. The CINCPAC agreed to help and created a joint task force, tasking it with finding and rescuing the missing Americans and rendering other disaster relief support to Western Samoa.

The result was an eighty-person joint task force (JTF) drawn primarily from the 25th Infantry Division. The "Tropic Lightning" Division provided the task force with two 500-kilowatt generators and engineers to run them, two OH–60 Black Hawk helicopters and crews, and soldiers from the division's support brigade. An Air Force search and rescue team and two naval construction engineers ("Seabees") also joined the force. The joint task force commander, an Army colonel, had a staff that included a judge advocate, Army Capt. Manuel E. F. Supervielle. *(Map 13)*

Captain Supervielle, having recently assumed duties as the operational law attorney at U.S. Army Western Command, received little advance notice that he was deploying to Western Samoa: he was told at 1 P.M. of his selection as JTF staff judge advocate and advised that he should be ready to deploy the next day at 5 A.M.[21]

After arriving in Western Samoa, Supervielle quickly discovered that living conditions would be Spartan. He slept the first night in a gutted hangar by the airfield. After that, he lived in a hotel otherwise vacant of guests in the aftermath of the typhoon. During the entire thirty-day deployment, Supervielle's diet consisted only of Meals, Ready-to-Eat. Only chemically treated water was safe to drink. Additionally, the lack of transportation and clerical support, combined with the hot climate, made for difficult working conditions. Captain Supervielle walked almost everywhere, including a mile and a half morning and evening trek between the hotel and the government building in Apia in which he worked.

Captain Supervielle, as the only U.S. government attorney in Western Samoa, found his legal skills in great demand. The chargé

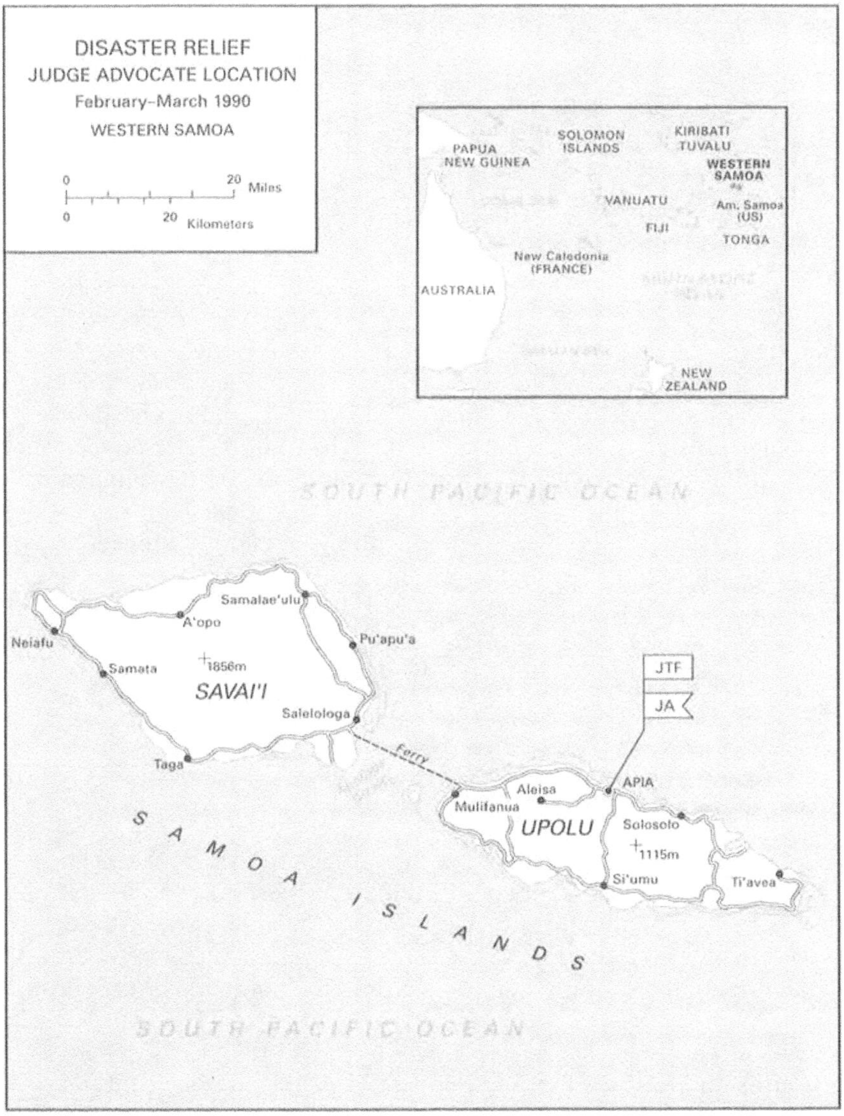

MAP 13

d'affaires looked to him for legal advice, as did the task force commander. Using a personal computer borrowed from Western Samoa's attorney general, Supervielle drafted a temporary Status of Forces Agreement within six days of his arrival. He negotiated a permanent Status of Forces Agreement by the third week. In both the temporary and the permanent agreements, Supervielle secured diplomatic status for all joint task force personnel equivalent to that afforded to the administrative officers and technical staff of a diplomatic mission. Under the 1961 Vienna Convention on Diplomatic Relations, this meant that all U.S. military personnel in Western Samoa would enjoy immunity from that country's criminal and civil proceedings while performing their official disaster relief duties.

Desiring a better understanding of the joint task force mission—which would enable him to better represent it in negotiations with Western Samoa—Supervielle also took time to perform some hands-on disaster relief. On the weekends, Captain Supervielle flew resupply missions as a Black Hawk helicopter crewman. This meant flying from the main island of Upolu to the isolated island of Saval'I with about 1,000 pounds of rice, flour, and dried milk. When Supervielle noticed that some villages over which he was flying were not getting supplied with food as often as others, he alerted the joint task force commander and chargé d'affaires, who then worked with the Western Samoans to reorder their priority of food deliveries.[22]

When Captain Supervielle returned to Hawaii in March 1990, he had a tremendous amount of practical experience in disaster relief—experience that would prove to be invaluable when he deployed as the sole judge advocate in a much larger disaster relief mission, Operation Sea Angel.

Operation Provide Comfort, 1991

On 28 February 1991, the cease-fire in the war against Iraq took effect. U.S. and coalition forces on Iraqi soil remained there for a time, but within weeks the redeployment of these forces was under way. As the allies withdrew from Iraq, however, an internal conflict began between the remaining Iraqi combat forces and Kurds in the north and Shiites in the south.

President Bush and his advisers, while sympathetic to the plight of the Kurds and Shi'ites battling Saddam Hussein's regime, did not wish to be drawn into an Iraqi civil war. An additional consideration was that both Saudi Arabia and Turkey counseled against military intervention, fearing that Iraq, if weakened further, might disintegrate. This would cause yet more turmoil and instability in the Persian Gulf region.[23]

On 5 April 1991, the UN Security Council condemned Iraq's repression of the Shi'ites and Kurds. Contemporaneously with this UN action, President Bush announced that, beginning 7 April, U.S.

Air Force transport planes would fly over northern Iraq dropping supplies of food, blankets, clothing, tents, and other relief-related items for Kurdish refugees. Within weeks, however, Operation PROVIDE COMFORT had expanded beyond airdropping supplies to building temporary shelters and providing protection for the Kurds. This expanded role led to the entry of military forces into northern Iraq, the establishment of temporary transit camps, and the return of most of the Kurdish displaced persons to their original homes within the security zone. At its peak, the combined task force, under the operational control of Headquarters, U.S. European Command (USEUCOM), involved more than 25,000 personnel. About 11,000 were Americans, with the remaining military personnel coming from the coalition countries, including France, Spain, Italy, The Netherlands, Germany, Belgium, Australia, Canada, Luxembourg, Turkey, and the United Kingdom. This made PROVIDE COMFORT the largest relief effort in modern military history.[24]

Army judge advocates participated in Operation PROVIDE COMFORT from the beginning. *(Map 14)* Col. Joseph L. Graves, Jr., staff judge advocate of the 21st Theater Area Army Command, deployed immediately to the Combined Task Force PROVIDE COMFORT headquarters at Incirlik Air Force Base, Turkey. Graves, along with Maj. Ralph L. Littlefield, a procurement and contract law expert, and Maj. Joseph A. Ridenhour, an operational law specialist, advised Lt. Gen. John M. Shalikashvili, the Combined Task Force PROVIDE COMFORT commander. Colonel Graves was replaced by Col. Quentin W. Richardson a short time later. *(Chart 14)*

Judge advocates initially serving in Army units of the combined task force included Capt. Alain C. Balmanno, the command judge advocate for the Combined Support Command, and Capt. Kevin H. Govern, the command judge advocate for Joint Task Force–A (10th Special Forces Group). Both were located in Silopi, Turkey, a small town approximately five kilometers north of the Iraqi-Turkish border. Across that border, in Zakhu, Iraq, Capt. Mitchell R. Chitwood served as command judge advocate for Joint Task Force–B (Security Operations). Other Army lawyers in the Zakhu area included Capt. William McQuade, who provided legal advice to the 3d Infantry Division's aviation brigade, and Capt. Corey L. Bradley, the judge advocate assigned to the 3d Battalion, 325th Infantry Regiment (Airborne Battalion Combat Team). Unlike his fellow judge advocates, who had deployed from Germany, Captain Bradley had flown to Iraq in July 1991 from his home station of Vicenza, Italy. Regardless of their normal place of duty, however, all found that living conditions in south-

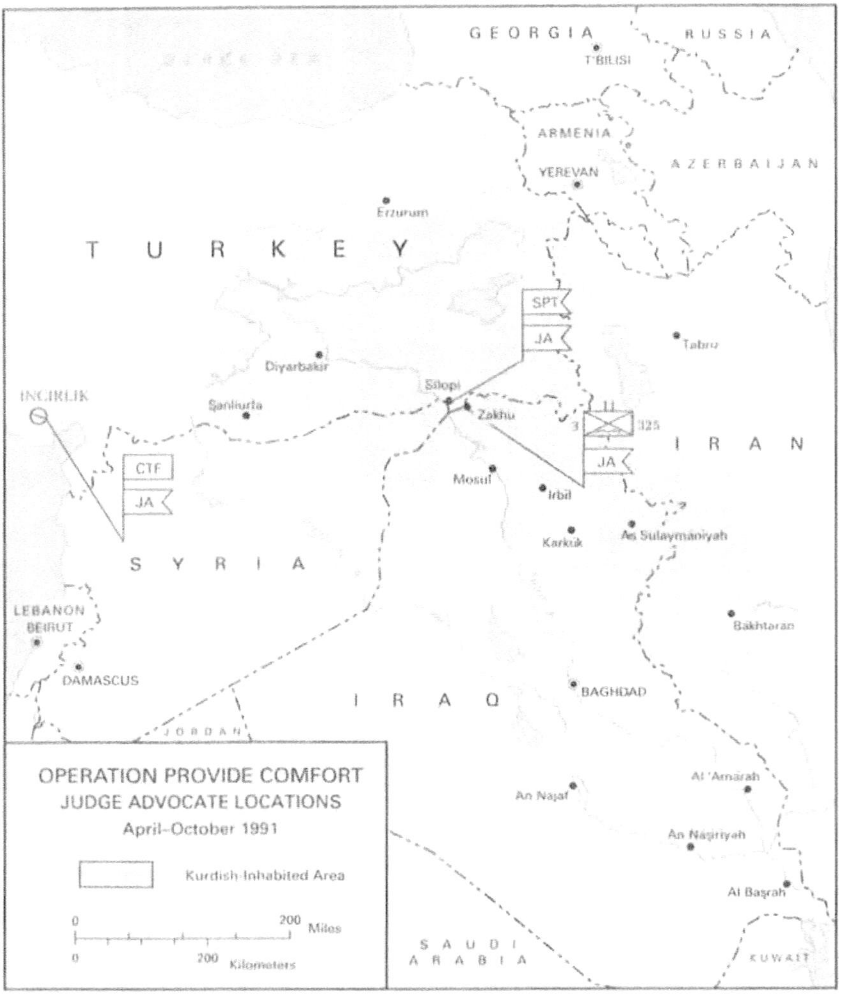

MAP 14

ern Turkey were extremely primitive. The weather was harsh—freezing conditions in the rugged mountains—and 110-degree summer days in the plains around Silopi and Zakhu.

As relief operations continued, other Army lawyers deployed to Turkey and Iraq, replacing those who had been part of the initial deployment. Thus, in mid-May 1991, Maj. Richard E. Gordon deployed to Silopi, replacing Captain Balmanno as the Combined Support Command's command judge advocate. Maj. Fred T. Pribble replaced

CHART 14—ORGANIZATION OF LEGAL SERVICES FOR COMBINED TASK FORCE PROVIDE COMFORT, MAY 1991

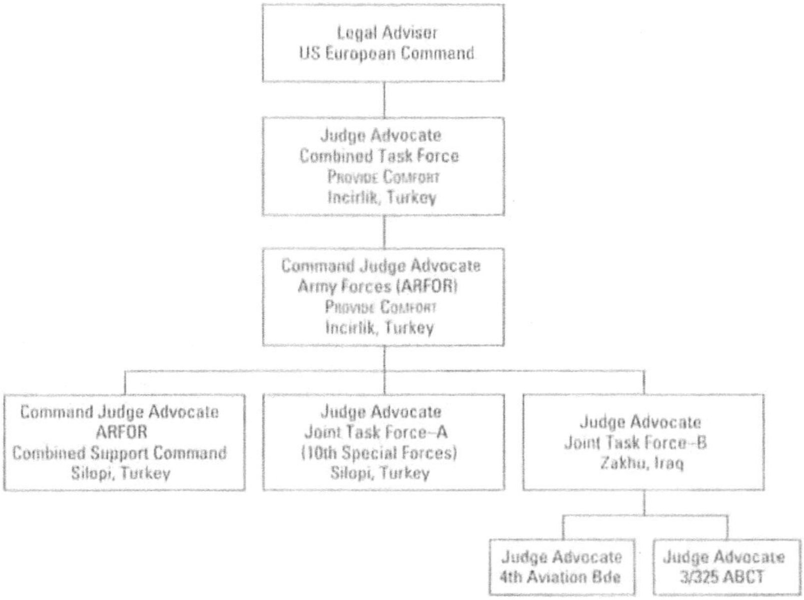

Gordon in late June, and Pribble, in turn, was replaced by Capt. Charles N. Pede in August.

The issues facing the judge advocates during PROVIDE COMFORT were varied. As had become common, drafting rules of engagement was the first order of business, even though PROVIDE COMFORT was a humanitarian operation. This resulted from the fact that the relief effort was not conducted at the invitation of the Iraqi government. Rather, PROVIDE COMFORT included a security mission—to separate the advancing Iraqi military forces from the personnel of the combined task force, international relief organization personnel, refugees, and other displaced persons. Additionally, once the Americans deployed into northern Iraq, the likelihood of armed conflict increased. Consequently, Colonel Richardson and his staff drafted rules that put into effect General Shalikashvili's guidance on the use of force. Applicable only to U.S. forces, and published on 1 May 1991, these rules of engagement stressed that Operation PROVIDE COMFORT was "a humanitarian assistance operation [and that] the multinational forces are not at war." Combined task force personnel, however, always had "the right to use force in self-defense when responding to attacks or threats of imminent attack against the multinational forces, humanitarian relief personnel and refugees." Finally, the rules of engagement specifically required task force personnel to "disarm and detain any Iraqi soldier or member of a paramilitary security force, or any civilian policeman . . . within a designated security area." In this regard, the minimum force necessary to disarm and detain was authorized, including the use of deadly force, if hostile intent was exhibited.

The rules remained in effect for the duration of PROVIDE COMFORT; however, as a result of the multinational character of the operation, the British, French, Dutch, Australian, and other military participants employed their own rules of engagement.[25] Some were markedly different. For example, some nations did not permit the use of deadly force in response to a demonstration of hostile intent, requiring instead that an individual or unit actually receive hostile fire before responding with fire. Fortunately, the existence of different rules of engagement never created problems in the combined task force, as the Iraqis never threatened the operation.[26]

Along with their work on rules of engagement, judge advocates also gave top priority to the drafting of General Order no. 1. Given the demonstrated success of a punitive general order in Operations DESERT SHIELD and DESERT STORM, General Shalikashvili sought a similar framework for PROVIDE COMFORT personnel. As had been the case during combat operations in Southwest Asia, the central prohibition in General Shalikashvili's general order was a ban on consuming alcoholic beverages. However, the presence of Turkish and Kurdish street vendors in Silopi and Zakhu, all of whom had large quantities of Turkish beer to sell, meant that violations of the ban on drinking alcohol were somewhat routine. Army lawyers advised that such breaches of discipline should be handled with nonjudicial punishment under Article 15 of the Uniform Code.[27]

Most legal issues—from administrative, fiscal, and contracting law to military justice and legal assistance—had been anticipated. The payment of foreign claims, however, was a different matter. Major Gordon and other judge advocates, for example, soon discovered that Kurdish farmers were filing claims not only for the loss of crops and animals caused by U.S. forces, but also for the depreciated value of their land. The land around Zakhu had been chosen for these camps because it had a sufficient amount of clean well water, was flat, and was readily accessible by road. The proximity of the Tigris River also made the land very fertile: wheat, barley, and rye grew during the fall months, while cucumbers, tomatoes, and watermelons were produced in the spring and summer. Consequently, the Kurdish farmers owning land upon which refugee camps were built claimed that the value of their land had been depreciated by these camps and demanded compensation for this damage.

For Army lawyers, the problem in adjudicating these claims resulted from the fact that no Kurdish claimant could produce documents or papers proving land ownership, since, to assure loyalty to his regime, Saddam Hussein had confiscated all land titles and deeds. Investigating land claims thus required a certain degree of ingenuity. The eventual solution involved calling all interested farmers and Kurdish leaders together and then, in a group discussion, reaching an agreement among all concerned as to which farmer owned a particular plot of land. For some judge advocates, this claims work resulted in memorable experiences. Major Gordon, for example, left Zakhu one afternoon in a military vehicle. Driving alone, his mission was to inspect the property lines of a piece of land on which U.S. engineers were planning to build a refugee camp. The land was flooded, and Gordon's vehicle became mired in the mud. He was in Iraq, some six miles away from the nearest combined task force element. Suddenly, an old farmer riding a donkey came by. The man stopped, got off his animal, pulled out a machete, and began cutting down stalks of grain to shove under the truck wheels. Within ten minutes the Kurd had cut a layer of grain, and Major Gordon was able to drive his vehicle back to the road. With a wave of his hand, the man got back on his donkey and rode off down the road.[28]

While Army judge advocates like Colonel Richardson and Major Gordon advised the commanders and staffs of large organizations, Army lawyers such as Captain Bradley had very different experiences. Arriving in late June 1991 to assume duties as the command judge advocate for a 1,200-man airborne battalion combat team headquartered in Vicenza, Italy, Captain Bradley discovered that his unit had deployed to Iraq in May as part of PROVIDE COMFORT. Within days, however, the battalion commander, Lt. Col. John P. Abizaid, directed that all deployable soldiers remaining in his unit in Italy deploy to Iraq as soon as possible. Bradley was faced with a challenge that few judge advocates experience, for, as the senior officer, he was in charge of deploying this group of infantrymen from Italy to Turkey. He was responsible not only for some thirty infantrymen—a lieutenant, two junior noncommissioned officers, three specialists, and two dozen privates, most of whom were fresh out of basic training—but also for their weapons. Arriving in Turkey, Captain Bradley not only discovered that there were no liaison personnel to assist the Americans through customs but that the Turks were refusing entry to some soldiers and attempting to confiscate their weapons. Getting his soldiers through customs "stands out as one of the most difficult and vital negotiations of my JAG career," remembers Bradley.[29]

After successfully passing through customs, Captain Bradley and his contingent boarded a chartered bus in Incirlik. The bus was to take the Americans to Silopi, but during the thirteen-hour ride, Captain Bradley often wondered whether they would make it to their destination. The Turkish bus driver had no map and spoke no English. In the end, however, Bradley and the paratroopers—and their weapons—arrived safely in Silopi and joined their unit in Iraq the next day.

Over the next month, Captain Bradley advised on a variety of operational law issues, including the formulation of a war trophies policy, legal aspects of security checkpoints, and humanitarian obligations toward Kurdish refugees. Bradley also provided legal assistance to American soldiers, processed military personnel and foreign claims, and advised Colonel Abizaid and his subordinate commanders on military justice matters. In addition to his busy legal work, Captain Bradley took part in airborne operations intended to demonstrate to Iraqi forces and Kurdish rebels that the Americans were both willing and able to project force in the region. In participating in this "show of force," Captain Bradley became one of the few Americans to parachute into Iraqi territory as part of a conventional military mission.[30]

Although Operation PROVIDE COMFORT did not end until 31 December 1996, direct Army Judge Advocate Corps participation ended with the departure of Captain Pede, who returned from Silopi to Germany in early October 1991. Between April and October, however, Army lawyers, along with Air Force, Navy, and Marine judge advocates, had fully supported task force commanders and had taken on a variety of challenges.

Operation SEA ANGEL, 1991

On the night of 29 April 1991, a killer cyclone with winds of 140 miles per hour, accompanied by a twenty- to thirty-foot-high tidal wave, struck Bangladesh. The destruction that the storm inflicted on this densely populated nation on the Indian subcontinent was immense. Within hours, the entire southeast countryside was under water. Trees were uprooted, villages leveled, and crops destroyed. Electricity and telephone services were eliminated, and roads washed away. Additionally, the bodies of some 139,000 men, women, and children and more than a million cattle were floating in water. The immediate problem was not repairing buildings, roads, and power grids. Rather, famine and disease now threatened some 11.2 million people in the areas of Bangladesh most devastated by the cyclone.[31]

The United States, along with the rest of the international community, responded quickly. Recognizing that humanitarian relief was imperative, American Ambassador William B. Milam requested immediate assistance from the Department of State and the Agency for International Development, particularly the former's Office of Foreign Disaster Assistance. Within days, Milam also queried U.S. Pacific Command as to whether Marine or Navy assets in the area might assist in relief operations.[32] This request for military aid resulted in the creation of Joint Task Force SEA ANGEL, and by 11 May, when President Bush formally issued the order dispatching American military personnel to Bangladesh, the planning for humanitarian relief efforts was well under way. *(Map 15)*

SEA ANGEL was primarily a Navy and Marine Corps operation, with significant Air Force participation in the form of C–5 Galaxy, C–130 Hercules, and C–141 Starlifter aircraft and crews. The Army contributed some UH–60 Black Hawk helicopters and Special Forces personnel. *(Chart 15)* Given the substantial sea service composition of Task Force SEA ANGEL and the fact that its commander was Marine Maj. Gen. Henry C. Stackpole III, it appeared to be somewhat unusual that Admiral Charles R. Larson, Commander in Chief, U.S. Pacific Command, selected an Army judge advocate, Maj. Manuel E. F. Supervielle, as the legal adviser for the task force. Not only did Supervielle become the only Army officer on Stackpole's staff, he was also the only lawyer. Central to Admiral Larson's decision, however, was his recognition that no other judge advocate in the area of operations was better qualified than Supervielle in view of his recent experiences with disaster relief operations in Western Samoa.

After arriving in Bangladesh by air in early May 1991, Major Supervielle set up his cot and sleeping bag in an empty house rented by the U.S. embassy. He conducted his legal work there or at the task force headquarters, both of which were located in the capital city of Dhaka. Supervielle quickly established a personal relationship with the ambassador, his deputy chief of mission, and others on the embassy staff. A close rapport with these officials proved to be the key to resolving a number of legal issues, as direct access to these decision makers often was critical to mission success.

Recognizing the importance of quickly establishing the legal status of all U.S. Pacific Command personnel present in Bangladesh, Major Supervielle spearheaded negotiations for a Status of Forces Agreement within days of arriving. On 14 May 1991, Ambassador Milam granted

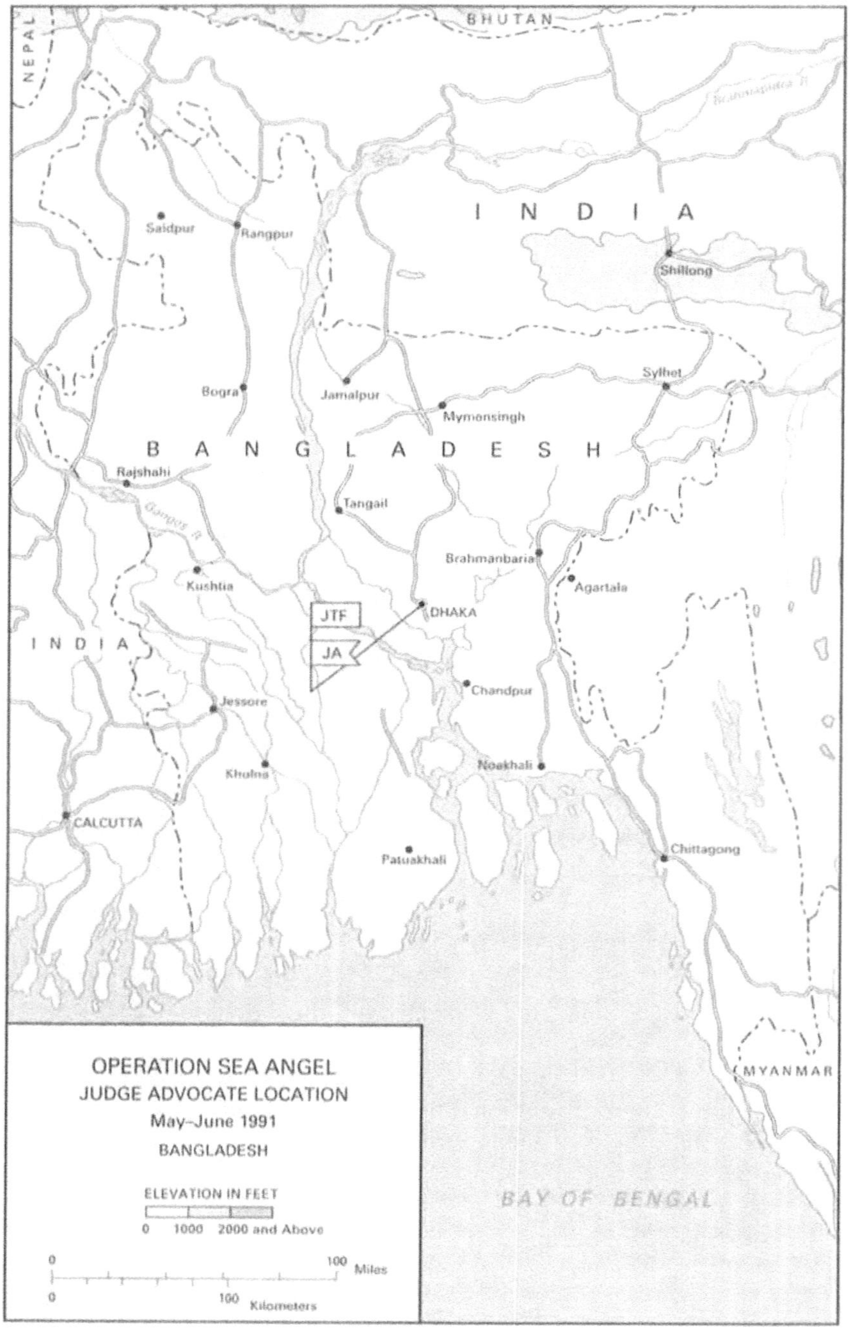
BHUTAN
NEPAL
INDIA
Saidpur
Rangpur
Shillong
Sylhet
Bogra
Jamalpur
Mymensingh
BANGLADESH
Rajshahi
Tangail
Brahmanbaria
Kushtia
Agartala
JTF
DHAKA
INDIA
JA
Chandpur
Jessore
Noakhali
Khulna
CALCUTTA
Patuakhali
Chittagong
MYANMAR
OPERATION SEA ANGEL
JUDGE ADVOCATE LOCATION
May–June 1991
BANGLADESH
ELEVATION IN FEET
0 1000 2000 and Above
0 100 Miles
0 100 Kilometers
BAY OF BENGAL

Map 15

Chart 15—Staff Organization for Operation Sea Angel

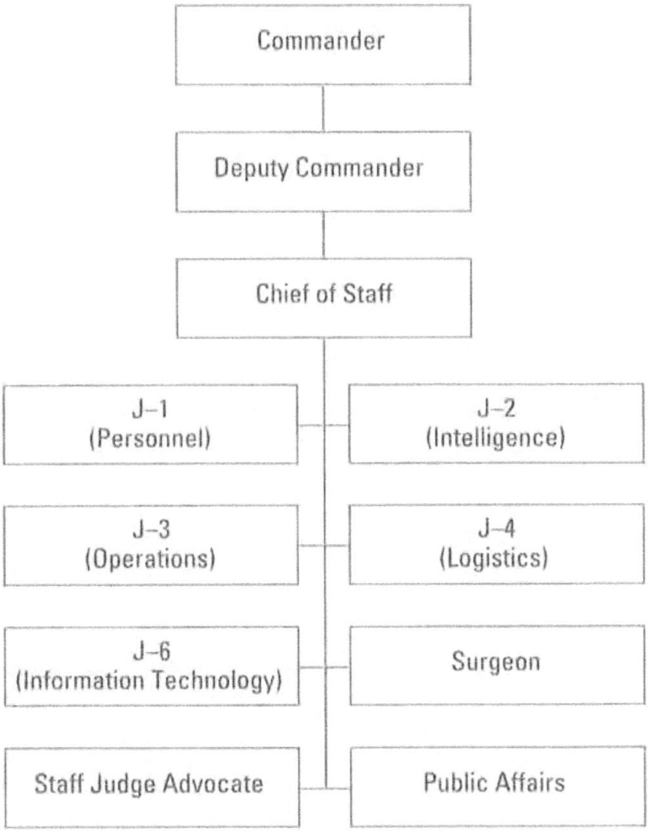

Major Supervielle the authority to negotiate and conclude a temporary Status of Forces Agreement, and, working closely with high-level Bangladeshi government lawyers in Dhaka, Supervielle reached an agreement on 18 May. The president of Bangladesh approved the wording of the agreement the next day, and Ambassador Milam and the Bangladeshi foreign secretary signed it the following day.

As he had done in Western Samoa, Major Supervielle negotiated a Status of Forces Agreement providing task force personnel with diplomatic status equivalent to that provided members of the administrative and technical staff of the U.S. embassy in Bangladesh. These personnel thus enjoyed immunity from the jurisdiction of the civil and criminal courts of Bangladesh while performing their official duties. They were also permitted to enter and leave Bangladesh upon presentation of their

U.S. Armed Forces identification and were exempt from customs and taxation. Additionally, U.S. aircraft landing in or leaving Bangladesh were exempt from customs inspections. This temporary Status of Forces Agreement remained in effect for the duration of SEA ANGEL.[33]

A second legal issue of critical importance to General Stackpole and the joint staff was the use of Department of Defense Operation and Maintenance (O&M) funds. Since the Department of State, through its Office of Foreign Disaster Assistance, and the U.S. Agency for International Development were primarily responsible for funding disaster relief overseas, Major Supervielle realized that using Defense Department funds for SEA ANGEL initiatives posed a problem. After some research, however, he concluded that the department's O&M funds could be used if General Stackpole, as the on-the-scene commander, determined that "time is of the essence and . . . humanitarian considerations make it advisable to do so." Consequently, Major Supervielle advised that O&M monies might be used when the commander determined that immediate aid was required to save lives and property. However, he noted, when SEA ANGEL relief efforts assumed the look of rehabilitation work normally expected of a civil government rather than emergency life-saving missions, the joint task force would be required to seek Office of Foreign Disaster Assistance funding.[34]

In addition to handling all legal issues associated with SEA ANGEL, Major Supervielle also assisted Marine Lt. Col. James R. Morris, the assistant chief of staff for personnel (J–1), in his work. This included recommending a course of action for the awarding of the Humanitarian Service Medal to disaster relief personnel and drafting and recommending a "liberty policy" for SEA ANGEL forces. The liberty policy covered cultural matters (conservative attire in recognition of the Islamic religion), safety issues (no individual could take liberty alone), shopping (must be particularly polite and patient), and other commonsense matters.

Major Supervielle left Dhaka's Zia International Airport, along with the rest of the joint task force command and staff, during the first week of June 1991. For him, SEA ANGEL's biggest lesson was the fact that a judge advocate must insert himself into humanitarian assistance and disaster relief planning from the very beginning of the operation. This was particularly important in the case of SEA ANGEL, as there had been no preexisting operations plan for disaster relief in Bangladesh. As a result, the plan was developed after the Americans were on the ground. Major Supervielle's early involvement in political issues, such as the legal status of U.S. forces, and fiscal law issues made his work as task force staff judge advocate critical to mission success.

Additionally, his ability to work in resolving nonlegal problems illustrated that judge advocates were well suited to handle a variety of tasks ordinarily handled by other staff principals.

Migrant Camp Operations at Guantanamo Bay Naval Base, 1991–1994

Violence and economic hardship in Haiti led thousands of Haitians to flee their country in flimsy wooden vessels. Most headed toward the United States. A picket line consisting primarily of U.S. Coast Guard cutters intercepted many of these "boat people," and after "on ship" asylum screening the vast majority were repatriated to Haiti. This high seas repatriation came to a halt, however, when human rights activists filed suit in the U.S. District Court for the Southern District of New York and obtained a restraining order against on-board screening and repatriation.

As a result of the federal judge's decision in *Haitian Community Council, Inc.* v. *McNary*, the Coast Guard began taking Haitian refugees intercepted at sea to Guantanamo Bay Naval Base in Cuba for immigration screening. While waiting for their immigration hearings, the men, women, and children lived in camps on the base. As the number of refugees grew into the thousands, additional military personnel were required to run the camps and maintain security. The U.S. Atlantic Command, responsible for operations in the geographic area that included Guantanamo, established Joint Task Force GITMO to oversee Haitian refugee operations. Given that its commander, Brig. Gen. G. H. Walls, Jr., was a marine, the bulk of the task force personnel were also marines. Soldiers from the XVIII Airborne Corps, however, also joined the task force, and sailors and Coast Guard personnel participated in the operation in addition to carrying out their complementary maritime activities on the high seas.

The first Army judge advocate involved in operations at Joint Task Force GITMO—a five-letter acronym for Guantanamo—was Maj. Allen K. Goshi. The operational law attorney at XVIII Airborne Corps, Goshi was part of a liaison team that traveled from Fort Bragg to Camp Lejeune, North Carolina, to coordinate Army participation in task force operations. A military police company from Fort Bragg was to provide security for the task force, and this Army involvement led the XVIII Airborne Corps staff judge advocate, Col. Fred E. Bryant, to recommend that a lawyer be included in the planning process. A judge advocate, he felt, would be particularly useful in formulating rules of engagement; writing guidance on the use of force became Major

Goshi's first task after he arrived in Cuba only a few days before Thanksgiving in 1991.[35]

Goshi drafted rules of engagement based on Department of Defense Civil Disturbance Plan GARDEN PLOT. After coordinating his efforts with joint task force operations (J–3), Major Goshi briefed General Walls, who approved the rules. Impressed with Goshi's work, Walls also decided to make him the joint task force staff judge advocate, and from that point on Major Goshi was fully integrated into the headquarters staff. During the roughly eighty days that he spent at Guantanamo Naval Base, Goshi dealt with a number of legal and nonlegal problems. He drafted the joint task force disciplinary policy, which left the responsibility to administer military justice to each component commander in the force, while reserving to General Walls the authority to handle a particular case in unusual circumstances. At Walls' direction, Major Goshi also drafted a general order prohibiting fraternization between task force personnel and refugees and authored a policy letter on the wearing of military uniform items in the joint task force, ordinarily a personnel (J–1) issue. Finally, he was the joint task force point of contact for liaison with all nongovernmental organizations. This meant that he worked closely with the United Nations high commissioner for refugees and international charity organizations with an interest in the Haitians. When the litigants involved in *Haitian Community Council, Inc.* traveled to Guantanamo to take depositions, Major Goshi was directly involved.

By December 1991, six camps, each surrounded by triple-strand concertina wire and each containing 1,500 refugees, had been built on an unused airfield on the American base. Most Haitians living in the camps were well behaved, but there were some troublemakers. Consequently, General Walls directed that Major Goshi write a disciplinary code to handle misdemeanor-level misconduct in the camps. Based on the commander's inherent authority over those under his control, and using the applicable punitive provisions of the Uniform Code of Military Justice as a guide, Goshi wrote a disciplinary code applicable to all those living in the camps. Its provisions ranged from requiring obedience to the commands of security personnel to waiting in line to receive food. A refugee who violated the code faced an administrative penalty that could include being separated from the general camp population and placed in a segregated area. Due process was satisfied through the use of a hearing at which the camp commander—a Marine Corps major—decided guilt or innocence and imposed the appropriate penalty. An appeals process through superior commanders was also established; however, Goshi was careful to preserve an administrative character for the disciplinary code so that there could be no contention that it was, in effect, a criminal code requiring additional due process.

In mid-December 1991, the refugees in one camp rioted and demanded immediate passage to the United States. Making weapons out of tent stakes and the metal frames of their cots, the Haitians used these to drive the military police out of their camp. The refugees were prevented from leaving the camp by the military police; however, their violent behavior continued until they were informed that they would receive no food until the violence abated. Over the next few days Major Goshi assisted the joint task force chief of staff and the UN high commissioner for refugees in negotiations with the Haitian agitators, putting his common sense to good use as a diplomat.[36]

During these December riots, more U.S. troops were rushed to Cuba from the XVIII Airborne Corps at Fort Bragg and the 10th Mountain Division at Fort Drum. When the 10th Mountain Division was tasked with providing a battalion task force for this security operation, Lt. Col. Gerard A. St. Amand, the division staff judge advocate, selected his operational law attorney, Capt. Norman F. J. Allen III, to accompany the task force. As a result, Allen was immediately involved in planning for the deployment of the task force from Fort Drum. He focused most of his efforts on the rules of engagement, briefing those soldiers of the 2d Battalion, 14th Infantry, selected for deployment, as well as the commanding general and staff. Since this was a security rather than a combat operation, the rules on the use of force focused on warning shots and graduated responses in using force. Deadly force was permissible only for self-defense or force protection. After arriving in Cuba and setting up his operation, Captain Allen provided legal assistance to the roughly 600-person battalion task force and also provided military justice and administrative law advice to Lt. Col. William C. David, the task force commander.[37]

Army judge advocates at Guantanamo Naval Base faced one other unanticipated challenge: the forcible repatriation of refugees. The Immigration and Naturalization Service determined that a large number of Haitians were not entitled to political asylum in the United States and should be returned to Haiti. Almost all of the refugees agreed to return voluntarily, but a few refused. Consequently, it was decided to forcibly repatriate them. As the Coast Guard wished to have military guards on board its cutters during the journey from Guantanamo to Port-au-Prince, Major Goshi advised that use-of-force rules were needed during these forcible repatriation operations. At General Walls' direction, Goshi drafted a rules of engagement annex to the repatriation operations plan, covering such matters as the amount of force permitted in carrying resisting refugees up a gangplank onto a ship, actions to be taken during any riot aboard a ship, and the circumstances under which deadly force could be used. Later, Goshi personally supervised the loading of refugees on each ship and briefed all troops and commanders on the rules of engagement. As events unfolded, it was not necessary to use force in repatriating Haitian refugees. However, the possibility had been anticipated, and a practical and lawful solution had been prepared.

After the departure of Captain Allen in January 1992 and of Major Goshi during the following month, other Army judge advocates followed. Maj. M. Tia Johnson, for example, deployed to Guantanamo as Joint Task Force GITMO's deputy staff judge advocate in May 1992. Until her return to the United States in August, she worked at the direction of the task force staff judge advocate, an Air Force colonel, in providing legal support to the continuing mission of caring for the Haitians. Major Johnson focused most of her energy, however, on supervising the military attorneys complying with a discovery demand for nearly 100,000 military documents. This request, arising out of a civil suit filed against the United States in the U.S. District Court for the Southern District of New York, required the reading and analysis of thousands of pages of documents.[38]

More than a year later, in May 1994, a new refugee crisis emerged after tens of thousands of Haitians and Cubans made their way to Florida in homemade rafts and rickety boats. The Haitians were fleeing the military dictatorship that ousted Jean Bertrand Aristide, while the Cubans were fleeing Fidel Castro's Communist regime. The Coast Guard, again manning a picket on the high seas, intercepted these refugees and took them to Guantanamo Bay. By the fall of 1994, some 21,000 Haitians and 29,000 Cubans were living in some of the same tent cities built for Haitian refugees in 1991. But while there had been roughly 10,000 refugees in 1991, there now were five times that number. As a result, new schools, churches, and basketball courts were erected, and telephones were installed. More than 8,000 soldiers, airmen, sailors, and marines participated in these new humanitarian relief operations, either as part of Operation SEA SIGNAL or as members of Joint Task Force 160, the task force established to oversee the refugee camps.

Joint Task Force 160, popularly known as Joint Task Force GITMO, had five judge advocates. The staff judge advocate, Marine Lt. Col. John McAdams, was joined by an Air Force major, a Navy lieutenant, and a Marine captain. Rounding out the legal staff was an Army judge advocate, Capt. Stewart A. Moneymaker. While serving as the chief of claims at the 1st Infantry Division, Fort Riley, Kansas, Moneymaker was selected to join the 81st Military Police Brigade deploying to Guantanamo from III Corps, Fort Hood, Texas. Captain Moneymaker arrived at Guantanamo in June 1994 and, until returning to the United States in November, was the sole Army judge advocate in Cuba.[39]

During his deployment, Moneymaker provided legal assistance to members of the joint task force and also advised the command on military justice matters, while arranging with his Navy judge advocate colleagues at Guantanamo for trial defense counsel for task force personnel who required the services of a defense attorney. He also acted as a liaison to Amnesty International when that organization's members arrived in Cuba to speak with Haitian and Cuban refugees and to inspect their living conditions.[40]

With Aristide's restoration to power following Operation UPHOLD DEMOCRACY, most Haitians were denied political asylum, and the last Haitian departed the Guantanamo camp for Haiti on 1 November 1995. Most of the Cubans, however, were granted entry into the United States, and when the last Cuban refugee left Guantanamo Naval Base for Florida on 31 January 1996, refugee operations ended.[41]

Los Angeles Civil Disturbance Operations, 1992

Late in the evening of 29 April 1992, following the acquittal of four white policemen charged with brutally beating a black motorist, Los Angeles erupted in violence. Governor Pete Wilson quickly ordered the mobilization and deployment of California National Guard troops. Despite a rapid National Guard response, however, the governor determined that escalating violence and disorder made a request for federal troops imperative. As a result of this request, President George Bush issued a proclamation on 1 May directing that all persons cease committing acts of violence and disorder. The president also signed an executive order providing for the federalization and use of National Guard troops and the deployment of active duty forces to Los Angeles. As a result, by 6 P.M. that day, Joint Task Force LOS ANGELES had begun operations at Los Alamitos Army Reserve Center. Commanded by Maj. Gen. Marvin L. Covault, the task force consisted of 9,800 members of the California National Guard's 40th Infantry Division (Mechanized) and its 49th Military Police Brigade; 1,500 marines from Camp Pendleton; and 1,700 soldiers from the 7th Infantry Division (Light), stationed at Fort Ord. At any given time between the 1st and 6th of May, some 2,600 National Guard troops, 440 marines, and 680 active duty soldiers patrolled the streets of Los Angeles. Their mission was to suppress violence and to restore law and order in the City and County of Los Angeles.[42]

Since General Covault, the task force commander, was also the commander of the 7th Infantry Division (Light), the nucleus for the task force staff was drawn from the division assault command post. This resulted in the division deputy staff judge advocate, Maj. Scott C. Black, becoming the initial JTF staff judge advocate. Anticipating that the task force would operate on a 24-hour-a-day basis, Major Black recognized that another judge advocate would be needed. Consequently, he selected Capt. Bradley D. Page, the chief of administrative and civil law at the 7th Infantry Division legal office, to join him in Los Angeles. Black and Page were joined a few days later by Marine Col. Richard G. Walls, the Marine Expeditionary Force staff judge advocate. As the senior ranking officer, Colonel Walls replaced Black as the task force staff judge advocate. Major Black thus became the deputy staff judge advocate, and Captain Page remained as the third judge advocate at the task force headquarters. A fourth judge advocate, Army Capt. Robert S. Bowers, also deployed from Fort Ord to Los Angeles, but did not join Walls, Black, and Page at the task force headquarters. Instead, he accompanied the division's brigade task force into the streets, providing both training and legal support.

From the beginning, these judge advocates confronted a number of difficult legal issues. These included the impact of the Posse Comitatus Act on task force operations, the lawfulness of loaning U.S. military property to state and local authorities, the effect on both military and civil law of federalizing the California National Guard, and the legality of arming orders and rules of engagement.

While the Posse Comitatus Act of 1878 generally prohibits federal troops from engaging or assisting in civilian law enforcement activities without a presidential mandate, the act does not restrict the president's authority to order U.S. troops to quell a civil disturbance. Consequently, Major Black immediately advised General Covault that there was no restriction on how his forces could be used. Thus, the Posse Comitatus Act never limited the decision-making process within the force headquarters, and U.S. troops did perform typical law enforcement functions. But while the law permitted soldiers and marines to be employed "piecemeal" under the control of individual members of the Los Angeles Police Department, General Covault determined that it was a wiser course of action to employ U.S. troops only under the control of their military leaders.

An issue related to Posse Comitatus, in the sense that it involved military assistance to law enforcement, was the legality of loaning and leasing U.S. military property to civilian government agencies. The Los Angeles Police Department, for example, requested the loan of twenty-five night vision goggles. Other similar requests for military property were also made to the joint task force during the first few days of its operation.

Using Army Regulations 500–51, *Support to Civilian Law Enforcement*, and 700–131, *Loan and Lease of Army Materiel*, as guidance, the judge advocates advised that military equipment fell into three loan or lease categories. Determining the applicable category was critical, as the level of approval authority differed for each. General Covault, for example, had the authority to approve the loan or lease of so-called Group Two military property ("riot control agents, concertina wire, and other like equipment"), but higher approval authority was required for other items. Given the nature of the task force mission and the need for prompt decisions, and given the danger to the citizens of Los Angeles, the judge advocates advised that requested property be classified in Group Two to the fullest extent possible. While this was expedient and necessary under the circumstances, the need for this action convinced the attorneys involved that the regulations governing requests for military equipment were ill-suited for civil disturbance situations.

A second significant legal determination in this area was the requirement, under AR 700–131, that some "loaning" of equipment be effected by lease. As a result, the requesting civilian agency, as the lessee, would be required to pay fair market value for the use of such equipment. This not only required a valuation of the property, but also meant that any lease must be accompanied by a surety bond equal to the value of the equipment provided. In the case of night vision goggles, the Los Angeles Police Department would be required to post a bond in excess of $100,000. This obligation created a significant problem, and by the time the Police Department's request for the goggles was successfully processed, the need for the devices no longer existed.

Although the federalization of the National Guard lasted but ten days, it also gave rise to a number of unanticipated legal issues. The most significant was that, in moving from a state status governed by Title 32, U.S. Code, to a federal status governed by Title 10, U.S. Code, National Guard personnel became subject to the Uniform Code of Military Justice. Thus, those soldiers who failed to report to their units after federalization were subject to prosecution only under the Uniform Code; they could not be disciplined under Title 32 or other state laws. Additionally, criminal offenses committed by guardsmen during the period of federal service required disposition under the Code. This new jurisdictional issue was an unanticipated—and unwelcome—consequence of federalization, particularly as the task force commander and his judge advocates had neither the assets nor the desire to pursue routine criminal proceedings against California National Guard soldiers. In any event, criminal misconduct by National Guard personnel would require quick action—either resolution of the matter prior to defederalization or reassignment of the soldier to a Regular Army unit for disposition of the pending charges.

Additionally, when California National Guard personnel transitioned to federal service, they obtained the rights and protections afforded by the Soldiers' and Sailors' Civil Relief Act. Accordingly, federalization resulted in increased legal assistance issues, ranging from limitations on interest rates and court appearances to reemployment and landlord-tenant rights.

At the direction of Major Black, Captain Page began drafting joint rules of engagement, using Department of Defense Civil Disturbance Plan GARDEN PLOT, existing rules of engagement from the 7th and 40th Infantry Divisions, and Army Regulation 190–14, *Carrying of Firearms and Use of Force for Law Enforcement and Security Duties*. Anticipating that street rioting would be the typical scenario facing task force personnel, yet concerned because the rioters would be American civilians, Captain Page's restrictions on the use of force included six levels of "arming orders"—features not ordinarily seen in rules of engagement.

These arming orders provided specific arming guidance for soldiers and marines in the task force, the idea being that any escalation in the level of force used would occur in a calculated and measured manner. Level One was the least, and Level Six the most, aggressive posture. This approach assisted the commander in maintaining a proper balance between the risk of an unnecessary discharge of a weapon and that of being caught unprepared in a dangerous situation. This was particularly true in the first few days in Los Angeles, when the level of tension was high.

Very early on the morning of 2 May 1992, General Covault approved the final draft of the rules of engagement. Key issues included the proper uses of lethal and nonlethal force, warning requirements, limits on automatic fire of weapons, changes to arming orders, the use of sniper teams, and the employment of riot control agents.[43]

General Covault implemented these rules by directing that 12,000 "helmet cards" be printed (a card small enough to fit under the elastic band holding the cloth camouflage cover on a helmet) and provided to all personnel. Additionally, Covault directed that all joint task force soldiers and marines be at Level One during normal circumstances. But the rules also provided the authority to the officer or noncommissioned officer in charge to determine the appropriate level of arming. The rationale for this was that first-line leaders were best able to determine the threat to their personnel and to direct the appropriate response. As a practical matter, however, some first-line supervisors directed increased arming order levels based on erroneous perceptions of a threat. Consequently, soldiers could be postured, at the same time, at Level One (rifle at sling, bayonet in scabbard, ammunition in magazine in pouch) through Level Six (rifle at port, bayonet fixed, ammunition in magazine, locked and loaded), with no well-defined reason for these differences. On the other hand, some first-line leaders incorrectly believed that they had to endure casualties before they could move beyond Level One. Colonel Walls and Major Black concluded that these problems could be solved in any future civil disturbance scenario with earlier dissemination of the rules of engagement and realistic threat training. This would produce a more consistent application of arming-order levels and result in a safer, more efficient military operation. In the end, however, the Task Force Los Angeles rules of engagement worked. The troops in the force carried some 350,000 rounds of ammunition and 3,700 tear gas grenades, yet they fired only twenty rounds during the course of the entire crisis.[44]

Although Joint Task Force Los Angeles operations lasted only ten days, their character as a tactical operation in a civil environment resulted in Army judge advocates' facing issues unlike those ordinarily seen in operations other than war. Whether advising on rules of engagement or processing requests to borrow military equipment, these attorneys always had to consider the unique domestic aspect of the mission—that it involved the potential use of deadly force against American citizens.

Hurricane Andrew, 1992

At 5 A.M. on 24 August 1992, Hurricane Andrew struck south Florida near Miami. Its sustained winds of over 145 miles per hour wreaked havoc, leaving more than 30,000 homeless and an eventual clean-up cost exceeding $30 billion. Federal Emergency Management Agency and Florida officials, after assessing the disaster, advised Governor Lawton Chiles that local resources were inadequate to respond to the emergency. Consequently, Chiles requested federal assistance, and on 24 August President Bush declared Dade, Monroe, and Broward Counties to be disaster areas. That same day, Secretary of the Army Michael Stone, as the executive agent for domestic disaster assistance for the Department of Defense, directed that the military prepare to provide support for humanitarian relief operations in south Florida.

On 25 August, Second U.S. Army commander Lt. Gen. Samuel Ebbesen deployed to Tallahassee, Florida, and linked up with civilian, federal, and state officials who were coordinating disaster relief. When on 27 August the president ordered an increased military role in the disaster zone, Joint Task Force ANDREW was created, with General Ebbesen as its commander. Its mission was to conduct humanitarian support and relief operations, and Ebbesen exercised operational control of all Department of Defense forces supporting hurricane relief operations. The task force eventually grew to more than 24,000 active duty U.S. military personnel, including soldiers from the XVIII Airborne Corps, 82d Airborne Division, 1st Corps Support Command, 10th Mountain Division, U.S. Army Materiel Command, and other active and reserve component units. Relief operations were also conducted by a significant number of Air Force, Navy, and Marine Corps personnel, while a contingent of Canadian military engineers gave the task force a multinational character. Though Florida National Guard personnel also conducted relief efforts at all levels, they were not federalized and consequently were not part of the task force.[45]

Col. Vahan Moushegian, Jr., then the legal adviser to the surgeon general of the Army, was selected as task force staff judge advocate. With continuous service as an Army lawyer since the Vietnam War, Moushegian was an experienced judge advocate. His legal staff included Army Majs. Steven T. Strong, an administrative law expert, and Steven A. Lamb, an expert in operational law. Within eight hours of notification, Moushegian, Strong, and Lamb were on a civilian aircraft en route to Miami.

Arriving on 31 August, they joined Army Maj. William M. Mayes, a veteran of Operation JUST CAUSE and now chief of administrative and civil law for the 101st Airborne Division (Air Assault). Mayes was already in south Florida, having deployed a day earlier. Moushegian, Strong, Lamb, and Mayes recognized that they must not only provide timely and accurate legal advice, but also ensure that General Ebbesen, his staff, and all subordinate component commands understood the relationship between the task force and the Federal Emergency Management Agency (FEMA). As the latter was the lead federal agency in the operation, the task force had no authority to engage in relief activities other than those directed by FEMA. Consequently, in advising on the lawfulness of a particular military activity, the judge advocates emphasized that the starting point for any discussion was the fact that the military could lawfully perform only those relief missions assigned them by FEMA. This was an important point: Joint Task Force ANDREW would be reimbursed for monies spent in performing missions directed by FEMA. Generally, there would be no reimbursement for military activities conducted by the joint task force on its own initiative. Consequently, while it was essential that the task force be aggressive in its hurricane relief activities, it was necessary that there be a link between resource expenditures and mission taskings.

Over the next few days, as elements of the Army component of the task force deployed, their judge advocates deployed with them. The deploying lawyers included Colonel Altenburg, staff judge advocate of the XVIII Airborne Corps; his administrative and civil law expert, Maj. Marjorie R. Mitchell; and Capts. Kurt D. Schmidt and Julie P. Schrank. Also deploying were Lt. Col. John M. Smith III, staff judge advocate of the 10th Mountain Division, and his deputy, Maj. Richard E. Gordon. Maj. Malinda E. Dunn, the deputy staff judge advocate for the 82d Airborne Division, headed the group of Army lawyers from that unit. Just as at task force headquarters, the operational tempo for judge advocates at the corps and divisions was high. In providing 24-hour-a-day legal support, however, these attorneys and their staffs maintained close contact with Army lawyers at the task force headquarters. Colonels Altenburg and Moushegian, for example, spoke daily to ensure that each was informed of current or potential problems, so that affirmative and coordinated action could be taken. Colonel Moushegian also conferred regularly with judge advocates assigned to other Army organizations having personnel in south Florida, such as Maj. Alfred L. Faustino of Army Materiel Command, thus ensuring that uniform legal advice was provided.[46]

As a variety of federal, state, and local laws regulate military actions that may be taken in disaster relief operations, Army lawyers advised on a range of diverse issues. Judge advocate guidance to the military police was critical during the early stages of disaster relief operations. The devastation caused by the hurricane left few traffic lights and signs standing. Many man-made landmarks used as navigation aids by drivers had been destroyed or damaged. As a result, traffic control points were needed in the area, and a number of them were manned by military police. Provided these traffic control points had a military purpose, such as maintaining a military supply route or controlling military convoy traffic, they constituted permissible law enforcement under the Posse Comitatus Act. However, manning these traffic control points merely to assist civilian law enforcement officials would violate the act. Additionally, active duty military police faced the practical question of how to deal with traffic violators, as these soldiers could not issue citations to civilians or otherwise enforce Florida state law. Consequently, while the judge advocates advised that active duty soldiers could always be used at traffic control points that served a military purpose, they recommended that Florida National Guard troops be used for this duty. As this state force had not been federalized, the Posse Comitatus Act did not apply to the National Guard personnel and they could provide unrestricted support to civilian law enforcement efforts.[47]

Similarly, as civilians living in the area welcomed the security that came with a military presence, they supported the idea of soldiers patrolling in their communities. The Posse Comitatus Act, however, prohibited "patrolling" for the sole purpose of providing neighborhood security. Consequently, the judge advocates advised that if the patrolling undertaken for the security of military forces or in the execution of relief missions had the incidental benefit of deterring lawbreakers in civilian neighborhoods, it was legal. Again, however, they counseled that, as the Posse Comitatus Act applied only to the activities of federal troops, National Guard soldiers were not restricted from patrolling in the civilian community. This judge advocate advice not only ensured compliance with the law, but also benefited the joint task force mission, as the efforts of active duty personnel could focus on the relief and recovery mission.

Initially, Joint Task Force ANDREW personnel did not have disaster relief rules of engagement, and subordinate units received no guidance as to whether troops should carry their individual weapons while performing disaster relief operations. Within a short time, however, the XVIII Airborne Corps developed rules of engagement that were subsequently adopted by the joint task force. While recognizing the inherent right of self-defense, the rules were defensive in nature. All troops who deployed with individual weapons carried them, but without ammunition. Generally, only the military police carried loaded weapons. Later, because of the potential for confrontations between task force personnel and armed gang members, the task force judge advocates prepared contingency rules of engagement for use if the force commander determined that ammunition should be issued to his personnel for their protection. These special rules were never implemented, however.

Another legal question involved state and local taxation. Task force military and civilian personnel residing in local hotels were subject to a 10 percent "transient rental" tax. Recognizing that this would ultimately be borne by the government, Major Strong spearheaded efforts to obtain tax relief available under state law. With the assistance of

Army Reserve judge advocates at the 174th Military Law Center in Miami, who provided quick and expert advice on state and local law, a tax exemption certificate was prepared and distributed. All members of Joint Task Force ANDREW thus received relief from state and local taxes on their accommodations.[48]

Finally, Army lawyers faced an unanticipated question involving local elections. Two primary elections were scheduled to be held in the area of the disaster relief operation. Because of the massive destruction in the area, the Dade County Election Commission requested that the joint task force provide tents and electric generators for eighty-six polling places. A federal statute, however, prohibited the positioning of troops "at any place where a general or special election is held." Could the requested support be provided, especially if this meant that military personnel would be near the sites providing support during the hours that voting occurred? Upon inquiry, the Department of Justice advised that providing Dade County with the tents and generators would not violate federal law, provided that the troops avoided any demonstration of military authority. As a result, Colonel Moushegian's staff prepared guidance for task force troops that ensured their absence from polling sites and minimized their presence in the surrounding areas. Thus, lawyers both ensured compliance with the applicable law and made possible the provision of badly needed support.[49]

When relief efforts ended in south Florida in mid-October 1992, nearly 900,000 meals had been served, 67,000 civilian patients had received medical care, and more than 6.2 million cubic yards of debris had been removed. Roughly thirty judge advocates had deployed to south Florida during the six weeks that the task force had operated there, and their efforts as part of the largest peacetime deployment of Defense Department forces in the continental United States contributed significantly to the accomplishment of the task force mission.

Operation SUPPORT HOPE, 1994

In March 1994, a surface-to-air missile struck an aircraft carrying the presidents of Rwanda and Burundi, killing all aboard. Earlier that same day, the two presidents had signed an agreement supporting the civil rights of Tutsi tribesmen living in their two countries. Extremists among the Hutus, the majority group in the Rwandan population, with a long history of antagonism toward the Tutsi, vigorously opposed the agreement.

The Rwandan militia, dominated by Hutu extremists, publicly blamed Tutsi rebels for the downing of the plane, and within days they began attacking and killing Tutsi citizens and politically moderate Hutus. By mid-April, over 500,000 men, women, and children were reported dead. A small group of UN peacekeepers in Rwanda, having failed to halt the bloodshed, was withdrawn.

Tutsi rebels living in nearby Uganda now struck south toward Kigali, the capital of Rwanda, in an attempt to stem the genocide. In a series of pitched battles, they defeated the Hutu militia and overthrew the Rwandan government that had tolerated the militia's excesses. Fearing a Tutsi reprisal if they remained, many of the Hutus fled their homes for sanctuary in southwest Rwanda, Zaire, Burundi, and Tanzania. In Zaire, the area around Goma alone soon had a refugee population of between two and one-half and five million. This huge influx of displaced persons completely overwhelmed the United Nations and other nongovernmental organizations operating in Central Africa. Camps were short of food, clean water, shelter material, and medicine. In June a cholera epidemic erupted, and by mid-July approximately 500 deaths per day were occurring in the camps around Goma.

The United Nations called on the international community for immediate aid, and the United States responded by sending a Department of State team to Zaire and alerting the Department of Defense that military assistance might be required. On 18 July 1994, U.S. European Command (USEUCOM) activated a joint task force at Kelly Barracks in Stuttgart, Germany, and it began planning for a possible humanitarian deployment.

Maj. Gen. Jack P. Nix, Jr., commander of the Army's Southern European Task Force in Vicenza, Italy, was named task force commander. Lt. Col. John P. Ley, Jr., his staff judge advocate, was selected as the force legal adviser. Ley, who had recently arrived from V Corps in Germany, had been in Italy for only eight days. Now, within four hours of learning of his designation as the force legal adviser, Colonel Ley found himself returning to Germany with ten other Southern European Task Force planners to join the USEUCOM crisis action team in planning for operations in Central Africa.

It was a daunting task. Just getting to Rwanda was challenging, as it was a small, landlocked country with few good roads and no continuous rail system. The one airport large enough to support jumbo-jet aircraft was not operating, as its administrators either had been killed or had fled. Internal conflict in Burundi and Zaire made any sustained operations in these two countries difficult and threatened the security of the refugees and anyone assisting them. Additionally, coordination with the new Rwandan government was difficult, as the United States had yet to grant it official recognition. Finally, and most important,

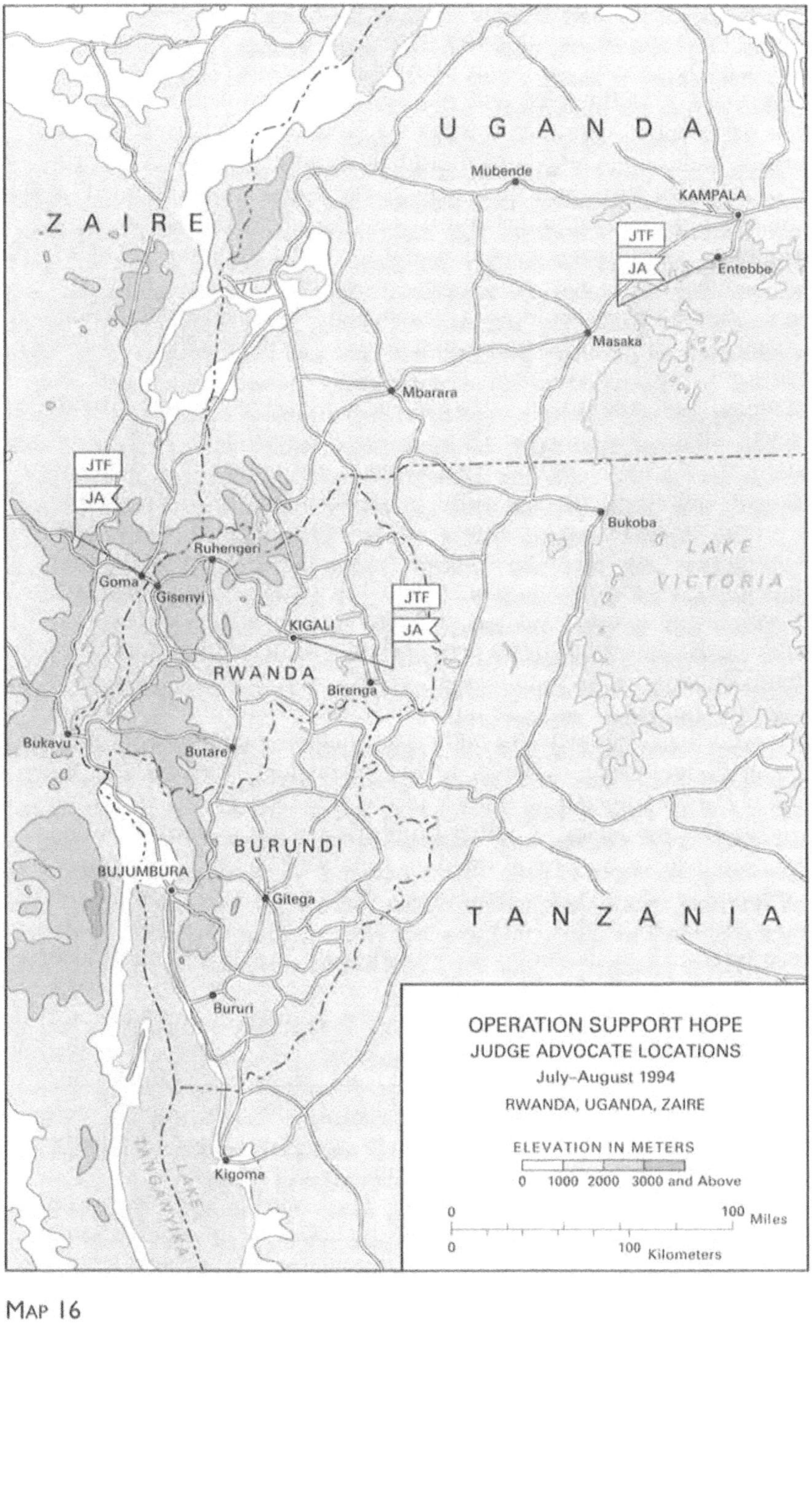
UGANDA
Mubende
KAMPALA
JTF
JA
Entebbe
ZAIRE
Masaka
Mbarara
JTF
JA
Bukoba
LAKE
VICTORIA
Ruhengeri
Goma
Gisenyi
JTF
KIGALI
JA
RWANDA
Birenga
Bukavu
Butare
BURUNDI
BUJUMBURA
Gitega
TANZANIA
Bururi
Kigoma
LAKE
TANGANYIKA
OPERATION SUPPORT HOPE
JUDGE ADVOCATE LOCATIONS
July–August 1994
RWANDA, UGANDA, ZAIRE
ELEVATION IN METERS
0 1000 2000 3000 and Above
0 100 Miles
0 100 Kilometers

Map 16

intelligence reports were confusing as to the extent of the disaster and the items necessary for relief purposes.

To better gauge the situation, Army General George A. Joulwan, the USEUCOM commander, directed General Nix to deploy to Africa with an assessment team. Nix traveled to Entebbe, Uganda, where the U.S. ambassador had gained a commitment from the Ugandan government to establish an intermediate staging base, and then continued on to Goma. After meeting with UN officials, Nix learned that the most immediate need was for fresh water and medicine to combat cholera. General Joulwan then determined that Nix would remain in Goma and begin executing local operations. Lt. Gen. Daniel R. Schroeder, deputy commander of U.S. Army, Europe, assumed command of what was now named Joint Task Force SUPPORT HOPE. *(Map 16)*

Colonel Ley, still at Kelly Barracks in Stuttgart, continued providing legal advice. He and the USEUCOM legal adviser, Army Col. Raymond C. Ruppert, drafted a legal annex for inclusion in the joint task force operations plan. Central to their work was promulgating rules of engagement and arranging for the printing of pocket cards containing these rules. Colonel Ley also planned realistic training scenarios on the use of force, scenarios which underscored the defensive nature of the task force rules of engagement. In one situation, for example, a soldier guarding a base perimeter at night observes an unidentified individual climbing over a barrier fence. When the soldier calls for the person to halt, the individual flees. The training scenario emphasized that force was not authorized once the unidentified individual no longer posed a threat to the perimeter. Consequently, no warning shot could be fired, nor could force be used to detain, injure, or kill the fleeing person. Another training scenario underscored that force could be used only to protect U.S. forces in Central Africa. Thus, if a Special Forces detachment observed a group of refugees preparing to ambush members of a rival clan, the detachment personnel could not use force to halt the ambush. But Colonels Ley and Ruppert also ensured that the overriding principle—that an American soldier always has the right of self-defense—was never diluted during rules of engagement training.[50]

Other legal issues addressed in the legal annex to the operations plan were the requirement to report and investigate allegations of violations of the Law of War and guidance on handling requests for political asylum and temporary refuge for humanitarian reasons.[51]

Colonel Ruppert, drawing upon the 1961 Vienna Convention on Diplomatic Relations, also requested that U.S. Joint Task Force SUPPORT HOPE personnel be accorded diplomatic status equivalent to that provided members of the administrative and technical staff of the U.S. embassies in Rwanda, Uganda, and Zaire. This would provide U.S. forces operating in these areas with immunity from civil and criminal jurisdiction while performing their official duties. Colonel Ley also drafted a General Order no. 1 for the force commander's signature. Approved on 27 July 1994, this punitive general order was intended to promote good order and discipline. Its chief provision was a ban on the consumption of alcoholic beverages.[52]

On 28 July 1994, Colonel Ley flew to Entebbe, Uganda, and was soon joined by Air Force judge advocate Capt. Gregory E. Lang and an Air Force legal clerk. While the Air Force attorney remained in Uganda handling mostly contract issues, Colonel Ley traveled to Kigali and Goma. Shortly thereafter, Maj. Richard Pregent, the deputy staff judge advocate of the 3d Infantry Division, deployed from Germany in support of Operation SUPPORT HOPE. Pregent dealt with a number of contract and command issues in Kigali, then returned to Entebbe before redeploying in September.

Colonel Ley spent the majority of his time in Goma, living and working in the rear of the general purpose, medium tent used for command briefings. He worked on a number of complex legal issues, including the legality of potentially using riot control chemical agents and pepper spray for crowd control purposes and the lawfulness of providing military security, transportation, and other forms of logistical support to nongovernmental organizations. This last issue resulted from the fact that there were some forty agencies in the area, all desiring support from the American military. Additionally, French soldiers and the forces of other North Atlantic Treaty Organization countries sought U.S. logistical support. Accordingly, Colonel Ley was repeatedly challenged to establish legal authority for such support. A final issue concerned the legitimate transfer of U.S. property to United Nations forces under presidential drawdown authority.

Overshadowing all of these legal questions was the larger question of "mission creep." Where did humanitarian relief end and nation building begin? Just as Colonel Supervielle had discovered in Operation SEA ANGEL, Operation and Maintenance monies could not be expended for the latter. Yet this presented a real-world problem, as virtually all of the economic infrastructure of Rwanda had been destroyed.[53]

Colonel Ley and Major Pregent redeployed with the rest of the joint task force in September, and Operation SUPPORT HOPE came to an end. There had been few U.S. troops involved, but the legal issues had been numerous and complicated.

Conclusion

This brief look at judge advocates in military operations other than war from POWER PACK to SUPPORT HOPE reflects—on a smaller scale—the same transformation in the role of the Army lawyer that occurred between Vietnam and UPHOLD DEMOCRACY. During POWER PACK in 1965 and 1966, judge advocates viewed their mission as providing legal services in the Dominican Republic in the same way that these services had been provided in garrison at Fort Bragg. While combat in and around Santo Domingo meant some involvement in Law of War matters, Army lawyers focused their energies on providing commanders and soldiers with the same form of legal support that they received in a peacetime environment. Judge advocates did not participate in the planning and execution of military operations, and they played no role in providing advice on legal issues impacting on the conduct of such operations.

By the mid-1980s, a number of factors, including the implementation of the Department of Defense Law of War program, the experiences of judge advocates such as Major Graham in the Multinational Force and Observers, and the emergence of operational law as a legal discipline, had transformed the role of the judge advocate in the Army. As a result, by the time Major Supervielle deployed as the lone Army lawyer to Western Samoa and Bangladesh, judge advocates were expected to, and did, use their legal training and soldier skills to resolve both legal and nonlegal issues associated with operations other than war. Advice on military justice matters and processing foreign claims remained important components of judge advocate work, but Army lawyers were now called upon to negotiate Status of Forces Agreements, draft and develop training for rules of engagement, formulate policies on awards and war trophies, resolve fiscal law matters, and assist high-level military and civilian decision makers in arriving at solutions for numerous other issues arising in an operational environment.

In sum, the metamorphosis in the role of judge advocates in the Army has not been restricted to one of resolving legal matters associated with conventional combat operations; the role of the judge advocate extends to operations other than war as well. The lack of sustained combat does not moderate the complexity of legal and nonlegal issues facing a judge advocate deployed on peacekeeping, humanitarian, and other forms of operations other than war. On the contrary, whether participating in civil disturbance operations in Los Angeles or in humanitarian relief efforts such as PROVIDE COMFORT, every Army lawyer has quickly learned that at times these missions pose even greater legal challenges than those arising during armed conflict. All available evidence indicates that the twenty-first century is unlikely to alter this state of affairs; the provision of timely legal support for operations other than war will continue to challenge Army judge advocates.

Notes

[1] Ltr, Lt Col Manuel E. F. Supervielle to author, 20 Jan 97, Historians files, OTJAG.

[2] Joint Publication 3–07, *Joint Doctrine for Military Operations Other Than War*, 16 Jun 95, p. I–1; Field Manual (FM) 100–23, *Peace Operations*, Dec 94, Glossary.

[3] Lawrence A. Yates, *Power Pack: U.S. Intervention in the Dominican Republic, 1965–1966* (Leavenworth, Kans.: Combat Studies Institute, 1988), pp. 61–71.

[4] Interv, author with Col Burnett H. Radosh, 15 Sep 97, Historians files, OTJAG.

[5] Interv, author with Col Raymond D. Cole, 22 Sep 97, Historians files, OTJAG; *United States* v. *Moore*, CM 413158, 36 CMR 531 (CMA 1966). On appeal, the Army Board of Military Review approved the findings but reduced the sentence to fifteen years. The Court of Military Appeals subsequently determined that the law officer's failure to instruct the court members that they could find Moore guilty of the lesser included offense of involuntary manslaughter required that the findings and sentence be set aside.

[6] *United States* v. *Darling*, CM 413433, 36 CMR 620 (1966).

[7] *United States* v. *Monday*, CM 413039, 36 CMR 711 (1966).

[8] *United States* v. *Wolfe*, CM 413647, 37 CMR 571 (1966).

[9] Interv, author with Cole, 22 Sep 97.

[10] Yates, *Power Pack*, p. 142.

[11] Interv, author with Col John R. Bozeman, 19 Aug 97, Historians files, OTJAG.

[12] Ltr, Col Daniel J. Dell'Orto to author, 31 Dec 96, Historians files, OTJAG. While there were no riots at Fort McCoy, Wisconsin, Army judge advocates like Capt. Bernard P. Ingold, who deployed from Fort Devens, Massachusetts, to Fort McCoy from June to July 1980, faced the same sort of issues handled by Captain Dell'Orto. These ranged from advising the provost marshal on the use of force to break up fights between refugees and permissible military assistance to civilian police in the area to the use of drug-sniffing dogs to detect marijuana in the refugee camp and administrative and legal options for handling refugee misconduct.

[13] For a thorough examination of Army Cuban refugee operations see U.S. Army Forces Command, *The Role of FORSCOM in the Reception and Care of Refugees from Cuba in the Continental United States* (Atlanta, Ga.: Headquarters, FORSCOM, 1984).

[14] Memo, Col David E. Graham for author, 16 Jan 97, sub: MFO, Historians files, OTJAG.

[15] Ibid.

[16] Ibid.

[17] Interv, author with Maj Catherine M. With, 7 Oct 97, Historians files, OTJAG.

[18] Memo, Capt Paul Fiorino for author, sub: Experiences in the MFO, 16 Sep 97, Historians files, OTJAG.

[19] Patrick Finnegan, "Operational Law: Plan and Execute," *Military Review* 76 (March–April 1996): 30; ltr, Finnegan to author, 29 Oct 97, Historians files, OTJAG.

[20] Ibid.

[21] Ltr, Supervielle to author, 20 Jan 97.

[22] Ibid.

[23] Philip A. Meek, "Operation PROVIDE COMFORT: A Case Study in Humanitarian Relief and Foreign Assistance," *Air Force Law Review* (1994): 225.

[24] Operation PROVIDE COMFORT After Action Report, Headquarters, US European Command/ECJ3, 29 Jan 92, Historians files, OTJAG.

[25] For the text of Operation PROVIDE COMFORT's rules of engagement, see Appendix B. They are also reprinted in John P. Cavanaugh, *Operation PROVIDE COMFORT: A Model for Future NATO Operations* (Fort Leavenworth, Kans.: School of Advanced Military Studies [Monograph], 1992).

[26] Meek, "Operation PROVIDE COMFORT: A Case Study," p. 236; PROVIDE COMFORT After Action Rpt, p. 3.

[27] Ltr, Lt Col Richard E. Gordon to author, 7 Nov 97, Historians files, OTJAG.

[28] Ibid.

[29] Interv, author with Capt Corey L. Bradley, 9 Jan 97, Historians files, OTJAG.

[30] Ibid.

[31] Charles R. Smith, *Angels from the Sea: Relief Operations in Bangladesh* (Washington, D.C.: Marine Corps History and Museums Division, 1995), p. 1.

[32] Ibid., p. 14.

[33] Info Paper, JTF SEA ANGEL SJA, 13 Jun 91, sub: Joint Task Force SEA ANGEL SJA After Action Report, Historians files, OTJAG.

[34] Info Paper, JTF SEA ANGEL SJA, 23 May 91, sub: Use of DOD O&M Funds for Disaster Relief Operations, Historians files, OTJAG.

[35] Interv, author with Lt Col Allen K. Goshi, 25 Nov 97, Historians files, OTJAG.

[36] Ibid.

[37] Interv, author with Capt Norman F. J. Allen III, 17 Mar 97, Historians files, OTJAG.

[38] Interv, author with Lt Col M. Tia Johnson, 8 Sep 97, Historians files, OTJAG.

[39] Interv, author with Capt Stewart A. Moneymaker, 8 Sep 97, Historians files, OTJAG.

[40] Ibid.

[41] "Last Cuban Refugees at Guantanamo Base Leave for U.S.," *Washington Post*, 1 Feb 96, p. A22. See also Department of Defense, U.S. Atlantic Command, *Migrant Camp Operations: The Guantanamo Experience* (Norfolk, Va.: U.S. Atlantic Command, 1996).

[42] International Law Note, "Civil Disturbance Rules of Engagement: Joint Task Force Los Angeles," *Army Lawyer* (September 1992): 32.

[43] For the text of the approved rules of engagement, see Appendix B.

[44] Harry Summers, "Fire Discipline All the Way to L.A.," *Washington Times*, 21 May 92, p. G–4.

[45] Memo, JTF ANDREW Legal Adviser for The Judge Advocate General, sub: Joint Task Force ANDREW—After Action Report, Oct 92, p. 2, Historians files, OTJAG.

[46] Ltr, Moushegian to author, 30 Dec 96; interv, author with Maj Gen John D. Altenburg, Jr., 3 Nov 97, both in Historians files, OTJAG.

[47] Memo, JTF ANDREW Legal Adviser for The Judge Advocate General, sub: Joint Task Force—After Action Report, pp. 9–10.

[48] Interv, author with Lt Col Steven T. Strong, 31 Oct 97, Historians files, OTJAG.

[49] Ltr, Moushegian to author, 30 Dec 96; interv, author with Altenburg, 3 Nov 97.

[50] Memo, JTF SUPPORT HOPE Rules of Engagement Training Scenarios, n.d., Historians files, OTJAG.

[51] Annex O to JTF SUPPORT HOPE OPLAN 94–002, Legal, undated, Historians files, OTJAG.

[52] Memo, EUCOM Operations Directorate, 30 Jul 94, sub: General Order no. 1, Historians files, OTJAG.

[53] Memo, Col John P. Ley, Jr., for author, 20 Oct 97, Historians files, OTJAG.

9

Conclusion

This history of judge advocates in military operations from Vietnam to Haiti records the experiences of a multitude of talented and dedicated men and women. It captures their individual stories and answers the questions "Who was there?" and "What did they do?" But the narrative also demonstrates that there was a metamorphosis in the role played by lawyers in the Army during this period. It was a remarkable transformation, for the old concept that the role of a deployed judge advocate was to support the mission by delivering the same legal services offered in a peacetime garrison environment was supplanted by a new idea: that, while a judge advocate participating in a military operation might still prosecute and defend at courts-martial, adjudicate claims, and provide legal assistance, an Army lawyer best enhanced mission success by integrating legal support into operations planning and execution at all levels. Since such integration meant that judge advocates identified and resolved a wide array of legal issues impacting directly on the conduct of an operation, they became increasingly important to mission success in the contingency-oriented Army of the late 1980s and early 1990s.

In 1959, when the first judge advocate arrived in Saigon for duty, he was expected to support the Military Assistance Advisory Group in the same manner as judge advocates would support a commander and staff at a U.S. Army installation during peacetime—by administering the military justice system, advising on administrative and civil law issues, and providing counsel to soldiers needing assistance with their personal legal problems. By 1996, however, when the last judge advocates returned from Haiti, Army lawyers were expected to advise on a broad range of legal issues and, schooled in a new legal discipline called "operational law," were routinely using their knowledge of domestic, foreign, and international law to enhance mission success. These judge advocates also were advising on a variety of nonlegal matters and were assuming nontraditional roles such as chief of staff in a division assault command post, public affairs officer, and adviser in political-military negotiations ordinarily the province of professional diplomats.

What caused the abandonment of the traditional perspective that judge advocates in combat simply continued with their same peacetime legal activities? What caused the Judge Advocate General's Corps' institutional recognition that judge advocates in a contingency-oriented Army must be fully prepared to advise on a broad range of legal and nonlegal issues? There were at least four causes. First, fundamental changes in the nature of warfare during the period from 1959 to 1996 naturally led to changes in U.S. Army doctrine and force structure, and this Army-wide evolution inexorably led to a need for a different role for Army lawyers. Second, starting with the war in Vietnam, an increasing number of judge advocates took *individual* initiatives to enhance mission success in ways not ordinarily considered to be part of judge advocate duties. Because these individual efforts proved that judge advocates could use their abilities to enhance mission success in non-traditional ways, an institutional change in the role of lawyers in the Army was shown to be both feasible and desirable. Third, the Army's experiences at My Lai and the resulting establishment of the Department of Defense Law of War Program directly altered the role of judge advocates in the Army. This was because, in complying with this new DOD program, Army lawyers were required to integrate themselves into military operations at all levels—a radical new role requiring knowledge of operational law. Fourth, and finally, were the experiences of judge advocates deploying to Grenada as part of Operation URGENT FURY in 1983, for this mission showed conclusively that Army lawyers must change the way they provided legal support if they were to meet the challenges of a contingency-based Army.

The first reason—that fundamental changes in the nature of warfare during the period 1959 to 1996 caused a change in the role of lawyers in the Army—is easily understood. The emergence of new technology on the battlefield, especially the explosive growth in the use of helicopters as weapons platforms and troop-carrying vehicles during Vietnam and after, naturally led to changes in U.S. Army doctrine and force structure. The end of the war in Vietnam also saw an end to "attrition" as the doctrine for combat success. In its place was a new AirLand Battle doctrine and a contingency-oriented Army characterized by an exceptionally high operational tempo. This new doctrine and force necessarily required uniformed lawyers who could do more than prosecute courts-martial, adjudicate claims, and instruct troops on the Law of War. Deployments for peacekeeping, humanitarian assistance, and civil disturbance operations—in the United States and overseas—required judge advocates who were as adept at drafting rules of engagement as they were at advising on political-military matters. These deployments also required Army lawyers who understood the legal basis for transferring U.S. property to United Nations forces and who recognized that certain U.S. fiscal and environmental laws might restrict ways to accomplish a mission. But, while changes in the nature of warfare and a corresponding reconfiguration of the Army effected changes in the role played by judge advocates in military operations, this was only a contributing influence.

Of greater importance than this first reason was a second factor: that, starting with the war in Vietnam, an increasing number of judge advocates took *individual* initiatives to enhance mission success in ways not ordinarily considered to be part of normal judge advocate duties. Of course, there always have been Army lawyers who enhanced mission success in nontraditional ways. Shortly before America's entry into World War I, Maj. Gen. Enoch H. Crowder, then the Judge Advocate General, was appointed Provost Marshal General. In that capacity, Crowder prepared the Selective Service Act of 1917 and supervised the induction of nearly three million men into the armed services. Another World War I Army lawyer, Col. Blanton Winship, demonstrated a similar versatility while serving as First Army judge advocate: he commanded two infantry regiments in combat. Some twenty years later, Maj. Gen. Allen W. Gullion, while serving as the Judge Advocate General, assumed additional duties as the Army's Provost Marshal General—and continued serving as the Army's top military police officer after retiring as the Army's top lawyer. Another World War II judge advocate, 1st Lt. Samuel A. Spitzer, was awarded the Silver Star medal after a daring act of personal courage that resulted in the capture of more than 500 German soldiers. Finally, during the Korean War Maj. Bruce C. Babbitt, then a judge advocate in the 2d Infantry Division, took command of a rear area perimeter defense after the front collapsed during a Chinese attack. As a direct result of his leadership, an enemy attack on the position was successfully repulsed. But, while these deeds certainly enhanced mission success, the prevailing view was that what Crowder, Winship, Gullion, Spitzer, and Babbitt did was unique—and outside the scope of a judge advocate's normal duties.

The Vietnam War, however, saw an increasing number of judge advocates assuming nontraditional roles—and addressing issues ordinarily handled by other staff principals—because these lawyers believed that the judge advocate's proper role in the Army was to use the law and their talents as judge advocates to enhance mission success in as many ways as possible.

From 1959 to 1962, while serving as the first judge advocates in Vietnam, Colonels Durbin and Eblen looked for ways in which the law could further the mission of their Military Assistance Advisory Group. Then Colonel Prugh, MACV staff judge advocate from 1964 to 1966, took even more far-reaching initiatives. Prugh led efforts to persuade the South Vietnamese military that their conflict with the Viet Cong and North Vietnamese was no longer an internal civil disorder. As a direct result of his work, the military—and later the government of South Vietnam—acceded to the American view that the insurgency was an armed conflict of an international character and that the benefits of the 1949 Geneva Prisoners of War Convention should be given to all captured Viet Cong and North Vietnamese soldiers. This was a public relations coup for the South Vietnamese. At the same time, applying the benefits of the Geneva Convention to those combat captives held in South Vietnam also enhanced the opportunity for survival of U.S. servicemen held by the Viet Cong and the North Vietnamese.

Colonel Prugh also reasoned that American lawyers under his authority could support U.S. military and political aims in Vietnam by helping to educate the Vietnamese about the beneficial effect of the rule of law in society. According to Prugh, "those who are familiar with the ways to combat insurgency have come to recognize that the law and lawyers have one of the most significant parts to play." That is, instilling a respect for law and order would support South Vietnam in its campaign against the terrorist activities of the Viet Cong and their North Vietnamese allies.

Finally, Prugh established a unique legal advisory program that monitored the real-world operation of South Vietnam's military criminal justice system. As a result, long after George Prugh's return to the United States, MACV judge advocate advisers used their legal talents to assist the Vietnamese military on issues ranging from desertion control, resources control, and security operations to obtaining transportation for Vietnamese judge advocates, providing storage for records of trial, and obtaining materiel for local prisons.

In the late 1960s and early 1970s, Army judge advocates continued to take individual initiatives in supporting combat operations in Vietnam. At MACV headquarters, Colonel Haughney and his staff promulgated the first procedural framework for classifying combat captives, using so-called Article 5 tribunals. While the MACV provost marshal was primarily responsible for advising the Vietnamese on prisoner of war issues, judge advocates spearheaded efforts in this area—

and also took the initiative in establishing a records system identifying and listing all prisoners of war. Similarly, while investigating and reporting war crimes was not a judge advocate responsibility, MACV lawyers took the lead in formulating guidance on investigating and reporting such crimes. By 1968 the Military Assistance Command, Vietnam, had decided, as a matter of policy, that judge advocates would be the primary focal point for all war crimes issues.

Judge advocates also enhanced mission success by providing legal support to decision makers outside the Army and the Department of Defense. Like his predecessors, Colonel Williams, MACV staff judge advocate from August 1969 to July 1970, provided legal advice to the U.S. ambassador and his staff. As the senior government lawyer in Vietnam, it was only natural for the MACV staff judge advocate to respond directly to inquiries from the top State Department officer in the country. In addition to meeting at least weekly with the U.S. ambassador, however, Colonel Williams expanded his role as an adviser and counselor while a member of the Irregular Practices Committee. This committee was composed of civilian representatives of the U.S. Overseas Mission and officers from MACV staff sections, including Colonel Williams as the MACV staff judge advocate. While officially tasked with coordinating the suppression of black marketing, currency manipulation, and other illegal activities affecting the Vietnamese economy, the committee's composition naturally made it a clearinghouse for a variety of policy issues—and a point of contact for Saigon government officials seeking assistance. As a result, by the time he departed Vietnam in 1970 Colonel Williams was conferring weekly with the Vietnamese minister of finance, director of customs, and minister of economy and with representatives of the U.S. Agency for International Development and the U.S. embassy.

Meanwhile, Army lawyers outside Saigon used their individual talents and abilities in a variety of nontraditional ways. At USARV headquarters in Long Binh, for example, Major Suter reviewed the Fire Support Base MARY ANN investigation, determined who was responsible for the disaster, and recommended an appropriate course of action. Lt. Gen. William J. McCaffrey, the USARV deputy commander, approved all of Suter's recommendations.

Army lawyers at brigades and divisions in the field took similar initiatives. At the 1st Cavalry Division, the Army's new airmobile experiment, Colonel Holdaway's innovative approach to practicing law enhanced mission success. With about 450 helicopters, the division had a very large area of operations, and this meant that Holdaway and his lawyers had to take their legal services to the field. As a result, the division's military attorneys were often airborne, flying out to base camps and fire bases on the "lawbird" to confer with and advise commanders—as well as provide personal legal assistance to their soldiers.

Finally, after the signing of the Paris Peace Accords, judge advocates serving on the Four Party Joint Military Commission and Four Party Joint Military Team between 1973 and 1975 did more than traditional lawyering. Thus, Captain Moushegian served as the U.S. delegation's expert on the peace treaty's provisions and also assumed the duties of principal liaison officer—meeting regularly with his Viet Cong, South Vietnamese, and North Vietnamese counterparts in what was essentially a diplomatic role. Similarly, Captain Scanlon, one of the last Army lawyers to serve in Vietnam, advised the chief of the U.S. delegation on the rights and obligations of all parties under the Paris Peace Accords. But Scanlon also assisted in gathering information on Americans still missing in action—which meant traveling to Hanoi, touring the infamous "Hanoi Hilton," and making contact with North Vietnamese government officials who might provide information about missing or dead Americans.

What was the reason for this significant number of individual initiatives? Certainly the nature of the Vietnam War itself encouraged nontraditional approaches to mission accomplishment. The unconventional nature of the guerrilla insurgency required responses that were novel, if not radical. The Army experimented with an airmobile division and created new combat units—Special Forces—adept at both combat and "winning hearts and minds." Seen from this perspective, efforts like Prugh's advisory program were a perfect complement to initiatives in the Army generally.

Another reason for increased individual initiative, however, certainly resulted from the reality that there were more lawyers in the Army than ever before. During World War II, for example, an armored division of 11,000 soldiers was authorized one judge advocate on its table of manpower. As other divisions were similarly structured, judge advocates participating in the fighting in Europe or the Pacific had little time for issues outside the established areas of military justice, claims, legal assistance, and administrative law. But as the Judge Advocate General's Corps increased in size during the Vietnam buildup—an expansion that accelerated when more lawyers were needed to satisfy the new requirements of the Military Justice Act of 1968—there simply were more judge advocates in the corps. Many were not content to adhere to the traditional concept of the role of lawyers in uniform. Better educated, exceptionally energetic, and unfettered by old approaches to lawyering, these judge advocates looked for new ways to serve.

The changing nature of warfare and the emergence of a contingency-oriented Army, combined with the multitude of individual judge advocate initiatives to enhance mission success in nontraditional ways, however, do not fully account for this metamorphosis in the role of the Army lawyer. There was a third factor: the promulgation of Department of Defense Directive 5100.77. Published in November 1974, this new directive tasked Army lawyers with ensuring that all U.S. military operations strictly complied with the Law of War.

This Department of Defense policy decision—and a subsequent and complementary Joint Chiefs of Staff directive requiring the chairman's legal counsel to review all operations plans—was a direct result of My Lai. After Lieutenant Calley and his men murdered more than a hundred Vietnamese men, women, and children in My Lai and the surrounding area, the Defense Department recognized that preventing similar incidents required a new approach to ensuring obedience to the Law of War. Requiring the Army's legal corps to take primary responsibility for implementing a DOD Law of War Program that would ensure that "the Armed Forces of the United States shall comply with the law of war in the conduct of military operations and related activities in armed conflict" was considered to be the best way to accomplish this goal.

A few perceptive Army lawyers realized that this meant judge advocates must review all operations plans, concept plans, rules of engagement, execution orders, deployment orders, policies, and directives to ensure compliance with the Law of War, as well as with domestic and international law. These same military lawyers also recognized that this could best be accomplished if judge advocates were integrated into operations at all levels and, while Army lawyers were not routinely to perform nonlegal duties, effective integration would sometimes require judge advocates to take on nonlegal tasks.

Interestingly, there was a direct link between the institutional change caused by the 1974 DOD directive and the individual initiatives of the Vietnam era. Colonel Prugh, who while serving as MACV staff judge advocate had developed new strategies for using the law to support the war effort in Vietnam, served as the Judge Advocate General from 1971 to 1975. It was during his tenure that Colonel Solf, then chief of the International Law Division, Office of the Judge Advocate General, proposed that the Army suggest that a DOD-level Law of War Program be created. Not surprisingly, General Prugh concurred, and Colonel Solf began the process which culminated in the new directive. Thus, Prugh's belief that the law and lawyers must support military operations in new ways now was influencing an institutional shift in the role of the judge advocate.

To sum up, the first two factors—the emergence of a contingency-oriented Army and individual initiatives to use the law and legal skills to enhance mission success in novel ways—along with a third factor, the Army's experiences at My Lai and the resulting DOD directive, combined to move the Judge Advocate General's Corps toward an institutional recognition that the role of the judge advocate must change. As a result, the Corps took some steps to integrate itself into military operations—but these were fragmented. Moreover, the pace of change between 1974 and 1983 was too slow—as evidenced by the "wake-up call" the Corps received in October 1983: Operation URGENT FURY. This deployment to Grenada was the fourth factor—and certainly the catalyst—causing the metamorphosis in the role of the Army lawyer.

URGENT FURY truly was a watershed event, for the Corps now recognized, as an institution, that it must reconfigure to meet the challenges of a contingency-oriented Army. Failure to do so would keep the Corps on the periphery of the Army, if not render it increasingly irrelevant in military operations. As Colonel Richardson, the 82d Airborne Division staff judge advocate during URGENT FURY, observed: "You can only tell the CO [commanding officer] that he can't shoot the prisoners so many times. You reach a certain point at which, when the boss has run out of beans and bullets, has certain equipment requirements, and has the locals clamoring to be paid for property damage, you have to be prepared to provide the best possible legal advice concerning these issues as well."[1]

Fortunately, in the last hours of planning for URGENT FURY, Colonel Richardson had prevailed on the division chief of staff to let him accompany the assault command post into Grenada. Consequently, he was at the right place at the right time to deal with a variety of nontraditional issues ranging from the status of Cuban diplomats and conditions in detainee camps to drafting rules of engagement and advising civilian authorities on the content of a preventive detention ordinance. Had Richardson not taken the individual initiative to secure a place on the lead aircraft deploying to Grenada, no Army lawyer would have been available to advise the commander on these extremely important matters.

Providing effective judge advocate support to military operations, however, could no longer be based on such individual initiative. As a result, the evolution in the role of the Army lawyer accelerated after 1983, with the Corps—as an institution—focusing its efforts on the development of operational law as a concept and a legal practice. This included a new recognition that, in order to be good lawyers giving good advice, judge advocates had to be located with commanders during a deployment. And to be with commanders, judge advocates had to be "operations smart"—able to understand and appreciate AirLand Battle and its maneuver warfare progeny. They also had to be able to use the law to enhance mission success at the tactical and operational levels of military operations.

Six years later, when U.S. forces deployed to Panama in Operation JUST CAUSE, Colonel Smith, the 82d Airborne Division staff judge advocate, parachuted into combat along with the rest of the division's assault command post. This put him on the ground with the command group from the very beginning and ensured that he was immediately available to give timely, accurate advice. When asked by the division logistics officer (G–4), for example, whether privately owned automobiles and trucks could be used to transport the 7,000 paratroopers now on the ground, Colonel Smith was able to correctly advise him that the Law of War permitted the immediate seizure of such property if necessary for mission success. Since the 82d Airborne Division needed these vehicles if it was to move rapidly in the next hours and days, seizing and using them were key to mission success—and therefore lawful.

Over the following days and weeks, Colonel Smith and his fellow judge advocates at the 7th Infantry Division, XVIII Airborne Corps, and Special Operations Command handled a variety of issues, including those involving detainees and prisoners of war, law and order, foreign claims, and military justice. Colonel Bozeman, the XVIII Airborne Corps staff judge advocate, was increasingly involved in political-military matters. For example, television news programming in the United States trumpeted that a "money for guns" program was under way in Panama—even though no such program yet existed. Although Bozeman thought the issue might be a corps logistics (G–4), or provost marshal, or even a joint task force intelligence (J–2) issue, he stepped into the vacuum and wrote a "guns for cash" policy for the corps and task force. By the end of JUST CAUSE, more than 8,800 weapons had been turned in at a cost of some $811,000. In addition to his work with guns for cash, Bozeman also drafted a war trophy policy for the joint task force. Again, while the issue was not a judge advocate responsibility, Colonel Bozeman authored a comprehensive joint policy, and before soldiers redeployed from Panama to the United States clear guidance had been published.

With the experiences of URGENT FURY and JUST CAUSE behind them, the members of the Judge Advocate General's Corps were ready for Operation DESERT SHIELD, the largest overseas deployment since Vietnam. The evolution of the role of the judge advocate in military operations was now virtually complete. The development of, and focus on, a comprehensive operational law program had paid tremendous dividends. The import of operational law was the fact that it was based on an incontrovertible premise: that judge advocates deploying on military operations are faced with a wide range of legal issues *uniquely* associated with the conduct of such operations. Judge advocates confront international law, administrative and civil law (including contract, fiscal, and environmental law), claims, and military justice issues that arise from, and impact on, the manner in which military activities are conducted across the entire operational spectrum. After Grenada and Panama, the Corps had identified these issues, collected and placed them under a common terminological umbrella, developed an extensive academic curriculum and training program focusing on them, and compiled comprehensive resource materials for use by judge advocates deploying on military operations. The result was that Army lawyers going to the field no longer needed to "reinvent the wheel." On the contrary, armed with operational law, the judge advocate became a key member of the commander's staff, and the Judge Advocate General's Corps, as a whole, emerged as a more relevant and essential organization in a contingency-oriented Army.

The Army's Law of War training program, an essential part of the operational law effort, also had taken hold. Commanders at all levels insisted that their operations be conducted in strict compliance with the requirements of the law. Thus, Army lawyers were integrated into operations at all levels—and were working more closely with commanders and operators than ever before. For example, CENTCOM judge advocates—along with their counterparts at VII Corps and XVIII Airborne Corps—took an active role in developing rules of engagement. Those planning operations were still responsible for formulating rules governing the use of force; however, they sought the involvement of lawyers from the beginning of the process. As a result, judge advocates at all levels did more than simply advise on and review rules of engagement—they also wrote them.

As DESERT SHIELD's buildup of men and materiel began, Army lawyers were called upon to use their talents in some very new ways. Thus, at the direction of General Schwarzkopf, judge advocates at U.S. Central Command drafted a general order regulating the behavior of all CENTCOM forces in the Persian Gulf. For the first time, certain activities lawful in the United States were prohibited in order to preserve good U.S.-host nation relations. Thus, as intoxicating beverages were illegal under Saudi law, judge advocates included a ban on alcohol consumption in the order. Similarly, Saudi cultural sensibilities in the area of sexually suggestive literature led to a regulatory prohibition on possessing these items. Both of these prohibitions were radical steps, as U.S. forces had never before had such restrictions placed upon their conduct. But subsequent events demonstrated that the general order drafted by Army lawyers and their other service colleagues greatly enhanced good order and discipline among U.S. troops. So successful was the alcohol ban in ensuring mission focus and safety of the force that it has been included in every similar general order promulgated since DESERT STORM.

Army lawyers used their operational law expertise in other areas as well. Because U.S. forces needed far more equipment and supplies than they brought with them, they negotiated a multitude of contracts for supplies and services. Colonel Hagan's contract and fiscal law experts at 22d Support Command, for example, provided advice and support on such matters. An early fiscal law issue was whether Army O&M funds could be used to fund the construction of a heliport—a critical operational need since the Saudis did not have adequate airfield space to accommodate the huge number of attack, scout, transport, and utility helicopters arriving daily. The problem was that a $200,000 limit on spending such monies for "unspecified minor construction" seemed to preclude building a heliport estimated to cost $1 million or more. The SUPCOM judge advocates researched and wrote a fiscal law opinion setting forth the facts and the law. They concluded that limits on spending O&M funds did not apply to real-world operations or to combat-related construction and advised General Pagonis, the SUPCOM commander, that there was no fiscal law bar to building the heliport. When the chief counsel of the Corps of Engineers concurred in this legal opinion, the authority for the heliport—and other similar projects—was established. Also settled was General Pagonis' confidence in the judge advocates in his command.

By the time Operation DESERT STORM started, judge advocates were combining their practice of law with building fighting positions, manning perimeters, and supervising soldiers pulling guard duty. In line with their new training as operational lawyers, they were also accompanying their commanders to the front to ensure that legal advice was consistently timely, accurate, and available. At VII Corps, for example, Colonel Huffman, the staff judge advocate, was in the tactical operations center and close to General Franks. Thus when, during the initial breaching of Iraqi defensive fortifications, Franks queried whether it was permissible under the Law of War for U.S. forces to use their combat earthmovers to bury Iraqi soldiers alive in their fighting positions, Colonel Huffman was able to assure General Franks that the breaching operations were entirely lawful.

Similarly, as VII Corps continued its advance, crushing Iraqi units and capturing thousands of the enemy, the volume of prisoners was so great that it threatened to slow the American advance. Because they did not want to give up vehicles to transport these Iraqis to rear holding areas or to assign soldiers to guard them, some recommended that the captives simply be given some food and water and instructed to start walking south. Colonel Huffman, over the protest of some combat commanders, answered that this was not possible under the Law of War and the 1949 Geneva Prisoners of War Convention. The VII Corps was responsible for the safety of the prisoners. There were minefields to the south, and Iraqi prisoners might be killed or injured if they inadvertently walked through such fields. Additionally, groups of Iraqi soldiers, traveling on foot behind the front lines, might be targeted by U.S. soldiers or aircraft. Were they killed or injured, VII Corps would bear responsibility for violating the Law of War. Colonel Huffman recognized that the large number of prisoners was hampering the forward progress of VII Corps units, but advised that the requirements of the law left no alternative. As a result of his timely and accurate advice, prisoners were safeguarded and moved under escort to the rear.

After the campaign in Southwest Asia there was much reflection and analysis about integrating judge advocates more fully into operations at all levels. Thus, at some Army units, like the 10th Mountain Division and XVIII Airborne Corps, a judge advocate with operational law experience sat at a desk alongside division or corps operations (G–3) personnel, rather than working in the Office of the Staff Judge Advocate.

This increasingly close working relationship between lawyers and operators paid big dividends when U.S. forces deployed to Somalia for Operation RESTORE HOPE in 1992. At the 10th Mountain Division, for example, Army lawyers developed rules of engagement only after close coordination with judge advocates at U.S. Central Command and the Unified Task Force. This practical approach ensured that no rules on the use of force were made in a vacuum and guaranteed smooth implementation of rules based on threatening conduct rather than status—a distinction critical to mission accomplishment in Somalia given the lack of an enemy and the essentially humanitarian nature of the deployment.

By the time 10th Mountain troopers boarded aircraft for Mogadishu, the division commander had such confidence in his legal support that he chose Colonel Smith, the staff judge advocate, to be the chief of staff for the 25-man advance party deploying to Somalia a few days before the main body of troops. Thus, in addition to providing legal advice to General Magruder, the assault command post commander, a judge advocate also supervised and coordinated the work of Magruder's principal staff (G–1 through G–5), as well as the activities of the engineers, communications cell, and security force.

During Operation RESTORE HOPE and the United Nations mission that followed, Army lawyers continued to enhance mission success in a variety of areas. Especially noteworthy was the leading role played in the area of law enforcement, particularly the formulation of a detainee policy. Somalia had no police and no prisons, much less a judicial system. Yet the establishment of law and order was required if RESTORE HOPE's humanitarian mission was to succeed. After researching the issue, Army judge advocates advised that Somalis who committed serious criminal offenses or attacked military personnel could be detained. Additionally, Somalis could be detained if their continued freedom likely would endanger UN and U.S. forces or innocent third parties. Using this legal basis, and recognizing that all detainee procedures must have minimum due process standards, judge advocates then drafted rules for detainee treatment. As a result, each Somali detainee lived in facilities, was provided food, and received medical care constituting a minimum standard of humane treatment. Later, when International Committee of the Red Cross representatives inspected these detainee facilities, they voiced approval of the operation.

Less than a year after U.S. forces returned from Somalia, judge advocates deployed to Haiti in Operation UPHOLD DEMOCRACY. During that operation and the United Nations mission that followed, these lawyers used the law and their legal talents in a number of novel areas. At the direction of General Shelton, the task force commander, Colonel Altenburg, the top lawyer at XVIII Airborne Corps, drafted a document for Shelton to use in convincing Haitian strongman Raoul Cedras that the Haitian military should accept a peaceful entry by U.S. forces into Haiti. But this did not end Altenburg's involvement, for General Shelton further directed Altenburg to sit with him at the face-to-face meeting with Cedras in Port-au-Prince; if a political-military issue arose, General Shelton wanted a judge advocate nearby who would be able to advise him. In addition to this unusual role as diplomatic adviser, Colonel Altenburg later also served as a public affairs officer for the task force when the officer organizing a conference for the international news media determined his skills as a judge advocate made him the best choice for answering questions about Operation UPHOLD DEMOCRACY.

By 1996, when the last Army lawyers returned from Haiti, the judge advocate's new role was firmly in place. The large-scale deployments to Panama, the Persian Gulf, Somalia, and Haiti, like the smaller military operations to Bangladesh, Western Samoa, and Rwanda, demonstrated that operational law was now the essence of the military legal practice. Judge advocates were using every source of law—international, foreign, and domestic—in the operational context. They were succeeding in this new role not only because they were good lawyers, but also because they were competent in military skills and understood the military unit and mission they were supporting.

The evolution of the role of the Army lawyer in military operations, however, is not at an end. On the contrary, this evolution continues. The International and Operational Law Department at the Judge Advocate General's School will continue to provide the intellectual foundation for operational law. Its work, combined with the greatly expanded role of the Center for Law and Military Operations located at the school, as well as realistic training at the Army's combat training centers, will ensure that Army lawyers continue developing their skills at the tactical and operational levels.

More significant developments, however, likely will occur at the strategic level. As Colonels Altenburg, Bozeman, Graham, and Supervielle's experiences in Haiti, Panama, the Sinai, Western Samoa, and Bangladesh demonstrate, judge advocates will be increasingly called upon to serve as political-military advisers. Whether serving as an adviser in political negotiations, creating a "money for muskets" program, drafting a turn-over document for the Multinational Force and Observers, or negotiating a Status of Forces Agreement, judge advocates will need the training and experience to be first-class soldiers, outstanding lawyers, and polished diplomats.

Judge advocates at the tactical level ensure that soldiers do not shoot prisoners. At the operational level, they use their talents and abilities to draft and implement rules of engagement. At the strategic level, they must use their skills as operational lawyers to enhance mission success as negotiators and advisers to those making political-military decisions. Given that judge advocates now are fully integrated into operations at the tactical and operational levels, it seems likely that the most significant future developments in the role of the Army lawyer will occur at the strategic level. Certainly this will mean a greater interest in training judge advocates in how to coordinate successfully with agencies and departments outside the Department of Defense. Thus, while current operational doctrine emphasizes "jointness" within the Defense Department, Army lawyers of the twenty-first century are likely to increasingly focus on interagency coordination and cooperation with "operators" at other government departments.

An examination of judge advocates in military operations from Vietnam to Haiti demonstrates that the role of lawyers in the Judge Advocate General's Corps has undergone a revolutionary transformation. As a result, there is today virtually no limit on what judge advocates may properly do—and are expected by commanders to do—to enhance mission success. Provided judge advocates continue their close working relationship with operators, they will continue to demonstrate that there are no defining limits on the role of the judge advocate. As the Army enters the twenty-first century, it is apparent that the Judge Advocate General's Corps will continue to be an integral part of the Army—providing important and valuable service to soldiers and the nation.

[1] As previously cited in Chapter 2, note 52.

APPENDIX A

Biographical Notes on Army Lawyers

While more than three hundred judge advocates are mentioned by name in this work, there are only about ninety biographical sketches in this appendix. As a general rule, the decision to include information on a particular individual was based on whether his or her experiences in a particular military operation were examined in the narrative; judge advocates mentioned in passing are not included.

Official personnel records maintained by the National Personnel Records Center, St. Louis, Missouri, along with data cards and personnel directories on file at the Personnel, Plans, and Training Office, Office of the Judge Advocate General, were the principal sources for biographical information on retired or deceased judge advocates. Department of the Army Officer Record Briefs provided the biographical data for judge advocates in the active and reserve components and the National Guard. While all information is believed accurate, any errors of commission or omission are the responsibility of the author.

Abbreviations:

Abn	Airborne
ACCA	Army Court of Criminal Appeals
ACMR	Army Court of Military Review
ACR	Armored Cavalry Regiment
Adm	Administrative
Adv	Adviser
AFB	Air Force Base
Affrs	Affairs
AJAG	Assistant Judge Advocate General
ARCENT	Army Central Command, Saudi Arabia
Arm	Armor
Armd	Armored

Arty	Artillery
Asslt	Assault
Asst	Assistant
Atty	Attorney
Aug	Augmentation
BA/S	Bachelor of Arts/Science
Bde	Brigade
Br	Branch
Brks	Barracks
CAC	Combined Arms Center, Fort Leavenworth, Kans.
CAD	Contract Appeals Division
Cav	Cavalry
C&GSC	Command and General Staff College, Fort Leavenworth
Cdr	Commander
Ch	Chief
Civ	Civil
CJA	Command Judge Advocate
CJCS	Chairman of the Joint Chiefs of Staff, Washington, D.C.
CLAMO	Center for Law and Military Operations, Charlottesville, Va.
Cmd	Command
Cmdt	Commandant
Commd	Commissioned
Commr	Commissioner
Cnsl	Counsel
Coll	College
CONUS	Continental United States
Ctr	Center
DA	Department of the Army
DAD	Defense Appellate Division, Falls Church, Va.
DC	Defense Counsel
Def	Defense
Dep	Deputy
Det	Detachment
Div	Division
DoD	Department of Defense
Dscpl	Disciplinary

Eng	Engineer
Enl	Enlisted
Exec	Executive
FECOM	Far East Command
FORSCOM	U.S. Forces Command, Fort McPherson, Ga.
Ft	Fort
GAD	Government Appellate Division, Falls Church, Va.
Grp	Group
IG	Inspector General
IMA	Individual Mobilization Augmentation
INSCOM	Intelligence and Security Command
Instr	Instructor
Intl	International
JA	Judge Advocate
JAGC/D	Judge Advocate General's Corps/Department
JAGO	Judge Advocate General's Office, Washington, D.C.
JCS	Joint Chiefs of Staff
JD	Juris Doctor
JFKSWC	John F. Kennedy Special Warfare Center, Fort Bragg, N.C.
JTF	Joint Task Force
Jus	Justice
Lgl	Legal
Lit	Litigation
LLB/M	Bachelor of Laws/Master of Laws
Log	Logistical
LSO	Legal Services Organization
MAAG	Military Assistance Advisory Group
MACV	Military Assistance Command, Vietnam
MDW	Military District of Washington, Washington, D.C.
M Ed	Master of Education
Med	Medical
MEDCOM	Medical Command
MFO	Multinational Force and Observers, Sinai, Egypt
Mil	Military

MJ	Military Judge
MA/S	Master of Arts/Science
Mgmt	Management
MP	Military Police
MPA	Master of Public Administration
MTMC	Military Traffic Management Command
Natl	National
NG	National Guard
OCLL	Office, Chief of Legislative Liaison, Department of the Army
Ofc	Office
Off	Officer
OGC	Office of the General Counsel
OIC	Officer in Charge
OJA	Office of the Judge Advocate, U.S. Army, Europe
Opnl	Operational
Opns	Operations
Org	Organization
OTJAG	Office of the Judge Advocate General, Washington, D.C.
OSJA	Office of Staff Judge Advocate
Pers	Personnel
PP&TO	Personnel, Plans, and Training Office, OTJAG
Rcrtg	Recruiting
RDC	Regional Defense Counsel
REFRAD	Released from active duty
Rgt	Regiment
Ret	Retired
Sch	School
Scty	Security
Svc	Service
SDC	Senior Defense Counsel
SETAF	Southern European Task Force
S&F	Staff and Faculty
SF	Special Forces
SJA	Staff Judge Advocate
Sp	Special

SOCOM	Special Operations Command
SOUTHCOM	U.S. Southern Command, Panama
Spt	Support
St	State
Strat	Strategic
SUPCOM	Support Command
TAACOM	Theater Army Area Command
TAJAG	The Assistant Judge Advocate General
TJAG	The Judge Advocate General
TJAGSA	The Judge Advocate General's School, U.S. Army
Tm	Team
Tng	Training
Trnsfd	Transferred
UN	United Nations
Univ	University
UNMIH	United Nations Mission in Haiti
UNOSOM II	United Nations Operation in Somalia II
USA	U.S. Army
USALSA	U.S. Army Legal Services Agency
USAREUR	U.S. Army, Europe
USARSO	U.S. Army, South, Panama
USARV	U.S. Army, Vietnam
USATDS	U.S. Army Trial Defense Service
USMA	U.S. Military Academy, West Point, N.Y.
USN	U.S. Navy

Ackerman, Mark S. (1951–). BA (1973) & JD (1977), St Univ N.Y. at Buffalo; LLM (1993), TJAGSA. Commd JAGC (1985). JA, 10th Mtn Div, Ft Drum (1985–89). DC, USATDS w/duty Ft Huachuca & Saudi Arabia (1989–92). Intl & Op Law Div, OTJAG (1993–96). Ch Lgl Adv, UNMIH (1995–96). OIC, USA Alaska, Ft Wainright (1996–98). OGC, Ballistic Missile Def Org, Washington, D.C. (1998–2000). CJA, ARCENT (Kuwait) (2000–present). Participation in DESERT SHIELD/STORM, UNMIH.

Allen, Norman F. J. III (1959–). BA (1981), Univ of the South; JD (1985), Univ of Ga., LLM (1996), TJAGSA. Commd JAGC (1988). DC, USATDS w/duty Ft Ord & JA, Ft Ord (1988–90). JA, 10th Mtn Div, Ft Drum (1991–93). SDC, USATDS w/duty 82d Abn Div, Ft Bragg (1993–95). S&F, TJAGSA (1996–99). DSJA, Task Force EAGLE (Tuzla)

(2000–present). Participation in JUST CAUSE, UPHOLD DEMOCRACY, deployed to Guantanamo Bay (1991–92).

Altenburg, John D., Jr. (1944–). BA (1966), Wayne St Univ; JD (1973), Univ of Cincinnati; MS (1986), C&GSC. Enl, 9th Inf Div, Vietnam (1969–70). DC & Instr, JFKSWC, Ft Bragg (1974). JA, 5th SF Grp (Abn), Ft Bragg (1974–78). JA, XVIII Abn Corps, Ft Bragg (1978). OIC, 3d Armd Div, USAREUR (1979–82). Crim Law Div & PP&TO, OTJAG (1982–85). Exec Off, OJA, USAREUR (1986–88). SJA, 1st Armd Div, USAREUR & Saudi Arabia (1988–91). SJA, XVIII Abn Corps, Ft Bragg (1992–95). AJAG for Mil Law & Opns (1995–97). TAJAG (1997–present). Participation in DESERT SHIELD/STORM, UPHOLD DEMOCRACY, deployed to Florida during Hurricane Andrew relief efforts.

Altieri, Michele H. (1956–). BA (1979), Creighton Univ; JD (1982), Wash Univ. Commd JAGC (1983). JA, 19th SUPCOM, Korea (1983–84). JA, USA Garrison, Ft Meade (1984–85). JA, Natl Scty Agency, Ft Meade (1985–88). Trnsfd to USAR (1988). JA, 2d COSCOM, Saudi Arabia (1991). Ministerial Assistance Tm, U.S. Embassy, Haiti (1995). Dep G–5, 1st Armd Div w/duty Bosnia (1996). Trnsfd to NY NG (1997). Participation in DESERT SHIELD/STORM, UPHOLD DEMOCRACY, JOINT ENDEAVOR.

Andrews, Howard R. (1942–1970). BS (1965), Ga Tech; JD (1968), Univ of Ala. Enl, 101st Abn Div, Vietnam (1969). Commd JAGC (1970). JA, 25th Inf Div, Vietnam (1970). Died in Vietnam (1970). Participation in Vietnam.

Babbitt, Bruce C. (1920–1999). BA (1941) & LLB (1947), Univ of Mont. Inf Off, 32d Inf Div w/duty Philippines & Pacific Theater (1942–45). Commd JAGC (1949). Mil Affrs Br, JAGO (1949–50). JA, 2d Inf Div, Ft Lewis (1950). JA, 3d Inf Div, Ft Benning (1957–58). SJA, 8th Inf Div, USAREUR (1959–61). Ch, GAD (1962–64). Exec Off, OTJAG (1966–68). SJA, MACV (1969–70). AJAG for Civ Law (1970–73). Ret, Brig Gen (1973). Participation in Vietnam.

Baker, John E. (1949–). BS (1972), USMA; JD (1979), Univ of Maine; LLM (1985), NY Univ. Arm Off, Ft Bragg & USAREUR (1972–76). Trnsfd to JAGC (1979). JA & SDC, USARSO (1979–82). Faculty, USMA (1983–86). JA, 8th Inf Div, USAREUR (1986–89). JA, XVIII Abn Corps, Ft Bragg (1989–90). DSJA, USMA (1990–93). MJ, USALSA (1993–94). CJA, Walter Reed Med Ctr, Washington, D.C. (1994–97). SJA, USA Arm Ctr, Ft Knox (1997–99). Ofc Genl Cnsl, Uniformed Services Univ. of Health Sciences (1999–present). Participation in JUST CAUSE.

Berry, J. Stevens (1938–). BA (1960), Stanford Univ; JD (1965),

Northwestern Univ. Inf Off, Ft Leonard Wood (1960–61). Commd JAGC (1968). Ch DC, II Field Force, Vietnam (1968–69). REFRAD (1969). Participation in Vietnam.

Black, Scott C. (1952–). BA (1974), Calif. Poly St Univ; JD (1980), Calif. Western Sch Law. Arm Off, 7th Inf Div, Ft Ord (1974–77). Trnsfd JAGC (1980). JA, USA Air Def Ctr, Ft Bliss (1980–84). Adm Law Div, OTJAG (1985–89). Asst Cnsl to President, White House (1989). DSJA, 7th Inf Div (1990–93). JA, OJA, USAREUR (1993–94). SJA, 3d & 1st Inf Div, USAREUR (1994–96). OCLL (1996–98). Chief, PP&TO, OTJAG (1999–2000). SJA, V Corps, USAREUR (2000–present). Participation in JTF Los Angeles.

Bozeman, John R. (1940–). BA (1963) & JD (1965), Univ of Ky.; LLM (1975) Univ of Calif.–Berkeley. Arty & SF Off, Ft Bragg, Korea, Thailand, & Vietnam (1965–72). Trnsfd to JAGC (1972). JA, 82d Abn Div, Ft Bragg (1972–74). SJA, 3d Armd Div, USAREUR (1981–83). Exec Off, OTJAG (1984–87). SJA, XVIII Abn Corps (1987–91), Ft Bragg, Panama, & Saudi Arabia. SJA, MTMC, Falls Church, Va. (1991–92). Ret, Col (1992). Participation in Just Cause, Desert Shield/Storm.

Bradley, Corey L. (1964–). BS (1986), Univ of So Calif.; JD (1989), Willamette Univ; LLM (1999), TJAGSA. Commd JAGC (1989). JA, 3d Armd Div, USAREUR (1990–91). CJA, 3/325th Inf Rgt (Abn), Vicenza, Italy, & Iraq (1991–93). JA, Tripler Med Ctr, Hawaii (1993–96). Lit Div, USALSA (1996–98). S&F, TJAGSA (1999–present). Participation in Desert Shield/Storm, Provide Comfort.

Brewer, Mark P. (1957–). BA (1979), Citadel; JD (1985), Miss Coll. Commd JAGC (1985). USATDS & JA, 82d Abn Div (1985–89), Ft Bragg. Trnsfd to USAR. JA, 199th JAG Det, St. Petersburg, Fla. (1989–present). Participation in Desert Shield/Storm.

Burrell, Robert A. (1956–). BA (1978), Hampton-Sydney Coll; JD (1981), Coll of Wm & Mary; LLM (1989), TJAGSA. Commd JAGC (1982). JA, 2d Armd Div, Ft Hood (1982–85). DAD & Rcrtg Off, USALSA (1985–88). JA, 25th Inf Div, Hawaii (1989–90; 91–92). JA, CENTCOM, Saudi Arabia (1990–91). S&F, TJAGSA (1992–95). DSJA, 101st Asslt Div, Ft Campbell (1996–98). SJA, 2d Inf Div, Korea (1998–99). S&F, TJAGSA (1999–present). Participation in Desert Shield/Storm.

Buyer, Stephen E. (1958–). BA (1980), Citadel; JD (1984), Valparaiso Univ. Commd JAGC (1984). JA, USA Transportation Ctr, Ft Eustis (1984–87). Trnsfd to USAR. JA, 22d SUPCOM, Saudi Arabia (1990–91). Member, U.S. House of Representatives (1993–present). Participation in Desert Shield/Storm.

Cole, Raymond D. (1937–). BS & JD (1962), S.Dak. St Univ. Commd JAGC (1962). JA, 5th Inf Div, Ft Carson (1962–64). JA, 82d Abn Div, Ft Bragg (1964–66). Bde JA, 173d Abn Bde, Vietnam (1966–67). SJA, 11th Army Terminal Cmd & 15th MP Bde, USAREUR (1968–71). JA, MACV (1972–73). CJA, US Dscpl Bks, Ft Leavenworth (1974–76). DSJA, V Corps, USAREUR (1976–80). SJA, 2d Inf Div, Korea (1980–81). MJ, USALSA (1981–92). Ret, Col (1992). Participation in POWER PACK, Vietnam, DESERT SHIELD/STORM.

Corrigan, Dennis M. (1940–). BA (1962), Fordham Univ; JD (1965), Rutgers Univ; MS (1978), C&GSC. Commd JAGC (1966). JA, USA Tng Ctr, Ft Ord (1967–69). JA, USA Spt Troops, U.S. Theater Army SUPCOM, USAREUR (1969–70). SJA, 1st Inf Div (Fwd), USAREUR (1970–72). JA, USARV, & MJ, USA Trial Judiciary, Vietnam (1972–73). S&F, TJAGSA (1974–77). SJA, 1st Inf Div (Fwd), USAREUR (1978–81). Lgl Adv, CJCS (1981–85). SJA, 21st SUPCOM, USAREUR (1986–90). MJ, ACMR (1990–91). Ch, PP&TO, OTJAG (1991–93). Mil Asst, OGC, DoD (1993–95). Ret, Col (1995). Participation in Vietnam.

Cottingham, Edward B. (1958–). BA (1980), Wofford Coll; MA & JD (1984), Univ of S.C. Commd JAGC (1985). JA & DC, 7th Inf Div, Ft Ord (1985–88). Trnsfd to USAR. JA, 187th JAG Det (1989–92). ARCENT, Saudi Arabia (1990–91). Participation in DESERT SHIELD/STORM.

Crowley, Leonard G. (1933–). LLB (1957), Boston Coll; LLM (1971), Geo Wash Univ. Enl, Ft Buchanan, P.R. (1957–58). Commd JAGC (1958). Lgl Assistance Div, OTJAG (1960–62). JA, USA Antilles Comd, USARSO, Ft Brooke, P.R. (1962–66). JA, Fourth USA, Ft Sam Houston (1967–69). Ch, Foreign Claims Div, USARV (1969–70). USA Procurement Agcy, USARV (1971–72). OGC, Army & Air Force Exchange Svc, USAREUR (1972–75). DSJA, USA SUPCOM Hawaii, Ft Shafter (1975–77). Ret, Lt Col (1977). Participation in Vietnam.

Cuthbert, Thomas R. (1938–). BS (1961), USMA; MS (1964), Princeton Univ; JD (1967), Harvard Univ; LLM (1971), Northwestern Univ. Eng Off w/duty Korea (1961–63). Trnsfd to JAGC (1967). Patent Atty, OTJAG (1967–69). DSJA, 101st Abn Div, Vietnam (1969–70). OIC (Nuremberg) & DSJA, 1st Armd Div, USAREUR (1971–74). SJA, USA Tng Ctr, Ft Leonard Wood (1974–76). Ch, Mil Pers Law Br, OTJAG (1977–81). Dir, Legislation & Lgl Policy, DoD (1982–86). SJA, VII Corps, USAREUR (1986–89). Ch, USATDS (1989–91). AJAG for Mil Law & Opns (1991–93). Cdr, USALSA & Ch Judge, ACCA (1993–95). Ret, Brig Gen (1996). Participation in Vietnam.

Dana, Matthew L. (1962–). BS (1985), Univ of San Francisco; JD (1988), Univ Calif.–Hastings. Commd JAGC (1989). JA, 1st Armd Div, USAREUR (1989–90). CJA, 2d ACR, Saudi Arabia (1990–91). JA, 3d Inf Div, USAREUR (1991–92). Post JA, TJAGSA (1992–94). Lit Div, USALSA (1995–97). Trnsfd to Calif. NG (1997). Participation in DESERT SHIELD/STORM.

Deline, Donald A. (1943–). BA (1965), JD (1968) & MA (1968), Univ of Miss. Commd JAGC (1968). JA, Tooele Army Depot (1968–69). JA, USARV (1969–70). JA, USAREUR (1970–74). OCLL (1975–78). SJA, 101st Abn Div (1978–81). SJA, 9th Inf Div, Ft Lewis (1982–85). Lit Div, OTJAG (1985–87). SJA, V Corps, USAREUR (1988–90). Legislative Cnsl to Secretary of Def, DoD (1990–93). Ret, Col (1993). General Cnsl, Senate Armed Services Committee (1993–96). Participation in Vietnam.

Dell'Orto, Daniel J. (1949–). BS (1971), Univ of Notre Dame; MBA (1975), Pepperdine Univ; JD (1980), St John's Univ; LLM (1992), Georgetown Univ. Arty Off, 25th Inf Div, Hawaii (1971–76). Trnsfd to JAGC (1979). JA, Ft Benning (1979–82). SDC, USATDS w/duty USAREUR (1983–86). OIC (Baumholder), 8th Inf Div, USAREUR (1986–87). Atty & Dep Ch, GAD (1988–92). MJ, USA Trial Judiciary (1992–94). Ch, USATDS (1994–96). Mil Asst, OGC, DoD (1996–98). Ret, Col (1998). Participation in Cuban refugee resettlement operations at Ft Chaffee (1980–81).

Dommer, Paul P. (1937–1975). BA (1959), Canisius Col; LLB (1962) & LLM (1964), Georgetown Law Ctr. Commd JAGC (1967). JA, USA Air Def Ctr, Ft Bliss (1967–69). JA, OJA, USAREUR (1969–72). JA, MACV (1972–73). Intl Law Div, OTJAG (1973–75). Lgl Adv, Ofc of JCS (1975). Died on active duty (1975). Participation in Vietnam; a principal author of *Law at War*.

Douglass, John J. (1922–). BA (1943), Univ of Nebr.; JD (1952) Univ of Mich.; LLM (1973), Univ of Va. Inf Off w/duty American Theater (1944–46). Commd JAGC (1953). JA, FECOM, Japan & Korea (1953–54). Lit Div, OTJAG (1959–62). Exec Off, OJA, USAREUR (1963–66). SJA, USA Garrison, Ft Riley (1966–68). SJA, USARV (1968–69). MJ, USA Trial Judiciary (1969–70). Cmdt, TJAGSA (1970–74). Ret, Col (1974). Participation in Vietnam.

Downes, Michael M. (1935–). BA (1957) & LLB (1959), Univ of Ga. Commd JAGC (1963). DSJA, 23d Inf Div (Americal), Vietnam (1969–70). SJA, 2d Inf Div (1976–77). Lgl Adv, DA IG (1978–81). SJA, XVIII Abn Corps, Ft Bragg (1981–85). SJA, USA Sig Ctr, Ft Gordon (1985–88). MJ, USA Trial Judiciary w/duty Ft Bragg (1988–90). Ret, Col (1990). Participation in Vietnam, URGENT FURY.

Durbin, Paul J. (1917–). BA (1938) & LLB (1941), Univ of Ky. Inf Off, w/duty France & Germany (1943–45). Commd JAGC (1948). JA, X Corps, FECOM, & 3d Log Cmd (1950). SJA, 7th Inf Div, Korea (1950–51). SJA, 1st Armd Div, Ft Hood (1952–53). SJA, 4th Armd Div, Ft Hood (1954). SJA, 82d Abn Div, Ft Bragg (1954–56). SJA, 101st Abn Div, Ft Campbell (1956–59). SJA, MAAG, Vietnam (1959–61). DSJA, USA Pacific, Ft Shafter (1962–65). DSJA, Fifth USA, Chicago (1965–66). SJA, II Field Force, Vietnam (1966). MJ, USA Trial Judiciary w/duty Vietnam (1966–67). Ret, Col (1968). Participation in Vietnam.

Eblen, George C. (1921–). BS (1942) & LLB (1948), Univ of Tenn. Enl & Inf Off w/duty France & Germany (1942–46). Commd JAGC (1949). SJA, MAAG, Vietnam, & MACV (1961–62). SJA, USA Missile Cmd, Redstone Arsenal (1963–66). Ret, Lt Col (1966). Participation in Vietnam.

Fair, Karen V. (1963–). BS (1985), USMA; JD (1991), Univ of Tenn.; LLM (1998), TJAGSA. Trnsfd to JAGC (1991). JA, III Corps, Ft Hood (1991–93). CJA, Task Force SUPCOM, Somalia (1993). Adm Law, OTJAG (1993–95). Lit Atty, USALSA (1995–97). OIC, 1st Inf Div, USAREUR (1998–2000). Participation in RESTORE HOPE.

Felder, Ned E. (1937–). BA (1959) & LLB (1961), S.C. St Coll. Finance Off, Korea & Ft Totten (1961–63). Commd JAGC (1963). JA, 4th Inf Div, Ft Lewis & Vietnam (1963–66). JA, II Field Force, Vietnam (1966–68). JA, VII Corps, USAREUR (1969–72). DSJA, Berlin Bde (1972–73). MJ, USALSA (1973–81, 1984–88). SJA, USA Garrison, Ft Meade (1981–84). Ret, Col (1988). Participation in Vietnam.

Finnegan, Patrick. (1949–). BS (1971), USMA; MPA (1973), Harvard Univ; JD (1979), Univ of Va. Trnsfd to JAGC (1979). JA, 8th Inf Div, USAREUR (1979–82). S&F, TJAGSA (1983–87). JA & DSJA, XVIII Abn Corps, Ft Bragg & Saudi Arabia (1988–91). Lgl Adv, Joint Sp Opns Cmd, Ft Bragg (1991–93). SJA, US SOCOM, MacDill AFB (1994–96). Lgl Adv, USEUCOM (1996–98). SJA, USMA (1998–99). Professor of Law, USMA (1999–present). Participation in HAWKEYE, DESERT SHIELD/STORM.

Fletcher, Douglas K. (1951–). BA (1973), Univ of Calif.–Davis; JD (1976), Univ of Calif.–San Francisco (1976). Commd JAGC (1977). JA & DC, 2d Armd Div, Ft Hood (1977–80). CJA, 38th Air Def Bde, Osan, Korea (1980–81). SDC, USATDS w/duty XVIII Abn Corps, Ft Bragg (1982–85). S&F, TJAGSA (1985–88). Asst SJA, CENTCOM (1988–91). CJA, JFKSWC & USA Sp Opns Cmd, Ft Bragg (1991–94). CJA, JTF–6, Ft Bliss (1994–97). S&F, Med Ctr & Sch, Ft Sam Houston (1997–99). Retd, Lt Col (1999). Participation in URGENT FURY, DESERT SHIELD/STORM.

Gingras, Mark J. (1962–). BA (1985), Univ of Mass.–Amherst; JD (1988), Univ of Va.; LLM (1996), TJAGSA. Commd JAGC (1989). JA, 82d Abn Div, Ft Bragg & Saudi Arabia (1988–92). JA, V Corps, USAREUR (1992–95). Adm Law & PP&TO, OTJAG (1996–99). Participation in MFO, DESERT SHIELD/STORM.

Gordon, Richard E. (1953–). BA & JD (1979), Fla. St Univ; LLM (1988), TJAGSA. Commd JAGC (1981). JA, 8th Inf Div, USAREUR (1981–84). DC & JA, 2d Armd Div, Ft Hood (1984–87). OIC, Rheinberg Lgl Ctr, 21st SUPCOM, USAREUR (1988–90). Asst Exec Off, OJA, USAREUR (1990–92). DSJA, 10th Mtn Div, Ft Drum (1992–95). SJA, JTF Somalia (1993). Politico-Mil Planner, JCS (1995–97). SJA, 4th Inf Div, Ft Hood (1998–99). SJA, USA Inf Ctr, Ft Benning (1999–present). Participation in PROVIDE COMFORT, RESTORE HOPE, UNOSOM II, UPHOLD DEMOCRACY, deployed to Florida during Hurricane Andrew relief efforts.

Goshi, Allen K. (1959–). BS (1981), USMA; JD (1989), Georgetown Univ; LLM (1993), TJAGSA. Trnsfd to JAGC (1989). Arm Off, w/duty USAREUR (1981–86). JA, XVIII Abn Corps, Ft Bragg & Saudi Arabia (1989–92). SJA, JTF, Guantanamo, Cuba (1991). SDC, USATDS w/duty Korea (1993–94). OIC (Katerbach), 3d Inf Div, USAREUR (1994–95). JA, SF Opns Det.–Delta, Ft Bragg (1996–98). DSJA, USA Garrison, Ft Sam Houston (1998–2000). JA, FORSCOM (2000–present). Participation in DESERT SHIELD/STORM, deployed to Guantanamo Bay during refugee rioting (1994).

Govern, Kevin H. (1962–). BA (1984) & JD (1987), Marquette Univ; LLM (1995), TJAGSA. Commd JAGC (1987). JA, XVIII Abn Corps, Ft Bragg & Saudi Arabia (1987–90). Grp JA, 10th SF Grp, Ft Devens (1990–92). SDC, USATDS, w/duty Ft Benning (1992–94). JA, 1st Armd Div, USAREUR (1995–97). JA, US SOCOM, MacDill AFB (1997–2000). JA, XVIII Abn Corps (2000–present). Participation in JUST CAUSE, HAWKEYE, DESERT SHIELD/STORM, PROVIDE COMFORT.

Graham, David E. (1944–). BA (1966), Tex. A&M Univ; MA (1968), George Washington Univ; JD (1971), Univ of Tex. Commd JAGC (1971). Faculty, TJAGSA (1971–75). JA, OJA, USAREUR (1975–78). Asst JA w/duty U.S. Embassy, Bonn (1978–79). Intl Law Div, OTJAG (1979–81, 1982–86). Force JA, MFO (1981–82). Ch, Intl Law Div, OTJAG (1986–90); Dir, CLAMO (1988–90); Dep Cmdt, TJAGSA (1989–90). SJA, US SOUTHCOM (1990–92). Ch, Strat Plans, OTJAG (1993–94). Ch, Intl & Opnl Law Div, OTJAG & Dir, CLAMO (1994–present). Participation in MFO.

Gray, Kenneth D. (1944–). BA (1966) W.Va. St Coll; JD (1969), Univ of W.Va. Commd JAGC (1969). DC, USA Tng Ctr, Ft Ord

(1969–70). JA & DC, USA SUPCOM, USARV (1970–71). Asst Mil Affrs Off, First USA, Ft Meade (1971–72). PP&TO, OTJAG (1972–74). S&F, TJAGSA (1975–78). DSJA, 1st Armd Div, USAREUR (1978–80). SJA, 2d Armd Div, Ft Hood (1981–84). Ch, PP&TO, OTJAG (1984–87). SJA, III Corps, Ft Hood (1988–89). Sp Asst to TJAG (1989–90). Cdr, USALSA & Ch Judge, ACMR (1991–93). TAJAG (1993–97). Ret, Maj Gen (1997). Participation in Vietnam.

Hagan, William R. (1947–). BA (1970), Ky. Wesleyan Coll; JD (1971), Univ of Ky. Commd JAGC (1972). JA, 2d Armd Div, Ft Hood (1972–73). JA & DC, 82d Abn Div, Ft Bragg (1973–76). DSJA, SETAF w/duty Vicenza, Italy (1977–80). S&F, USMA (1980–83). DSJA, 24th Inf Div, Ft Stewart (1984–87). SJA, 25th Inf Div, Hawaii (1987–90). SJA, 22d SUPCOM, Saudi Arabia (1990–91). MJ, ACMR (1991–92). Lgl Adv, DA IG (1992–94). Ret, Col (1994). Participation in DESERT SHIELD/STORM.

Ham, Patricia. See Martindale, Patricia A.

Hamelin, Normand J. (1935–). BS (1964), Bryant Coll; JD (1970), St. Louis Univ. Enl, U.S. Air Force w/duty Germany & Korea (1952–61). Commd JAGC (1970). JA, OJA, USAREUR (1972–75). Intl Affrs, OTJAG (1976–79). Liaison Off, U.S. Embassy, Bonn (1979–83). JA, JFKSWC, Ft Bragg (1983–84). Ret, Lt Col (1984). Participation in JUST CAUSE.

Hamlin, Guy A. (1923–1992). LLB (1950), Duke Univ. Enl & Inf Off w/duty France & Germany (1945–46). Commd JAGC (1950). JA, Okinawa Eng District, FECOM-Japan (1951–53). JA, Third USA, Ft McPherson (1953–57). JA, Seventh Army, USAREUR (1958–62). SJA, 82d Abn Div, Ft Bragg & Dominican Republic (1963–66). JA, MACV, Vietnam (1966–67). SJA, 12th Support Bde, Ft Bragg (1967). Ret, Lt Col (1967). Participation in POWER PACK, Vietnam.

Haughney, Edward W. (1917–). LLB (1949), St. Johns Univ; MS (1966), George Washington Univ. Arty Off, France & Germany (1942–47). Commd JAGC (1949). JA, FECOM (1950–53). S&F, TJAGSA (1954–58). Ch, Intl Affrs Br, OJA, USAREUR (1958–62). Asst Ch & Ch, Intl Affrs Div, OTJAG (1963–65). SJA, MACV (1966–67). Ch, Intl Affrs, OTJAG (1967–69). Lgl Adv, USEUCOM (1969–72). Ret, Col (1972). Participation in Vietnam.

Hodson, Kenneth J. (1913–1995). BA (1935) & LLB (1937), Univ of Kans. Trnsfd to JAGD (1942). JA, Trinidad Base Sector & Base Cmd, Trinidad, W. Indies (1942–44). JA, Western Tactical Tng Cmd (Army Air Corps) (1944). JA, 52d Medium Port, Ft Hamilton & Le Havre, France (1944–45). Asst JA, Normandy Base Section, France

(1945). SJA, Chanor Base Section, France (1946). Asst SJA, Exec Off, & SJA, Western Base Section, France (1946–47). Asst SJA, U.S. Constabulary, Paris, France (1947). Asst SJA & SJA, American Graves Registration Cmd (1947). Ch, Sp Projects Div & Asst to Board of Review No. 1 (1948–51). S&F, TJAGSA (1951–53). Asst SJA, Army Forces, Far East/Eighth USA (Rear) (1954–57). Ch, Mil Pers Div & Mil Jus Div & Exec Off, OTJAG (1958–62). AJAG for Mil Jus (1962–67). TJAG (1967–71). Ret & recalled (1971). Ch Judge, ACMR (1971–74). Ret, Maj Gen (1974).

Holdaway, Ronald M. (1934–). BA (1957) & LLB (1959), Univ of Wyo. Commd JAGC (1959). JA, 4th Inf Div, Ft Lewis (1960–63). JA, USA Hawaii (1963–66). S&F, TJAGSA (1967–69). SJA, 1st Cav Div, Vietnam (1969–70). Dep & Ch, GAD, USALSA (1970–75). Ch, PP&TO (1975–77). SJA, VII Corps, USAREUR (1978–80). Exec Off & AJAG for Civ Law (1980–83). JA, OJA, USAREUR (1983–87). Cdr, USALSA & Ch Judge, ACMR (1987–89). Ret, Brig Gen (1989). Participation in Vietnam.

Horton, Victor L. (1953–). BA (1975) & JD (1978), Univ of Ark. Commd JAGC (1978). JA & DSJA, USA Garrison, Ft McPherson (1978–81). JA, USA Arm Ctr, Ft Knox (1981–84). Post JA, 6th Inf Div w/duty Ft Greely (1985–87). Adm Law Div, OTJAG (1987–91). DSJA, 6th Inf Div, Ft Richardson (1991–93). SJA, UN Comd, Somalia (1993). Asst Exec Off, OTJAG (1993–96). SJA, USA Eng Cmd, Ft Leonard Wood (1996–98). SJA, INSCOM, Ft Belvoir (1998–present). Participation in UNOSOM II.

Hudson, William A., Jr. (1957–). BA (1979), Citadel; JD (1982), Univ of S.C.; LLM (1991), TJAGSA. Commd JAGC (1984). JA, USA 1st Sp Opns Cmd & SF Grp, Ft Bragg (1984–87). Instr, USA Inf Sch, Ft Benning & SDC, USATDS w/duty Ft Benning (1987–90). S&F, TJAGSA (1991–94). DSJA, Army Sp Opns Cmd, Ft Bragg (1994–96). UN Force Lgl Adv, UNMIH (1995). JA, CENTCOM, MacDill AFB (1996–97). DSJA, III Corps, Ft Hood (1998–99). SJA, Spec Forces Cmd, Ft Bragg (2000–present). Participation in UNMIH.

Huffman, Walter B. (1944–). BS (1967), M Ed (1968), & JD (1977), Tex. Tech Univ. Arty Off, Ft Sill; 5th Inf Div & XXIV Corps, Vietnam; III Corps, Ft Hood (1968–74). Trnsfd to JAGC (1977). JA, USA Garrison, Ft Ord (1977–79). S&F, TJAGSA (1980–83). SJA, 2d Armd Div (Fwd), USAREUR (1983–85). PP&TO, OTJAG (1985–88). SJA, VII Corps, USAREUR, & Saudi Arabia (1989–91). SJA, CENTCOM (1991–93). Exec Off, OTJAG (1993). AJAG for Mil Law & Opns (1993–95). AJAG for Civ Law & Lit (1995–97). TJAG (1997–present). Participation in DESERT SHIELD/STORM.

Jackson, Richard B. (1954–). BS (1976), USMA; JD (1983), Georgetown Univ; LLM (1987), TJAGSA. Trnsfd to JAGC (1983). Inf Off w/duty Panama & Ft Benning (1976–80). Grp JA, 5th SF Grp, Ft Bragg (1984–85). JA, USA 1st Sp Opns Cmd, Ft Bragg (1985–86). SDC, USATDS w/duty Panama (1987–90). Lgl Adv, JTF–6 & JA, USA Air Def Ctr, Ft Bliss (1990–93). JA & DSJA, U.S. Atlantic Cmd, Norfolk (1994–96). S&F, TJAGSA (1996–98). SJA, USA Sp Opns Cmd, Ft Bragg (1998–2000). SJA, 25th Inf Div, Schofield Bks, Hawaii (2000–present). Participation in JUST CAUSE.

Johnson, M. Tia (1959–). BA (1980), Hampton Univ; JD (1983), Temple Univ; LLM (1991), TJAGSA. Commd JAGC (1984). JA, USA Tng Ctr, Ft Dix (1984–88). JA, Eighth USA, Korea (1988–90). JA, III Corps, Ft Hood (1991–94). DSJA, JTF, Guantanamo Bay, Cuba (1992). Intl & Opnl Law, OTJAG (1994–96). PP&TO, OTJAG (1997–99). DSJA, US Army South, Ft Buchanan, P.R. (1999–2000). S&F, TJAGSA (2000–present). Deployed to Guantanamo Bay (1992).

Johnson, Robin L. (1961–). BA (1984), Univ of Wisc.; JD (1987), Haline Univ; LLM (1997), TJAGSA. Commd JAGC (1987). DC, USATDS w/duty Ft Ord & MFO (1987–89). JA, USA Garrison, Ft Ord (1989–90). CJA, 23d Support Grp & 2d Inf Div, Korea (1990–92). JA, 6th Inf Div, Ft Wainwright (1992–93). JA, USA Test & Evaluation Cmd, Aberdeen Proving Gnd (1993–94). S&F, C&GSC (1994–96). DSJA, Natl Tng Ctr, Ft Irwin (1997–99). Intl & Opnl Law, OTJAG (1999–present). Participation in MFO, JUST CAUSE.

Kent, Irvin M. (1920–). BA (1940), Syracuse Univ; LLB (1947), Harvard Univ. Inf Off w/duty France & Germany (1943–45). Prosecution Staff, Nuremberg War Crimes Trials (1947). Commd JAGC (1948). JA, 1st Inf Div, EUCOM (1948–51). GAD, OTJAG (1951–54). SJA, First USA, Governors Island, N.Y. (1956–58). JA, 3d Log Cmd, USAREUR (1958–63). JA, USA Combat Development Cmd, Fts Gordon & Belvoir (1963–66). SJA, USA Air Def Cmd, Ent AFB, Colo. (1966–68). SJA, II Field Force & JA, MACV (1968–69). MJ, USA Trial Judiciary w/duty Ft Carson (1969–71). Ret, Col (1971). Participation in Vietnam.

Lamberth, Royce C. (1943–). BA (1965) & JD (1967), Univ of Tex. Enl, USA (1967). Commd JAGC (1968). JA, XVIII Abn Corps (1968–69). JA & DC, 1st Cav Div & Saigon SUPCOM, Vietnam (1969–70). Lit Div, OTJAG (1970–74). REFRAD (1974). Judge, U.S. District Court for District of Columbia (1987–present). Participation in Vietnam.

Larson, Michael L. (1955–). BA (1977) & JD (1980), Univ of Ky. Commd JAGC (1981). JA, White Sands Missile Range (1981–85).

JA, USA Garrison, Okinawa, Japan (1985–88). JA, XVIII Abn Corps, Ft Bragg & Saudi Arabia (1989–91). Trnsfd to USAR (1991). Participation in DESERT SHIELD/STORM, UPHOLD DEMOCRACY.

Ley, John P., Jr. (1951–). BA (1973), Pa. St Univ; JD (1976), Chicago-Kent Coll Law. Commd JAGC (1977). JA, DC & OIC (Augsburg), VII Corps, USAREUR (1977–80). OIC, Claims Off (Brussels), USA Claims Svc Europe (1980–82). S&F, TJAGSA (1983–85). JA, 1st Cav Div & DSJA, 2d Armd Div, Ft Hood (1987–90). RDC, USATDS w/duty Korea (1990–92). DSJA, V Corps, USAREUR (1992–94). SJA, SETAF (1994–96). DJA, OJA, USAREUR (1996–97). Ch, USATDS (1997–2000). SJA, 8th Army, Korea (2000–present). Participation in SUPPORT HOPE.

McNamee, Alfred A. (1931–). BA (1958), Univ of Fla.; LLB (1961), Wake Forest Univ. Enl & Inf Off w/duty USAREUR, Ft Benning, & Ft Bragg (1949–63). Trnsfd to JAGC (1962). JA, 101st Abn Div, Ft Campbell (1963–64). SJA, USA SUPCOM, Vietnam (1964–65). S&F, TJAGSA (1965–67). SJA, 5th Inf Div, Ft Carson (1967–68). DSJA & SJA, USA Arm Ctr, Ft Knox (1968–71). S&F, C&GSC, Ft Leavenworth (1972–75). Lgl Adv, Allied Forces Europe, South, Italy (1975–78). Ret, Col (1978). Participation in Vietnam.

Martindale (later Ham), Patricia A. (1960–). BS (1982), Syracuse Univ; JD (1985), Temple Univ; LLM (1998), TJAGSA. Commd JAGC (1990). JA, XVIII Abn Corps, Ft Bragg & Saudi Arabia (1990–91). JA, I Corps, Ft Lewis (1991–94). DC, USATDS w/duty USAREUR (1994–96). Instr, Intl Tng Det, Naval Justice Sch, Newport, R.I. (1996–97). GAD, USALSA (1998–2000). SDC, III Corps, Ft Hood (2000–present). Participation in DESERT SHIELD/STORM.

Miller, Hubert E. (1918–2000). LLB (1941), Albany Law Sch; MA (1964), George Washington Univ. Inf Off w/duty France & Germany (1943–45) & Korea (1950–52). Trnsfd to JAGC (1953). S&F, TJAGSA (1956–57). DSJA, 101st Abn Div, Ft Campbell (1957–59). SJA, 1st Cav Div, Korea (1959–60). Mil Jus Br, OTJAG (1960–63). DSJA, USA Pacific, Hawaii (1964–66). SJA, 1st Log Cmd, Vietnam (1966–67). MJ, USA Judiciary, OTJAG (1967–68). S&F, War Coll (1971–72). SJA, USA Air Def Cmd, Ent AFB (1972–74). SJA, USA Air Def Ctr, Ft Bliss (1974–75). Ret, Col (1975). Participation in Vietnam.

Miranda, Ralph G. (1938–). BA (1959), Tex. West Coll; LLB (1962), Univ of San Francisco. Commd JAGC (1964). JA, USA Pacific, Okinawa, Japan (1964–65). JA, USA Garrison, Ft Ord (1966–67). JA, VII Corps, USAREUR (1967–70). JA, USARV (1971–72). DSJA, USA Arty Ctr, Ft Sill (1972–73). Trnsfd to USAR (1973). JA, 22d JAG Det, Austin, Tex. (1977–87). Ret, Lt Col (1987).

Moushegian, Vahan, Jr. (1942–). BA (1965), Vanderbilt; LLB (1968), Univ of Va. Trnsfd to JAGC (1972). JA, USARV/MACV (1972–73). JA, V Corps & OJA, USAREUR (1973–76). Adm Law Div, OTJAG (1977–82). DSJA, 1st Inf Div (Mech) (1983–86). Adm Law Div, OTJAG (1986–88). SJA, 24th Inf Div (1988–91). Lgl Adv, USA Surgeon General (1991–94). SJA, First USA, Ft Meade (1994–95). Ret, Col (1995). Participation in Vietnam, deployed to Florida during Hurricane Andrew relief efforts (1992).

Myers, William G. (1924–). BS , (1944) Southwest La. Institute; MA (1948), Univ of Ark.; JD (1955), Univ of Mich. Enl & Off, USN (1943–49). Commd JAGC (1956). JA, USAREUR (1956–59). JA, MACV (1963–64). SJA, 4th Log Cmd, Ft Lee (1965–67). Ret, Lt Col (1970). Participation in Vietnam.

Page, Bradley D. (1960–). BA (1983), Univ of Mich.; JD (1989), Univ of Wisc.; LLM (1994), TJAGSA. Chem Off w/duty 25th Inf Div, Hawaii (1983–86). Trnsfd to JAGC (1989). JA, USA Eng Ctr, Ft Leonard Wood (1989–91). JA, 7th Inf Div, Ft Ord (1991–93). Adm Law Div, OTJAG (1994–96). Trnsfd to USAR (1996). JA, 91st LSO, Ft Sheridan (1996–98). IMA, Adm Law Div, OTJAG (1998–present). Participation in JTF LOS ANGELES.

Passar, Arthur L. (1949–). BA (1971), Univ of N.C.; JD (1975), Univ of Calif.–Berkeley; LLM (1988), George Washington Univ. Commd JAGC (1977). JA, 3d Inf Div, USAREUR (1977–79). GAD, OTJAG (1981–83). S&F, USMA (1984–87). Contract Law Atty, FORSCOM, Ft McPherson (1988–92). JA, USA Missile Cmd, Redstone Arsenal (1992–95). SJA, Joint Log SUPCOM, Haiti (1994–95). Contract Law Div, OTJAG (1995–98). Ret, Lt Col (1999). Participation in UPHOLD DEMOCRACY.

Pede, Charles N. (1961–). BA (1984) & JD (1987), Univ of Va.; LLM (1996), TJAGSA. Commd JAGC (1988). DC, USATDS w/duty USAREUR (1988–90). JA, 21st TAACOM, USAREUR (1990–92). JA, 10th Mtn Div, Ft Drum & Somalia (1992–95). S&F, TJAGSA (1996–98). PP&TO, OTJAG (1999–2000). Asst Exec Off, OTJAG (2000–present). Participation in PROVIDE COMFORT, RESTORE HOPE.

Pfeiffer, Bernard A. (1950–). BA (1972), Western Md Coll; JD (1975), Dickinson. Commd JAGC (1976). JA, USA Inf Ctr, Ft Benning (1976–80). Trnsfd to USAR (1980). JA, 200th JAG Det, Columbus, Ga., & Third USA, Ft McPherson, Ga. (1980–90). JA, ARCENT, Saudi Arabia (1990–91). DSJA, Third USA (1991–93). SJA, 200th JAG Det (1993–94). SJA, 87th Tng Div, Birmingham, Ala. (1997–99). SJA, 81st Regional Spt Cmd, Birmingham, Ala. (1999–present). Participation in DESERT SHIELD/STORM.

Prugh, George S. (1920–). BA (1941), Univ of Calif.–Berkeley; LLB (1948), Hastings Coll, Univ of Calif.; MA (1963), George Washington Univ. Enl, Calif. NG & Coast Art Off w/duty Philippines & Pacific Theater (1939–44). Commd JAGC (1947). Mil Jus, Claims & Lit Divs, OTJAG (1949–50). JA, Wetzlar Mil Post, USAREUR (1950–51). Exec Off & SJA, Rhine Mil Post, USAREUR (1951–53). Board of Review & Mil Jus Div, OTJAG (1953–56). DSJA, Eighth USA (1957–58). DSJA & Asst Exec Reserve Affrs, Sixth USA, Presidio of San Francisco (1958–61). Ch, Career Mgmt Div & Exec Off, OTJAG (1962–64). SJA, MACV (1964–66). Lgl Adv, USEUCOM w/duty France & Germany (1966–69). JA, OJA, USAREUR (1969–71). TJAG (1971–75). Ret, Maj Gen (1975). Participation in Vietnam.

Radosh, Burnett H. (1935–). BA (1953), Univ of Chicago; LLB (1956), N.Y. Univ; LLM (1960), Georgetown Univ. Enl w/duty Ft Benning (1958–59). Commd JAGC (1959). DAD, OTJAG (1959–61). CJA, 4th Log Cmd & 1st Log Cmd, France (1961–62). 4th Log Cmd, France (1962–63). 3d Log Cmd, France (1963–64). DSJA, 82d Abn Div, Ft Bragg & Dominican Republic (1965–66). JA, 1st Log Cmd, Vietnam (1966–67). CAD, OTJAG (1967–70). SJA, 25th Inf Div, Vietnam (1970–71). CAD, USALSA (1971–74). Ch, Mil Pers Br, Lit Div, OTJAG (1974–75). Adm Judge, Armed Svc Board Contract Appeals (1975–78). SJA, USA Tng Ctr, Ft Dix (1978–79). SJA, DA Materiel Development & Readiness Cmd (1979–80). Ret, Col (1980). Participation in POWER PACK, Vietnam.

Richardson, Quentin W. (1942–). BA (1970) & JD (1974), Univ of Wyo.; LLM (1980), George Washington Univ. Commd JAGC (1974). JA, 82d Abn Div, Ft Bragg (1975–78). Lit Div, OTJAG (1978–82). SJA, 82d Abn Div, Ft Bragg & Grenada (1982–85). Dep Lgl Adv, Ofc JCS (1985–88). Senior RDC, USATDS w/duty USAREUR (1988–90). Dep JA, OJA, USAREUR (1990–91). SJA, USA Garrison & 4th Inf Div, Ft Carson (1991–94). Ret, Col (1994). Participation in URGENT FURY, PROVIDE COMFORT.

Robinson, George R. (1925–1989). LLB (1950), Univ of Okla. Enl, USN w/duty Atlantic (1943–45) & Korea (1948–50). Commd JAGC (1950). JA, 82d Abn Div, Ft Bragg (1959–63). JA, 101st Abn Div, Ft Campbell (1963–64). JA, MACV (1964–65). JA, DSJA & SJA, USA Arty Ctr, Ft Sill (1965–68). MJ, USA Trial Judiciary w/duty Ft Sill & Ft Bragg (1968–72). SJA, USA Arty Ctr, Ft Sill (1972–73). Ret, Col (1973). Participation in Vietnam.

Ruppert, Raymond C. (1946–). BS (1968), Fordham Univ; JD (1976), Univ of Okla. Mil Intelligence Off w/duty Ft Bragg, Vietnam,

& USAREUR (1968–73). Trnsfd to JAGC (1976). JA & SDC, XVIII Abn Corps, Ft Bragg (1976–79). Crim Law Div, OTJAG (1980–81). Dep Ch, DAD, USALSA (1981–82). PP&TO, OTJAG (1983–85). SJA, 8th Inf Div, USAREUR (1985–88). SJA, CENTCOM (1989–91). Ch, Intl & Opnl Law, OTJAG (1991–94). SJA, USEUCOM (1994–96). Ch, Standards of Conduct, OTJAG (1996–98). Ret, Col (1998). Participation in DESERT SHIELD/STORM.

Russell, James S. (1948–). BA (1970) & JD (1978), Univ of Me. Inf Off, Ft Bragg (1971–75). Commd JAGC (1978). JA & DC, 25th Inf Div, Hawaii (1978–81). OIC (Butzbach), 3d Armd Div, USAREUR (1982–85). Adm Law Div, OTJAG (1985–89). DSJA & SJA, USARSO, Ft Clayton (1989–90). Adm Law Div, OTJAG (1990–91). SJA, USA Garrison, Ft Belvoir (1992–93). MJ, ACCA (1993–97). Ret, Col (1997). Participation in JUST CAUSE.

Sawi, Tarek (1958–). BS (1980), St Univ of N.Y.; JD (1983), Vanderbilt Univ. Commd JAGC (1984). GAD, USALSA (1984–87). DC, USATDS w/duty Korea (1987–88). JA, Mil Review Board Agcy, Washington, D.C. (1988–89). Trnsfd to USAR (1989). JA, 22d SUPCOM, Saudi Arabia (1990–91). Participation in DESERT SHIELD/STORM.

Scanlon, Jerome W., Jr. (1940–). BA (1961), St. Bonaventure Univ; JD (1964), Fordham Univ. Arty Off w/duty USAREUR (1966–69) Trnsfd to JAGC (1969). JA, USA Tng Ctr, Ft Dix (1969–72). Lgl Off, Def Attaché Ofc w/duty Saigon, Vietnam (1973–74). DSJA, USA Arty Ctr, Ft Sill (1974–77). S&F, USMA (1977–81). SJA, USA Garrison, Ft Sam Houston (1981–85). Ret, Lt Col (1985). Participation in Vietnam.

Sherwood, John T., Jr. (1935–1995). BS (1960) & JD (1962), American Univ; LLM (1975), Univ of Mich. Commd JAGC (1963). JA, 82d Abn Div, Ft Bragg (1963–65). JA, MACV (1965–66). S&F, USMA (1966–70). OIC (Nuremberg), VII Corps, USAREUR (1970–73). DSJA, USMA (1977–80). DSJA, Sixth USA, Presidio of San Francisco (1980–83). SJA, 19th SUPCOM, Korea (1983–85). SJA, Third USA, Ft McPherson (1985–86). Ret, Col (1986). Participation in Vietnam.

Shull, David A. (1946–). JD (1971), Univ of Fla. Commd JAGC (1972). JA, 3d Inf Div, USAREUR (1980–82). JA, OJA, USAREUR (1982–84). Opns Off, USATDS (1984–87). SJA, 7th Inf Div, Ft Ord (1987–90). Lgl Policy, DoD (1990–93). SJA, MTMC (1994–97). OGC, Ofc Natl Drug Control (1997–99). Ret, Col (2000). Participation in JUST CAUSE.

Smith, James J. (1944–). BA (1967) & JD (1973), Creighton Univ. Arm Off, USAREUR & Vietnam (1968–71). Commd JAGC

(1974). DC & JA, 172d Inf Bde, Ft Wainwright (1974–77). Crim Law Div, OJTAG (1978–81). JA, XVIII Abn Corps, Ft Bragg (1981–85). DSJA, USA Arty Ctr, Ft Sill (1985–88). SJA, 82d Abn Div w/duty Ft Bragg, Panama, & Saudi Arabia (1988–91). Ch, Professional Conduct, OTJAG (1991–92). Lgl Cnsl, DA OGC (1993–94). SJA, I Corps, Ft Lewis (1994–96). MJ, USA Trial Judiciary w/duty Korea & Ft Bragg (1996–2000). Ret, Col (2000). Participation in JUST CAUSE, DESERT SHIELD/STORM.

Smith, John M., III (1950–). BS (1972), USMA; JD (1979), Hastings Coll, Univ of Calif. MP Off, 25th Inf Div, Hawaii (1973–76). Trnsfd to JAGC (1979). JA, 82d Abn Div, Ft Bragg (1979–81). DC, USATDS w/duty Ft Bragg & MFO (1981–82). SDC, USATDS w/duty USAREUR (1983–87). CJA, 1st SF Opns Det–Delta & DSJA, XVIII Abn Corps, Ft Bragg (1987–92). SJA, 10th Mtn Div, Ft Drum & Somalia (1992–94). Ch, GAD, USALSA (1994–97). Ret, Col (1997). Participation in MFO, RESTORE HOPE, deployed to Florida during Hurricane Andrew relief efforts.

Solf, Waldemar A. (1913–1987). LLB (1937), Univ of Chicago. Arty Off (1942–46) w/duty France & Germany. Trnsfd to JAGD (1946). MJ, USA Trial Judiciary w/duty Korea & Ft Meade (1959–62). SJA, Eighth USA, Korea & U.S. Strat Cmd (1962–65). Ch, Mil Jus Div, OTJAG (1966–68). Ret, Col (1968). DA Civilian, Intl Affrs Div, OTJAG (1970–71). Ch, Intl Affrs Div, OTJAG (1971–77). Sp Asst to TJAG (1977–79). Delegate, Diplomatic Conference on the Reaffirmation and Development of Humanitarian Law in Armed Conflict (1977 Protocols to Geneva Conventions) (1975–77). Ret, DA Civilian (1979).

Supervielle, Manuel E. F. (1956–). BA (1978), St. Mary's Univ; JD (1981), Univ of Tex.; LLM (1988), TJAGSA; LLM (1993), Univ of Miami. Commd JAGC (1982). JA, 24th Inf Div, Ft Stewart & DC, USATDS w/duty Ft Stewart (1982–85). JA, 19th SUPCOM, Korea (1985–87). Ch, Intl Law, USA Western Cmd, Ft Shafter (1988–90). Asst SJA, U.S. Pacific Cmd, Hawaii (1990–92). JA, Intl & Opnl Law Div, OTJAG (1993–96). SJA, USARSO (1996–98). S&F, TJAGSA (1998–2000). SJA, SOUTHCOM, Miami, Fla. (2000–present). Participation in relief operations in Western Samoa, SEA ANGEL.

Suter, William K. (1937–). AB (1959), Trinity Univ; LLB (1962), Tulane Univ. Commd JAGC (1962). JA, USA Alaska (1963–66). S&F, TJAGSA (1967–70). Asst SJA, USA SUPCOM, Thailand (1970). DSJA, USARV (1971). OTJAG (1971–74). SJA, 101st Abn Div, Ft Campbell (1975–77). Ch, PP&TO, OTJAG (1977–79). Cmdt, TJAGSA (1981–84). Cdr, USALSA & Ch Judge,

ACMR (1984–85). TAJAG (1985–89). Acting TJAG (1989–91). Retired, Maj Gen (1991). Clerk of Court, U.S. Supreme Court. Participation in Vietnam.

Taylor, Arthur H. (1930–). AA & LLB (1954), Boston Univ. Trnsfd to JAGC (1960). JA, USA Tng Ctr, Ft Dix (1960–62). JA, USA Support Grp, Vietnam (1962–63). JA, XVIII Abn Corps w/duty Ft Bragg & Dominican Republic (1964–66). SJA, JFKSWC, Ft Bragg (1967–68). JA, US Air Def Cmd, Ent AFB (1968–69). JA, Eighth USA, Korea, & SJA, Korea SUPCOM (1971–72). Exec Reserve Affrs, FORSCOM, Ft McPherson (1972–74). Ret, Lt Col (1974). Participation in Vietnam, POWER PACK.

Tenhet, Joseph N., Jr. (1924–). LLB (1949), Univ of Richmond; LLM (1950), Duke Univ. Inf Off w/duty France & Germany (1943–46). Commd JAGC (1953). JA, Mil Affrs Div, OTJAG (1953–55). JA, 23d Inf Div, USA Caribbean, Panama (1955–58). JA, Fourth USA, Ft Sam Houston (1959–62). JA, Eighth USA, Korea (1963–64). Mil Affrs Div, OTJAG (1964–69). SJA, 3d Inf Div, USAREUR (1969–70). SJA, V Corps, USAREUR (1970–71). SJA, USARV & MACV (1972–73). Ch, Adm Law Div, OTJAG (1973–75). AJAG for Mil Law (1975–78). Ret, Brig Gen (1978). Participation in Vietnam.

Trant, Charles E. (1951–). BA (1973) & JD (1975), Suffolk Univ; LLM (1981), Georgetown Univ. Commd JAGC (1976). DC & JA, 5th Inf Div, Ft Polk (1976–79). DAD & Commr, USALSA (1979–82). MJ, USA Trial Judiciary w/duty USAREUR (1983–86). Crim Law Div, OTJAG (1986–89). Mil Crim Law Adv, Dept State (1987–88). JA & DSJA, VII Corps, USAREUR & Saudi Arabia (1989–91). MJ, USA Trial Judiciary w/duty USAREUR (1991–94). Ch, Crim Law Div, OTJAG (1994–97). MJ, ACCA (1997–2000). Ret, Col (2000). Participation in DESERT SHIELD/STORM.

Walsh, Gary L. (1951–). BA (1973) & JD (1980), Creighton Univ; LLM (1988), TJAGSA. Inf Off w/duty Ft Bragg (1973–77). Commd JAGC (1980). JA, 82d Abn Div, Ft Bragg & Grenada (1980–84). Grp JA, SF Grp, Ft Lewis (1984–87). S&F, TJAGSA (1988–91). CJA, SF Opns Det–Delta, Ft Bragg & Somalia (1991–94). SJA, 4th Inf Div, Ft Carson (1995–96). DSJA, USA Garrison, Ft Carson (1996–97). SJA, USA Sp Opns Cmd, Ft Bragg (1997). Ret, Lt Col (1998). Participation in URGENT FURY, RESTORE HOPE.

Warner, Karl K. (1952–). BS (1974), USMA; JD (1980) W.Va. Univ. Arty Off w/duty USAREUR (1974–77). Trnsfd to JAGC (1980). JA, USA Arm Ctr, Ft Knox (1980–82). JA, 1st Inf Div (Fwd), USAREUR (1983–84). SDC, USATDS w/duty USAREUR (1984–86). Lit Atty, USALSA (1987–89). S&F, TJAGSA (1989–92). DSJA, XVIII

Abn Corps, Ft Bragg (1993–94). SJA, 10th Mtn Div, Ft Drum (1994–96). Lgl Cnsl, CJCS (1996–98). SJA, U.S. SOCOM (1998–present). Participation in UPHOLD DEMOCRACY.

Warren, Marc L. (1957–). BA (1978) & JD (1981), Univ of Fla.; LLM (1993), TJAGSA. Commd JAGC (1981). JA, XVIII Abn Corps, Ft Bragg & Grenada (1981–84). Trnsfd to USAR (1984). Returned to active duty (1988). JA, V Corps, USAREUR (1988–90). Rgt JA, 11th Armd Cav Rgt, USAREUR & Saudi Arabia (1990–92). S&F, TJAGSA (1993–95). Lgl Adv, Joint Sp Opns Cmd (1996–99). SJA, 101st Asslt Div, Ft Campbell (1999–present). Participation in URGENT FURY, DESERT STORM.

Weaver, Robert K. (1920–). BS (1943) & LLB (1947), Univ of S.Dak. Commd JAGC (1949). JA, EUCOM (1950–53). S&F, TJAGSA (1954–58). Asst SJA, VII Corps, USAREUR (1958–61). S&F, USMA (1961–65). SJA, MAAG, Taiwan (1965–67). Ch, Lit Div, OTJAG (1967–70). SJA, MACV (1970–71). Ch, Procurement Law Div, OTJAG (1971–74). Ret, Col (1974). Participation in Vietnam.

Westerman, George F. (1916–1985). BS (1939) & LLB (1947), Univ of Wisc. Signal Off w/duty France & Germany (1941–45). Commd JAGC (1947). Asst SJA, MDW (1947–50). JA, Eighth USA & Japan Log Cmd w/duty Korea (1950–52). Ch, Patents Div, OTJAG (1952–56). Patent Adv, U.S. Mission to NATO, Paris (1957–60). Ch, Patents Div, OTJAG (1961–62). SJA, MACV (1962–63). Ch, Mil Affrs Div & Ch, Intl Affrs Div, OTJAG (1963–67). MJ & Ch Judge, ACMR (1968–70). Ret, Col (1971). Participation in Vietnam.

Williams, Lawrence H. (1922–1999). BS (1947), Univ of Minn.; JD (1948), Univ of Colo. Aviation Off w/duty No. Africa, France, & Germany (1942–46). Commd JAGC (1948). Mil Affrs Div, OTJAG (1948–52). S&F, TJAGSA (1952–53). JA, USA Carribean, Panama (1953–56). Asst Lgl Adv, Ofc Dep Ch Staff Logistics, Washington, D.C. (1956–57). Mil Affrs Div, OTJAG (1957–61). SJA & Asst Ch Staff (G–1), 3d Armd Div, USAREUR (1961–64). Ch, Mil Affrs Div, OTJAG (1964–66). SJA, III Corps, Ft Hood (1967–69). SJA, MACV (1969–70). Ch, Mil Affrs Div, OTJAG (1970–71). AJAG for Mil Law (1971–75). TAJAG (1975–79). Ret, Maj Gen (1979). Participation in Vietnam.

Winand, Daniel L. (1959–). BA (1981), Penn. St Univ; JD (1984), Dickinson Univ. Commd JAGC (1984). JA, USA Inf Ctr, Ft Benning (1984–86). DC, USATDS w/duty Korea & Ft Stewart (1986–89). Trnsfd to USAR (1989). JA, Third USA (Reserve Aug), Ft McPherson (1989–91). JA, ARCENT, Saudi Arabia (1990–91). Participation in DESERT SHIELD/STORM.

With, Catherine M. (1963–). BA (1985) & MA (1985), Coll of St. Rose; JD (1988), Albany Law Sch; LLM (1997), TJAGSA. Commd JAGC (1989). JA, 7th Inf Div, Ft Ord (1990–91). CJA, MFO (1991). JA, USARSO (1991–94). JA, 25th Inf Div, Hawaii & Haiti (1994–96). CJA, U.S. Forces Haiti (1995). Lgl Cnsl, CJCS (1997–99). Lit Atty, USALSA (1999–present). Participation in MFO, Uphold Democracy, UNMIH.

Woodling, Dale N. (1953–). BA (1975), Miami Univ; JD (1978), Univ of Akron (1978). Commd JAGC (1979). JA, 4th Inf Div, Ft Carson (1979–81). JA, 5th Inf Div, Ft Polk (1981–84). Adm Law Div, OTJAG (1985–89). JA & OIC (Hanau), 3d Armd Div, USAREUR (1989–92). Asst Exec, OTJAG (1992–94). SJA, U.S. Forces Somalia (1993). Dep Ch, Adm Law Div, OTJAG (1994–96). SJA, 25th Inf Div, Hawaii (1996–98). SJA, I Corps, Ft Lewis (1998–present). Participation in UNOSOM II.

APPENDIX B

Rules of Engagement

CUSTOMS AND COURTESIES

(Given to VII Corps Troops in Southwest Asia)

DO:

1. BE FRIENDLY AND COURTEOUS. A handshake accompanied with the phrase Al-Salaama' Alaykum (Peace be upon you) is the most common form of greeting.

2. IF YOU SMOKE (Most Arab men do). Offer to share cigarettes with those present.

3. SIT PROPERLY IN CHAIRS, UPRIGHT WITH FEET ON THE GROUND.

4. WHEN IN DOUBT, OBSERVE LOCALS AND IMITATE THEIR BEHAVIOR.

5. AVOID CONTACT WITH ARAB WOMEN. If introduced, be polite but do not stare or engage in any lengthy conversation.

6. REMAIN AT THE SCENE IF INVOLVED IN AN ACCIDENT. Be polite but say nothing about the circumstances until US authorities arrive. Host Nation police determine liability at the scene, and an offer to pay the other driver is taken as an admission that you are at fault.

DON'T:

1. MAKE CRITICAL COMPARISONS of your religion to Islam. Most Arabs speak English and understand what you are saying.

2. ASK AN ARAB NOT TO SMOKE.

3. POINT YOUR FINGER OR USE YOUR INDEX FINGER TO BECKON PEOPLE: it is considered demeaning.

4. USE ALCOHOL.

5. POSSESS OR USE PORNOGRAPHIC OR SEXUALLY EXPLICIT MATERIAL.

6. USE YOUR LEFT HAND TO EAT OR PASS FOOD.

PEACE TIME RULES OF ENGAGEMENT

(For VII Corps Troops During Operation DESERT SHIELD)

A. THE UNITED STATES IS NOT AT WAR.

B. YOU HAVE THE RIGHT TO USE FORCE TO DEFEND YOURSELF against attacks or threats of attack.

C. COMMANDERS are authorized to use force in self-defense when responding to attacks or threats of imminent attack against US or host nation forces, citizens, property, or commercial assets.

D. Unless directed by your chain of command, YOU MAY NOT ENTER THE LAND, SEA, OR AIRSPACE OF OTHER COUNTRIES besides the Host nation.

E. IF YOU ACCIDENTALLY ENTER the land, sea, or airspace of another country, withdraw quickly and use force only to prevent harm to yourself during the withdrawal.

F. You may NOT SEIZE THE PROPERTY OF OTHERS to accomplish your mission.

G. You must FOLLOW NORMAL CONTRACTING PROCEDURES to secure supplies to accomplish your mission.

H. TREAT ALL PEOPLE AND PROPERTY WITH DIGNITY AND RESPECT.

I. FOLLOW HOST NATION LAW and obey Host nation officials.

J. IF HOSTILITIES BEGIN, follow the War Time Rules of Engagement.

WAR TIME RULES OF ENGAGEMENT
(For VII Corps Troops During Operation DESERT STORM)

ATTACK all enemy soldiers, vehicles, positions, supplies, and equipment using these rules:

A. DO NOT ATTACK anyone who has surrendered, is out of battle due to sickness or wounds, is shipwrecked, or is an aircrew member parachuting from a disabled aircraft.

B. DO NOT HARM CIVILIANS unless it is necessary to save US lives. Do not fire into civilian populated areas or buildings which are not being defended or being used for a military purpose.

C. THE USE OF BOOBY TRAPS IS LIMITED. Traps that are likely to kill or injure unsuspecting people performing an innocent act are prohibited. Devices that protect friendly positions or slow down enemy forces are authorized. Booby traps may not be used on civilian personal property. All traps must be recovered or destroyed when they are no longer needed.

D. DO NOT ATTACK CHURCHES, Mosques, Shrines, Schools, Museums, National Monuments, or any other historical or cultural site except in self-defense.

E. HOSPITALS will be given special protection. Do not attack hospitals unless the enemy uses the hospital to commit acts harmful to US forces, and then only after giving a warning and allowing a reasonable time to pass before engaging, if the tactical situation permits.

F. LOOTING AND TAKING WAR TROPHIES ARE PROHIBITED.

G. DO NOT ATTACK CIVILIAN PROPERTY unless it is necessary to protect US forces or the civilian property is being used by the enemy for military purposes and the accomplishment of the mission depends on it.

H. TREAT ALL CIVILIANS AND THEIR PROPERTY WITH RESPECT AND DIGNITY. Before using privately owned property, see if government owned property can substitute. Do not requisition civilian property, including vehicles, without permission of a commander and without giving a receipt. If an ordering officer can contract the property, then do not requisition it.

I. Treat all PRISONERS and DETAINEES humanely and with respect and dignity. You are responsible for their protection.

J. FIRE ONLY THE MINIMUM amount and type of ammunition necessary to accomplish the mission.

K. PREVENT UNNECESSARY SUFFERING and restrict destruction to what the mission requires.

REMEMBER

1. FIGHT ONLY COMBATANTS AND ATTACK ONLY MILITARY TARGETS.
2. SPARE CIVILIAN PERSONS AND OBJECTS.
3. RESTRICT DESTRUCTION TO WHAT YOUR MISSION REQUIRES.
4. REPORT VIOLATIONS OF THESE RULES OF ENGAGEMENT THROUGH YOUR CHAIN OF COMMAND.

Rules of Engagement
Operation RESTORE HOPE

JTF for Somalia Relief Operation
Ground Forces Rules of Engagement

Nothing in these rules of engagement limits your right to take appropriate action to defend yourself and your unit.

A. You have the right to use force to defend yourself against attacks or threats of attack.

B. Hostile fire may be returned effectively and promptly to stop a hostile act.

C. When U.S. forces are attacked by unarmed hostile elements, mobs, and/or rioters, U.S. forces should use the minimum force necessary under the circumstances and proportional to the threat.

D. You may not seize the property of others to accomplish your mission.

E. Detention of civilians is authorized for security reasons or in self-defense.

REMEMBER

1. The United States is not at war.

2. Treat all persons with dignity and respect.

3. Use minimum force to carry out mission.

4. Always be prepared to act in self-defense.

Appendix B

Rules of Engagement
United Nations Operation in Somalia II

General Bir's approved rules of engagement, published in the UNOSOM II Operations Plan were:

1. UNOSOM PERSONNEL MAY USE DEADLY FORCE:

a. To defend themselves, other U.N. personnel, or persons and areas under their protection against hostile acts or hostile intent.

b. To resist attempts by forceful means to prevent the Force from discharging its duties.

2. CHALLENGING.

a. Whenever practicable, a challenge should be given before using deadly force.

b. Challenging is done by:

(1) Shouting in English: "U.N., STOP OR I WILL FIRE," or;

(2) Shouting in Somali: "U.N., KA HANAGA JOOGO AMA WAA GUBAN," or;

(3) firing warning shots in the air.

3. PRINCIPLES FOR USE OF FORCE. When it becomes necessary to use force, the following principles apply:

a. Action which may reasonably be expected to cause excessive collateral damage is prohibited.

b. Reprisals are prohibited.

c. Minimum force is to be used at all times.

4. SPECIFIC RULES.

a. UNOSOM Forces may use deadly force in response to a hostile act or when there is clear evidence of hostile intent.

b. Crew-served weapons are considered a threat to UNOSOM Forces and the relief effort whether or not the crew demonstrates hostile intent. Commanders are authorized to use all necessary force to confiscate and demilitarize crew-served weapons in their area of operations.

c. Within those areas under the control of UNOSOM Forces armed individuals may be considered a threat to UNOSOM and the relief efforts whether or not the individual demonstrates hostile intent. Commanders are authorized to use all necessary force to disarm and demilitarize groups or individuals in those areas under the control of UNOSOM. Absent a hostile or criminal act, individuals and associated vehicles will be released after any weapons are removed/demilitarized.

d. If UNOSOM Forces are attacked or threatened by unarmed hostile elements, mobs and/or rioters, UNOSOM Forces are authorized to employ reasonable minimum force to repel the attacks or threats. UNOSOM Forces may also employ the following procedures: verbal warnings to demonstrators, shows of force including use of riot control formations, and warning shots.

e. UNATTENDED MEANS OF FORCE. Unattended means of force, including booby traps, mines, and trip guns, are not authorized.

f. DETENTION OF PERSONNEL. Personnel who interfere with the accomplishment of the mission or who otherwise use or threaten deadly force against UNOSOM, U.N. or relief material, distribution sites, or convoys may be detained. Persons who commit criminal acts in areas under the control of U.N. Forces may likewise be detained. Detained personnel will be evacuated to a designated location for turnover to military police.

5. DEFINITIONS. The following definitions are used:

a. SELF DEFENSE. Action to protect oneself or one's unit against a hostile act or hostile intent.

b. HOSTILE ACT. The use of force against UNOSOM personnel or mission-essential property, or against personnel in an area under UNOSOM responsibility.

c. HOSTILE INTENT. The threat of imminent use of force against UNOSOM Forces or other persons in those areas under the control of UNOSOM.

d. MINIMUM FORCE. The minimum authorized degree of force which is necessary, reasonable and lawful in the circumstances.

6. Only the Force Commander, UNOSOM, may approve changes to these ROE.

The complete Rules of Engagement published and distributed on the ROE card were:

[front of card]

NOTHING IN THESE ROE LIMITS YOUR RIGHT TO TAKE ALL NECESSARY AND APPROPRIATE ACTION TO DEFEND YOURSELF AND YOUR UNIT.

1. Use all necessary force, including deadly force:

a. To defend yourself, other U.N. personnel, or persons and area under your protection against the use of force or clear evidence of intent to use force.

b. To confiscate and demilitarize crew-served weapons.

c. To disarm and demilitarize individuals in areas under UNOSOM control.

2. Always use the minimum force necessary under the circumstances and proportional to the threat.

3. If the tactical situation permits, you should give a challenge before using deadly force. Challenge by:

a. Shouting in English: "U.N., STOP OR I WILL FIRE,"or

b. Shouting in Somali: "U.N., KA HANAGA JOOGO AMA WAA GUBAN," or

c. Firing warning shots in the air.

[reverse of card]

4. Unattended weapons, such as booby traps, mines, and trip guns, are prohibited.

5. Detain individuals who interfere with your mission, who use or clearly threaten deadly force, or who commit criminal acts in areas under UNOSOM control. Evacuate detainees to a designated location for turnover to military police. Treat all detainees humanely.

6. Do not seize civilian property without your commander's authorization.

7. Treat all persons with dignity and respect.

Human Rights Guidelines
Operation UPHOLD DEMOCRACY

Ten Commandments of Human Rights for Soldiers

THOU SHALL:

1. Honor the spirit of the Universal Declaration of Human Rights.

2. Give and obey only lawful orders.

3. Report crimes and human rights violations to proper authorities.

4. Respect individual integrity and human dignity.

5. Abide by the military code of honor, be professional, and tell the whole truth in human rights investigations.

6. Spread the word: order depends on respect for human rights.

THOU SHALL NOT COMMIT, NOR TOLERATE:

7. Murder, rape, torture, or the excessive use of force.

8. Disappearances.

9. The unnecessary destruction of property.

10. Extrajudicial punishment.

Appendix B

Rules of Engagement
Operation Uphold Democracy

Each soldier in the Combined Joint Task Force was issued an ROE card containing the following:

Combined JTF Haiti
Rules of Engagement (ROE) Card 1
6 September 1994

Nothing in the ROE limits your right to use necessary force to defend yourself, your fellow service members, your unit, other JTF personnel, key facilities, and property designated by your commander.

1. Repel hostile acts with necessary force, including deadly force. Use only the amount of force needed to protect lives/property and accomplish mission. Engage targets with observed, direct, deliberately aimed fire.

2. Do not hesitate to respond with force against hostile acts and signs of hostile intent.

3. You may use necessary force to stop, disarm, and detain members of the Haitian military, police, other armed persons, or other persons committing hostile acts or showing hostile intent. Stop and detain other persons who interfere with your mission. Evacuate detainees to a designated location for release to proper authorities. Treat all detainees humanely.

4. When a tactical situation permits, you should give a challenge before using deadly force. Challenge by:

a. Shouting in English "U.S. STOP OR I WILL FIRE!"
b. Shouting in Creole "U.S. KANPE OUBIEN MAP TIRE!"
Phonetic: "U.S. kaHJnpey oobeeEH(n) mahp tEErey!
c. Fire warning shots into the air.

5. Treat all persons with dignity and respect.

6. Do not take private property without commander's permission.

7. Remember: no force has been declared hostile, including the Haitian Army and police. Use of deadly force must be based on hostile acts or clear indicators of hostile intent.

The modified rules of engagement follow:

COMBINED JTF HAITI
PEACETIME RULES OF ENGAGEMENT
CIVIL MILITARY OPERATIONS IN HAITI

NOTHING IN THESE ROE LIMITS YOUR OBLIGATION TO TAKE ALL NECESSARY AND APPROPRIATE ACTION TO DEFEND YOURSELF AND YOUR UNIT
21 September 1994

1. No forces have been declared hostile. Offensive military operations (raids, assaults, etc.) require CJTF 190 approval.

2. Treat all persons with dignity and respect.

3. Use all necessary force, up to and including deadly force, to defend U.S. forces, U.S. citizens, or designated foreign nationals against an attack or threat of imminent attack. When deadly force is employed, engage targets with observed, deliberately aimed fire.

4. Members of the military, police, or other armed persons may be stopped, detained, and if necessary, disarmed if they appear to threaten essential civic order.

5. Civilians may be stopped if they appear to be a threat to U.S. forces, protected persons, key facilities, or property designated mission essential by CJTF 190. If determined to be a threat, they may be further detained. If not, they will be released.

6. Necessary and proportional force is authorized to control disturbances and disperse crowds threatening essential civic order.

7. Persons observed committing serious criminal acts will be detained using minimal force necessary up to and including deadly force. Serious criminal acts include: homicide, aggravated assault, rape, arson and robbery. Non-lethal force is authorized to detain persons observed committing burglary or larceny. Release persons suspected of serious criminal acts to Haitian law enforcement officers/other appropriate authorities as soon as possible.

8. Civilian vehicles may be stopped and their occupants' identities checked for security purposes. If a civilian vehicle does not stop on

order and is approaching a check point or security perimeter, you may fire to disable the vehicle.

9. Do not enter the Dominican Republic without permission from CINC-USACOM.

10. Deadly force is not authorized to disarm Haitians, enforce curfews, or stop looting, unless those individuals involved engage in hostile acts or demonstrate hostile intent.

11. Possession of a weapon in public by any individual does not, by itself, constitute a hostile act or demonstrate hostile intent.

12. U.S. forces are not authorized to grant political asylum. Temporary refuge will be granted only if necessary to protect human life.

13. Respect diplomatic personnel, residences, facilities, and property. Do not enter diplomatic residences/facilities unless invited by appropriate diplomatic officials or approved by CINCUSACOM.

Rules of Engagement
Operation PROVIDE COMFORT

Each coalition soldier received a pocket-size card printed with the following:

RULES OF ENGAGEMENT

1. All military operations will be conducted in accordance with the Law of War.

2. The use of armed force will be utilized as a measure of last resort only.

3. Nothing in these rules negates or otherwise overrides a commander's obligation to take all necessary and appropriate actions for his unit's self-defense.

4. U.S. forces will not fire unless fired upon, unless there is clear evidence of hostile intent.

a. HOSTILE INTENT. The threat of imminent use of force by an Iraqi or other foreign force, terrorist group, or individual against the United States, U.S. forces, U.S. citizens, or Kurdish or other refugees located within a U.S. or allied safe haven refugee area. When the on-scene commander determines, based on convincing evidence, that HOSTILE INTENT is present, the right exists to use proportional force to deter or neutralize the threat.

b. HOSTILE ACT. Includes armed force directly to preclude or impede the missions and/or duties of U.S. or allied forces.

5. Response to hostile fire directly threatening U.S. or allied care shall be rapid and directed at the source of fire using only the force necessary to eliminate the threat. Other foreign forces (such as reconnaissance aircraft) that have shown an active integration with the attacking force may be engaged. Use minimum amount of force necessary to control the situation.

6. You may fire into Iraqi territory in response to hostile fire.

7. You may fire into another nation's territory in response to hostile fire only if the cognizant government is unable or unwilling to stop that force's hostile acts effectively or promptly.

8. Surface-to-air missiles will engage hostile aircraft flying north of the 36th Parallel.

9. Surface-to-air missiles will engage hostile aircraft south of the 36th Parallel only when they demonstrate hostile intent or commit hostile acts. Except in cases of self-defense, authorization for such engagement rests with the designated air defense commander. Warning bursts may be fired ahead of foreign aircraft to deter hostile acts.

10. In the event U.S. forces are attacked or threatened by unarmed hostile elements, mobs, or rioters, the responsibility for the protection of U.S. force rests with the U.S. commanding officer. On-scene commander will employ the following measures to overcome the threat:

a. Warning to demonstrators.

b. Show of force, including use of riot control formations.

c. Warning shots fired over the heads of hostile elements.

d. Other reasonable use of force necessary under the circumstances and proportional to the threat.

11. Use the following guidelines when applying these rules:

a. Use of force only to protect lives.

b. Use of minimum force necessary.

c. Pursuit will not be taken to retaliate; however, immediate pursuit may begin and continue for as long as there is an immediate threat to U.S. forces. In the absence of JCS approval, U.S. forces should not pursue any hostile force into another nation's territory.

d. If necessary and proportional, use all available weapons to deter, neutralize, or destroy the threat as required.

Rules of Engagement

Los Angeles Civil Disturbance
Joint Task Force L.A.
(as of 020100 May 1992)

A. Every serviceman has the right under law to use reasonable and necessary force to defend himself against violent and dangerous personal attack. The limitations described below are not intended to infringe on this right, but to prevent indiscriminate use of force.

B. Force will never be used unless necessary, and then only the minimum force will be used.

(1) Use nondeadly force to:

(a) control the disturbance;

(b) prevent crimes;

(c) apprehend or detain persons who have committed crimes.

(2) Use deadly force only when:

(a) lesser means of force exhausted or unavailable;

(b) risk of death or serious bodily harm to innocent persons is not significantly increased by the use; and

(c) purpose of use

1- self defense, to avoid death or serious bodily harm;

2- prevention of crime involving death or serious bodily harm;

3- prevention of destruction of public utilities that have been determined vital by the TF commander; or

4-detention, or prevention of escape, of persons who present a clear threat of loss of life.

(3) When possible, the use of deadly force should be preceded by a clear warning that such force is contemplated or imminent.

(4) Warning shots will not be used.

(5) When firing, shots will be aimed to wound, if possible, rather than kill.

(6) Weapons will not be fired on automatic.

(7) When possible, let civilian police arrest lawbreakers.

(8) Allow properly identified news reporters freedom of movement, so long as they do not interfere with your mission.

(9) Do not talk about this operation, or pass on information or

rumors about it, to unauthorized persons; refer them to your commander.

(10) The JTF commander withholds authority for use of riot control agents and sniper teams.

C. Arming orders:

Arming Order	*Rifle*	*Bayonet/ Scabbard*	*Bayonet*	*Pistol*	*Baton*	*Magazine/ Chamber*	*Control*
AO–1	Sling	On Belt	Scabbard	Holstered	Belt	In Pouch/ Empty	OIC/ NCOIC
AO–2	Port	On Belt	Scabbard	Holstered	Belt	In Pouch/ Empty	OIC/ NCOIC
AO–3	Sling	On Belt	Fixed	Holstered	Hand	In Pouch/ Empty	OIC/ NCOIC
AO–4	Port	On Belt	Fixed	Holstered	Hand	In Pouch/ Empty	OIC/ NCOIC
AO–5	Port	On Belt	Fixed	Holstered	Hand	Weapon/ Empty	OIC/ NCOIC
AO–6	Port	On Belt	Fixed	In Hand	Belt	Weapon/ Locked & Loaded	OIC

Bibliography

The location of various unpublished sources is indicated by the following abbreviations:

NARA National Archives and Records Administration, College Park, Maryland, and St. Louis, Missouri

OTJAG Office of the Judge Advocate General, Washington, D.C.

TJAGSA The Judge Advocate General's School, U.S. Army, Charlottesville, Virginia

USALSA U.S. Army Legal Services Agency, Arlington, Virginia

Primary Sources

Official Records

Agreement on Ending the War and Restoring Peace in Viet-Nam [Paris Peace Accords]. 27 January 1973. Published in U.S. Treaties and Other International Acts 7542 (1973), page 1.

Annual Report of the Code Committee on Military Justice Pursuant to the Uniform Code of Military Justice, 1983–96. Washington, D.C.: Government Printing Office, 1984–97.

Annual Report of the United States Court of Military Appeals and The Judge Advocates General of the Armed Forces and the General Counsel of the Department of the Treasury Pursuant to the Uniform Code of Military Justice, 1952–66. Washington, D.C.: Government Printing Office, 1954–67.

Annual Report of the U.S. Court of Military Appeals and The Judge Advocates General of the Armed Forces and the General Counsel of the Department of Transportation Pursuant to the Uniform Code of Military Justice, 1967–83. Washington, D.C.: Government Printing Office, 1968–83.

Department of Defense. *Conduct of the Persian Gulf War: Final Report to Congress*. Washington, D.C.: Government Printing Office, 1992.

Foreign Claims Act. Title 10, United States Code, Section 2734.
General Accounting Office. Report to the Honorable Fred Thompson. *Operation DESERT STORM: Investigation of a Fratricide Incident*. April 1995.
Geneva Convention for Amelioration of the Condition of the Wounded and Sick in Armed Forces in the Field. 12 August 1949. Published in U.S. Treaties and Other International Acts Series 3326 and in DA Pamphlet 27–1, *Treaties Governing Land Warfare* (Washington, D.C.: Department of the Army, 1956).
Geneva Convention Relative to the Protection of Civilian Persons in Time of War. 12 August 1949. Published in U.S. Treaties and Other International Acts Series 3356 and in DA Pamphlet 27–1, *Treaties Governing Land Warfare* (Washington, D.C.: Department of the Army, 1956).
Geneva Convention Relative to the Treatment of Prisoners of War. 12 August 1949. Published in U.S. Treaties and Other International Acts Series 3364 and in DA Pamphlet 27–1, *Treaties Governing Land Warfare* (Washington, D.C.: Department of the Army, 1956).
Hague Convention No. IV Respecting the Laws and Customs of War on Land and Annex Thereto Embodying Regulations Respecting the Laws and Customs of War on Land. 18 October 1907. U.S. Statutes 36: 2277, reprinted in DA Pamphlet 27–1, *Treaties Governing Land Warfare* (Washington, D.C.: Department of the Army, 1956).
Joint Chiefs of Staff. Joint Pub 3–07, *Joint Doctrine for Military Operations Other Than War*. Washington, D.C.: Joint Chiefs of Staff, June 1995.
———. Joint Pub 3–07.3, *Joint Tactics, Techniques and Procedures for Peacekeeping Operations*. Washington, D.C.: Joint Chiefs of Staff, April 1994.
Joint Task Force South. Memo, sub: Disposition of Captured Enemy Property. 4 January 1990. Historians files, OTJAG.
———. Memo, sub: Search of Alleged Diplomatic Private Residence. Undated [January 1990]. Historians files, OTJAG.
Joint Task Force SUPPORT HOPE. Annex O to JTF SUPPORT HOPE OPLAN 94–002, Legal. Undated [July 1994]. Historians files, OTJAG.
———. General Order no. 1. 27 July 1994. Historians files, OTJAG.
Judge Advocate General, U.S. Army, Office of the. *JAGC Personnel and Activity Directory*. Washington, D.C.: Office of the Judge Advocate General, 1962–95.
———. DESERT STORM Assessment Team Report to the Judge Advocate General of the Army. 22 April 1992. TJAGSA.

———. Report to the Judge Advocate General by the Wartime Legislation Team. September 1983. TJAGSA.

Judge Advocate General's School, U.S. Army. *Law and Military Operations in Haiti 1994–1995: Lessons Learned for the Judge Advocate*. Charlottesville, Va.: The Judge Advocate General's School, 1995.

———. *Operational Law Handbook*. Charlottesville, Va.: The Judge Advocate General's School, 1996.

Military Assistance Advisory Group (MAAG), Vietnam, Headquarters. MAAG Vietnam Judge Advocate Responsibilities, Appendix I to Annex L (Legal), MAAG Vietnam OPLAN 61–61. 10 February 1961. Record Group (RG) 334, NARA.

———. Memorandum no. 10, Missions and Functions of MAAG Advisers. 27 January 1961. RG 334, NARA.

Military Assistance Command, Vietnam (MACV), Headquarters. Memorandum, sub: Resources Control—Arrest, Search, and Seizure Laws in the Republic of Vietnam. 21 January 1965. Historians files, OTJAG.

———. Memorandum, sub: RVNAF [Republic of Vietnam Armed Forces] Military Justice. 12 February 1965. Historians files, OTJAG.

———. Memorandum, sub: Summary and Outline of RVN Judicial and Police Functions. 24 February 1965. Historians files, OTJAG.

———. Memorandum for Record, sub: Status of Forces in Thailand. 17 July 1962. Historians files, OTJAG.

———. Staff Study, "Purpose: To examine the role of the Civil Law in counterinsurgency in Vietnam and to make pertinent recommendations." 31 May 1965. Historians files, OTJAG.

Military Justice Act of 1968. Public Law no. 90–632. U.S. Statutes at Large 82: 1335.

Mutual Defense Assistance in Indochina [Pentalateral Agreement]. 23 December 1950. Reprinted in Prugh, *Law at War*, Appendix I.

1977 Protocols Additional to the Geneva Conventions. 12 December 1977. Published in International Legal Materials 16: 1391 and in Department of the Army Pamphlet 27–1–1, *Protocols to the Geneva Conventions of 12 August 1949* (Washington, D.C.: Department of the Army, 1979).

Posse Comitatus Act. Title 18, United States Code, Section 1385.

U.S. Army. *Report of the Investigation (Volume I), Report to the Department of the Army, Review of the Preliminary Investigation into the My Lai Incident [Peers Report]*. 14 March 1970. Reprinted in Joseph Goldstein, Burke Marshall, and Jack Schwartz. *The My*

Lai Massacre and Its Cover-up: Beyond the Reach of the Law? (New York: Free Press, 1976).

U.S. Army, Europe (USAREUR), and Seventh Army. Commander in Chief, USAREUR, Deployment Order no. 22. 10 November 1990. Reprinted in Stephen P. Gehring, *From the Fulda Gap to Kuwait* (Washington, D.C.: Department of the Army, 1998).

U.S. Atlantic Command. *Migrant Camp Operations: The Guantanamo Experience*. Norfolk, Va.: U.S. Atlantic Command, 1996.

U.S. Central Command (CENTCOM). General Order no. 1, *Prohibited Activities for U.S. Personnel Serving in the USCENTCOM AOR*. 30 August 1990. Historians files, OTJAG.

Vienna Convention on Diplomatic Relations. 18 April 1961. Published in United States Treaties 3: 3227 and in International Legal Materials 8: 679 (1969).

After Action and Command Reports

XVIII Airborne Corps. Office of the Staff Judge Advocate. *After Action Report: Operation DESERT SHIELD/DESERT STORM*. Undated [1991]. Historians files, OTJAG.

———. *After Action Report: Operation UPHOLD DEMOCRACY*. Undated [1994]. Historians files, OTJAG.

82d Airborne Division. Office of the Staff Judge Advocate. Memo for SJA, XVIII Airborne Corps, sub: After Action Report—URGENT FURY. 9 November 1983. Historians files, OTJAG.

———. Memo for SJA, XVIII Airborne Corps, sub: After Action Report—Operation URGENT FURY. 14 February 1984. Historians files, OTJAG.

———. Memo, sub: After Action Report—Operation JUST CAUSE. Undated. Historians files, OTJAG.

Joint Task Force SEA ANGEL. *SJA After Action Report*. 13 June 1991. Historians files, OTJAG.

VII Corps. *Command Report, Operation DESERT SHIELD/STORM*, Tab L (Legal). Undated [May 1991]. Historians files, OTJAG.

10th Mountain Division. Office of the Staff Judge Advocate. *After Action Report, sub: U.S. Army Legal Operations, Operation RESTORE HOPE, 5 December 1991–5 May 1992*. Undated [June 1992]. Historians files, OTJAG.

———. *After Action Report, sub: Operation UPHOLD DEMOCRACY*. March 1995. Historians files, OTJAG.

22d Support Command. Office of the Staff Judge Advocate, Contract Law Branch. *After Action Report*. March 1991. Historians files, OTJAG.

22d Support Command. *DESERT SHIELD Command Report*. 23 March 1991. Historians files, OTJAG.

———. *DESERT STORM Command Report*. 5 April 1991. Historians files, OTJAG.

U.S. Army Trial Judiciary. Memo for Chief Trial Judge, U.S. Army Trial Judiciary, sub: Trip Report on Saudi Arabia. Undated [April 1990]. Historians files, OTJAG.

U.S. European Command. *After Action Report, Operation PROVIDE COMFORT*. 29 January 1992. Historians files, OTJAG.

Directives, Instructions, Manuals, Pamphlets, and Regulations

Army Regulation (AR) 15–6, *Procedures for Investigating Officers and Boards of Officers*. 11 May 1988.

AR 15–185, *Army Board for Correction of Military Records*. 18 May 1977.

AR 27–1, *Judge Advocate Legal Services*. 15 September 1989.

AR 27–10, *Military Justice*. 22 December 1989.

AR 27–3, *Legal Assistance*. 30 September 1992.

AR 27–20, *Claims*. 22 February 1990.

AR 27–50, *Status of Forces Policies, Procedures, and Information*. 15 December 1989.

AR 27–52, *Consular Protection of Foreign Nationals Subject to the Uniform Code of Military Justice*. 5 November 1968.

AR 27–53, *Review of Legality of Weapons Under International Law*. 1 January 1979.

AR 190–8, *Enemy Prisoners of War—Administration, Employment, and Compensation*. 2 December 1985.

AR 190–14, *Carrying of Firearms and Use of Force for Law Enforcement and Security Duties*. 23 September 1988.

AR 190–57, *Civilian Internees—Administration, Employment, and Compensation*. 4 March 1987.

AR 350–216, *Training: The Geneva Conventions of 1949 and Hague Convention no. IV of 1907*. 28 May 1970.

AR 381–26, *Army Foreign Materiel Exploitation Program*. 27 May 1991.

AR 500–50, *Civil Disturbances*. 1 June 1972.

AR 500–51, *Support to Civilian Law Enforcement*. 1 August 1983.

AR 500–60, *Disaster Relief*. 1 September 1981.

AR 550–1, *Procedures for Handling Requests for Political Asylum and Temporary Refuge*. 1 November 1981.

AR 600–8–22, *Military Awards*. 25 February 1995.

AR 600–20, *Command Policy*. 20 August 1986, 30 March 1988.

AR 600–43, *Personnel—General: Conscientious Objection*. 7 August 1987.

AR 608–4, *Control and Registration of War Trophies and War Trophy Firearms*. 28 August 1969.

AR 635–200, *Personnel Separations, Enlisted Personnel*. 15 July 1966.

AR 672–5–1, *Military Awards*. 1 October 1990.

AR 700–99, *Acquisition, Accounting, Control, and Disposal of Captured Enemy Equipment and Foreign Materiel*. 15 July 1976.

AR 700–131, *Loan and Lease of Army Materiel*. 4 September 1987.

AR 700–137, *Logistics Civil Augmentation Program (LOGCAP)*. 16 December 1985.

AR 755–2, *Disposal of Excess, Surplus, Foreign Excess, Captured, and Unwanted Materiel*. 24 July 1970.

AR 870–20, *Museums and Historical Artifacts*. 9 January 1987.

Department of the Army (DA) Field Manual (FM) 19–40, *Enemy Prisoners of War, Civilian Internees, and Detained Persons*. February 1976.

FM 27–2, *Your Conduct Under the Law of War*. November 1984.

FM 27–10, *The Law of Land Warfare*. 1956.

FM 27–100, *Legal Operations*. September 1991.

FM 41–10, *Civil Affairs Operations*. January 1993.

FM 71–100–2, *Infantry Division Operations: Tactics, Techniques, Procedures*. August 1993.

FM 100–5, *Operations*. June 1993.

FM 100–8, *The Army in Multinational Operations*. November 1997.

FM 100–20, *Military Operations in Low Intensity Conflict*. December 1989.

FM 100–23, *Peace Operations*. December 1994.

Department of the Army Pamphlet (DA Pam) 27–1, *Treaties Governing Land Warfare*. December 1956.

DA Pam 27–1–1, *Protocols to the Geneva Conventions of 12 August 1949*. September 1979.

DA Pam 27–24, *Selected International Agreements*, Vol. II. December 1976.

DA Pam 27–161–1, *Law of Peace*, Vol. I. September 1979.

DA Pam 27–161–2, *International Law*, Vol. II. October 1962.

DA Pam 27–162, *Claims*. December 1989.

DA Pam 27–166, *Soldiers' and Sailors' Civil Relief Act*. August 1981.

DA Pam 715–16, *Contractor Deployment Guide*. February 1996.

Department of the Army. Office of the Chief of Staff. Chief of Staff Regulation 11–2, *Army Programs: Implementation of DoD Law of War Program*. 7 May 1975.

Department of Defense (DoD). DoD Directive 2000.11, *Procedures for Handling Requests for Political Asylum and Temporary Refuge*. 3 March 1972.

DoD Directive 5100.69, *DoD Enemy PW/Detainee Program*. 27 December 1972.

DoD Directive 5100.77, *DoD Program for the Implementation of the Law of War*. 5 November 1974, 10 July 1979.

DoD Directive 5530.3, *International Agreements*. 11 June 1987.

Joint Chiefs of Staff. Office of the Chairman. Instruction 5810.01, *Implementation of the DoD Law of War Program*. 12 August 1996.

Manual for Courts-Martial, U.S. Army, 1928, A. Washington, D.C.: Government Printing Office, 1927.

Manual for Courts-Martial, United States, 1949. Washington, D.C.: Government Printing Office, 1948.

Manual for Courts-Martial, United States, 1951. Washington, D.C.: Government Printing Office, 1951.

Manual for Courts-Martial, United States, 1969. Washington, D.C.: Government Printing Office, 1969.

Manual for Courts-Martial, United States, 1984. Washington, D.C.: Government Printing Office, 1984.

Manual for Courts-Martial, United States (1994 ed.). Washington, D.C.: Government Printing Office, 1994.

Military Assistance Command, Vietnam (MACV). MACV Directive 20–4, *Inspections and Investigations of War Crimes*. 18 May 1968. Reprinted in Prugh, *Law at War*, Appendix F.

MACV Directive 20–5, *Inspections and Investigations: Prisoners of War—Determination of Eligibility*. 17 May 1966, 21 September 1966.

MACV Directive 25–1, *Claims*. 14 May 1965.

MACV Directive 27–1, *Litigation: Legal Services and Litigation*. 16 April 1965.

MACV Directive 27–4, *Legal Services: Foreign Jurisdiction Procedures and Information*. 2 November 1967.

MACV Directive 27–5, *Legal Services: War Crimes and Other Prohibited Acts*. 2 November 1967.

MACV Directive 27–6, *Legal Services and Legal Obligations in Vietnam*. 16 June 1965, 14 September 1968.

MACV Directive 190–3, *Military Police: Enemy Prisoners of War*. 6 April 1967.

MACV Directive 335–1, *Reports of Serious Crimes or Other Incidents*. 5 January 1966.

MACV Directive 381–46, *Military Intelligence: Combined Screening of Detainees*. 27 December 1967.

U.S. Army, Vietnam. Supplement 1 to AR 27–10, *Military Justice*. 15 June 1970.

U.S. Central Command (CENTCOM). CENTCOM Regulation 27–1, *Legal Services: Law of War Program*. 3 January 1989.

CENTCOM Regulation 27–25, *Reporting and Documentation of Alleged War Crimes*. 3 January 1989.

Cases

Haitian Centers Council, Inc. v. *McNary*, Federal Supplement (F. Supp.) 789 (1992): 541, Federal Reporter 2d Series (F. 2d.) 969 (1992): 1350; see also *Sale* v. *Haitian Centers Council, Inc.*, United States Reports (U.S.) 509 (1993): 155.

Morrison v. *United States*, F. Supp. 316 (1970): 78.

United States v. *Brewer*, Army Court of Criminal Appeals (Army) 9401039, unpublished opinion 23 May 95. USALSA.

United States v. *Brown*, Army Court of Military Review (ACMR) 9100880, Military Justice Reporter (MJ) 41: 504 (Army Court of Criminal Appeals [Army] 1994).

United States v. *Bryan*, ACMR 9002397 (1990). USALSA.

United States v. *Bryant*, ACMR 9101419, MJ 35: 739 (ACMR 1992).

United States v. *Calley*, Court Martial (CM) 426402, Court-Martial Reports (CMR) 46: 1131 (ACMR 1973), affirmed CMR 48: 19 (U.S. Court of Military Appeals [CMA]); see also *Calley* v. *Calloway*, F. Supp. 382 (1974): 650, *Calley* v. *Hoffmann*, F. 2d 519 (1975): 814, certiorari (cert.) denied, U.S. 425 (1976): 911.

United States v. *Conde*, Navy-Marine Court Martial 95–01136 (1995).

United States v. *Darling*, CM 413433, CMR 36: 620 (Army Board of Military Review [ABMR] 1966).

United States v. *Duffy*, CM 424795, CMR 47: 658 (ACMR 1973).

United States v. *Finsel*, ACMR 9001841, MJ 33: 739 (ACMR 1991).

United States v. *C. Garcia*, CM 416159, CMR 38: 625 (ABMR 1967).

United States v. *J. Garcia*, CM 416160 (1967). USALSA.

United States v. *Gargaro*, ACMR 9201846, MJ 45: 99 (Court of Appeals for the Armed Forces [CAAF] 1996).

United States v. *Gervase*, CM 416161 (1967). USALSA.

United States v. *Goldman*, CM 420332, CMR 43: 77 (ACMR 1970).

United States v. *Green*, ACMR 9100972 (1991). USALSA.

United States v. *Gussen*, ACMR 9001229, MJ 33: 746 (ACMR 1991).

United States v. *Henderson*, CM 428589 (1972). USALSA.

United States v. *Hodges*, CM 420341 (1969). USALSA.

United States v. *Huet-Vaughn*, ACMR 9101873, MJ 39: 545 (ACMR 1994), MJ 43: 105 (CMA 1995), cert. denied U.S. 516: 1118 (1996).

United States v. *Johnson*, ACMR 9100315, MJ 37: 982 (ACMR 1993).
United States v. *Levy*, CM 416463, CMR 39: 672 (ABMR 1968); *Levy* v. *Resor*, 37 CMR 399 (CMA 1968).
United States v. *Lund, United States* v. *Francis*, CM 420181 (1969). USALSA.
United States v. *McGee*, CM 422412 (1969). USALSA.
United States v. *McMonagle*, MJ 34: 852 (ACMR 1992), MJ 38: 53 (CMA 1993).
United States v. *Medina*, CM 427162 (1971). USALSA.
United States v. *Monday*, CM 413039, CMR 36: 711 (ABMR 1966).
United States v. *Moore*, CM 413158, CMR 36: 531 (ABMR 1966).
United States v. *Mowris*, General Court Martial No. 68 (Fort Carson and 4th Infantry Division [Mechanized], 1 July 1993). USALSA.
United States v. *Noriega*, F. Supp. 808 (1992): 791.
United States v. *Pacheco*, Army 9500002, affirmed ACCA, 11 September 1998 (unpublished). USALSA.
United States v. *Presbury*, ACMR 9100658 (1991). USALSA.
United States v. *Rockwood*, Army 9500872, MJ 48: 501 (Army 1998).
United States v. *Rowland*, CM 421750 (1969). USALSA.
United States v. *Thomas*, CM 416162, CMR 38: 655 (ABMR 1968).
United States v. *Walker*, ACMR 9191209, MJ 37: 892 (ACMR 1993).
United States v. *Wiggins*, ACMR 9100982 (Army, memorandum opinion 30 July 1993). USALSA.
United States v. *Wiley*, ACMR 9101188, MJ 37: 885 (ACMR 1993).
United States v. *Williams*, CM 419872 (1968). USALSA.
United States v. *Wolfe*, CM 413647, CMR 37: 571 (ABMR 1966).
United States v. *Woods*, CM 416803 (1966). USALSA.

Interviews

The author conducted interviews with the following individuals:

Maj. Mark S. Ackerman, Maj. Ida F. Agamy, Mr. Irvin M. Allen, Capt. Norman F. J. Allen III, Maj. Gen. John D. Altenburg, Jr., Capt. Thomas N. Auble, Brig. Gen. Bruce C. Babbitt, Col. John E. Baker, Mr. J. Stevens Berry, Col. Scott C. Black, Col. John R. Bozeman, Capt. Steven E. Buyer, Capt. Corey L. Bradley, Maj. Mark P. Brewer, Mr. Samuel T. Brick, Jr., Col. Fred Bright, Jr., Lt. Col. David H. Brunjes, Maj. Robert A. Burrell, Col. Raymond D. Cole, Col. Dennis M. Corrigan, Capt. Edward B. Cottingham, Maj. Mark Cremin, Col. Jack H. Crouchet, Lt. Col. Leonard G. Crowley, Brig. Gen. Thomas R. Cuthbert, Maj. Matthew L. Dana, Col. Donald A. Deline, Col. Daniel J. Dell'Orto, Col. Dean R. Dort, Col. John J. Douglass, Col. Joseph Dudzik, Col. Paul J. Durbin, Lt. Col. George C. Eblen, Col. William G. Eckhardt, Col. John

T. Edwards, Maj. Karen V. Fair, Col. Gilpin R. Fegley, Col. Ned E. Felder, Col. Patrick Finnegan, Capt. Paul Fiorino, Lt. Col. Douglas K. Fletcher, Col. Leroy F. Foreman, Hon. Herman F. Gierke, Col. Francis A. Gilligan, Maj. Mark J. Gingras, Lt. Col. Richard E. Gordon, Lt. Col. Allen K. Goshi, Maj. Kevin H. Govern, Col. David E. Graham, Maj. Gen. Kenneth D. Gray, Col. Fred K. Green, Col. William R. Hagan, Brig. Gen. Donald W. Hansen, Col. Edward W. Haughney, Brig. Gen. Ronald M. Holdaway, Col. Victor L. Horton, Lt. Col. William A. Hudson, Jr., Maj. Gen. Walter B. Huffman, Col. Dennis R. Hunt, Col. Robert H. Ivey, Lt. Col. Richard B. Jackson, Maj. M. Tia Johnson, Maj. Robin L. Johnson, Maj. Paul E. Kantwill, Hon. Royce C. Lamberth, Capt. Michael L. Larson, Col. John P. Ley, Jr., Lt. Col. Philip W. Lindley, Capt. Charles D. Luckey, Capt. Patricia A. Martindale, Lt. Col. William M. Mayes, Col. Alfred E. McNamee, Maj. Stewart A. Moneymaker, Col. Vahan Moushegian, Lt. Col. William G. Myers, Col. Hubert E. Miller, Maj. Bradley D. Page, Lt. Col. Arthur L. Passar, Maj. Charles N. Pede, Maj. Gen. Wilton B. Persons, Lt. Col. Bernard A. Pfeiffer, Maj. Gen. George S. Prugh, Maj. Mark C. Prugh, Col. Burnett H. Radosh, Col. Quentin Richardson, Col. Raymond C. Ruppert, Col. James S. Russell, Maj. Tarek Sawi, Lt. Col. Jerome W. Scanlon, Col. David A. Shull, Col. James R. Smith, Col. John M. Smith III, Maj. Michael E. Smith, Col. James O. Smyser, Maj. Bradley P. Stai, Col. Barry P. Steinberg, Lt. Col. Steven T. Strong, Lt. Col. Manuel E. F. Supervielle, Maj. Gen. William K. Suter, Col. Robert L. Swann, Maj. Clyde J. Tate II, Col. Arthur H. Taylor, Brig. Gen. Joseph N. Tenhet, Maj. Jack T. Tomarchio, Col. Charles E. Trant, Lt. Col. Gary L. Walsh, Lt. Col. Marc L. Warren, Col. Benjamin H. White, Col. Charles A. White, Col. Barry E. Willey, Capt. Daniel L. Winand, Lt. Col. John I. Winn, Maj. Catherine M. With, Col. Dale N. Woodling, Maj. Daria P. Wollschlaeger.

Military Personnel Records, National Personnel Records Center, NARA

The author consulted the personnel records of the following individuals:

Brig. Gen. Bruce C. Babbitt, Col. Fred Bright, Jr., Col. Raymond D. Cole, Lt. Col. Leonard G. Crowley, Col. Dean R. Dort, Col. John J. Douglass, Col. Paul J. Durbin, Col. Ned E. Felder, Lt. Col. George C. Eblen, Col. Leroy F. Foreman, Col. Francis A. Gilligan, Col. Herbert J. Green, Col. William R. Hagan, Lt. Col. Normand J. Hamelin, Lt. Col. Guy A. Hamlin, Col. Edward W. Haughney, Brig. Gen. Ronald M. Holdaway, Col. Irvin M. Kent, Lt. Col. John N. Kulish, Col. Alfred A. McNamee, Maj. Ralph G. Miranda, Lt. Col. William G. Myers, Col. George R. Robinson, Lt. Col. Jerome W. Scanlon, Jr., Col. John T. Sherwood, Jr., Col. Arthur H. Taylor, Brig. Gen. Joseph N. Tenhet, Jr.,

Col. Robert K. Weaver, Lt. Col. John P. Weber, Col. George F. Westerman, Col. Charles A. White, Jr.

Secondary Sources

Books

Adkin, Mark. *URGENT FURY: The Battle for Grenada*. Lexington, Mass.: D.C. Heath and Co., 1989.

Allard, Kenneth. *Somalia Operations: Lessons Learned*. Washington, D.C.: National Defense University Press, 1995.

Anderson, David L., ed. *Facing My Lai: Moving Beyond the Massacre*. Lawrence, Kans.: University Press of Kansas, 1998.

Atkinson, Rick. *Crusade: The Untold Story of the Persian Gulf War*. Boston: Houghton Mifflin Co., 1993.

Ballard, John R. *Upholding Democracy: The United States Military Campaign in Haiti, 1994–1997*. Westport, Conn.: Praeger Publishers, 1998.

Berry, J. Stevens. *Those Gallant Men*. Novato, Calif.: Presidio Press, 1984.

Bilton, Michael, and Kevin Sim. *Four Hours in My Lai*. New York: Viking Press, 1972.

Bin, Alberto, Richard Hill, and Archer Jones. *DESERT STORM: A Forgotten War*. Westport, Conn.: Praeger Publishers, 1998.

Bowden, Mark. *Black Hawk Down: A Story of Modern War*. New York: Atlantic Monthly Press, 1999.

Briggs, Clarence E. *Operation JUST CAUSE: Panama, December 1989*. Harrisburg, Pa.: Stackpole Books, 1990.

Brinkerhoff, John R. *Waging the War and Winning the Peace: Civil Affairs in the War with Iraq*. Washington, D.C.: Office of the Chief of the Army Reserve, 1991.

———, Ted Silva, and John Seitz. *United States Army Reserve in Operation DESERT STORM—Enemy Prisoner of War Operations: The 800th Military Police Brigade*. Washington, D.C.: Office of the Chief, Army Reserve, 1992.

Burkett, B. G., and Glenna Whitley. *Stolen Valor*. Dallas, Tex.: Verity Press, 1998.

Cavanaugh, John P. *Operation PROVIDE COMFORT: A Model for Future NATO Operations*. Fort Leavenworth, Kans.: School of Advanced Military Studies, 1992.

Clancy, Tom, with Frederick M. Franks, Jr. *Into the Storm: A Study in Command*. New York: Putnam Publishing Group, 1997.

Clarke, Jeffrey J. *Advice and Support: The Final Years, 1965–1973*. U.S. Army in Vietnam. Washington, D.C.: U.S. Army Center of Military History, 1987.

Claudio, Arnaldo, and Ronald Dutton. *Somalia: Background Materials for Planners*. Fort Leavenworth, Kans.: U.S. Army Combined Arms Command, 1992.

Cole, Ronald H. *Operation Just Cause: The Planning and Execution of Joint Operations in Panama*. Washington, D.C.: Office of the Chairman of the Joint Chiefs of Staff, 1995.

Crouchet, Jack H. *Vietnam Stories: A Judge's Memoir*. Niwot, Colo.: University Press of Colorado, 1997.

DeLong, Kent, and Steven Tuckey. *Mogadishu: Heroism and Tragedy*. Westport, Conn.: Praeger Press, 1994.

Donnelly, Thomas, Margaret Roth, and Caleb Baker. *Operation Just Cause*. New York: Lexington Books, 1991.

Donovan, David. *Once a Warrior King*. New York: McGraw-Hill, 1985.

Eckhardt, George S. *Command and Control, 1950–1969*. Vietnam Studies. Washington, D.C.: Department of the Army, 1974.

Fishel, John T. *Civil Military Operations in the New World*. Westport, Conn.: Praeger Publishing, 1997.

Gehring, Stephen P. *From the Fulda Gap to Kuwait: U.S. Army, Europe, and the Gulf War*. Washington, D.C.: Department of the Army, 1998.

Generous, William T., Jr. *Swords and Scales: The Development of the Uniform Code of Military Justice*. New York: Kennikat Press, 1973.

Gordon, Michael R., and Bernard E. Trainor. *The General's War: The Inside Story of the Conflict in the Gulf*. Boston: Little, Brown and Co., 1995.

Greenberg, Lawrence M. *United States Army Unilateral and Coalition Operations in the 1965 Dominican Republic Intervention*. Washington, D.C.: U.S. Army Center of Military History, 1987.

Griffith, Robert K., Jr. *The U.S. Army's Transition to the All-Volunteer Force, 1968–1974*. Washington, D.C.: U.S. Army Center of Military History, 1997.

Hammond, William H. *Public Affairs: The Military and the Media, 1962–1968*. U.S. Army in Vietnam. Washington, D.C.: U.S. Army Center of Military History, 1988.

———. *Public Affairs: The Military and the Media, 1968–1973*. U.S. Army in Vietnam. Washington, D.C.: U.S. Army Center of Military History, 1996.

Hayden, Cynthia L. *JTF 180 Operation Uphold Democracy Oral History Interviews*. Fort Bragg, N.C.: XVIII Airborne Corps Training Support Center, 1995.

Hersh, Seymour. *My Lai 4: A Report on the Massacre and Its Aftermath*. New York: Random House, 1970.

House, Jonathan M. *The United States Army in Joint Operations, 1950–1983*. Washington, D.C.: U.S. Army Center of Military History, 1992.

Judge Advocate General's Corps, U.S. Army. *The Army Lawyer: A History of the Judge Advocate General's Corps, 1775–1975*. Washington, D.C.: Government Printing Office, 1975.

Krepinevich, Andrew F. *The Army and Vietnam*. Baltimore: Johns Hopkins University Press, 1986.

Lang, Daniel. *Casualties of War*. New York: McGraw, 1969.

Le Gro, William E. *Vietnam from Cease-Fire to Capitulation*. Washington, D.C.: Government Printing Office, 1981.

Lewy, Guenter. *America in Vietnam*. New York: Oxford University Press, 1978.

Lurie, Jonathan. *Pursuing Military Justice*. Princeton, N.J.: Princeton University Press, 1998.

McDonnell, Janet. *Supporting the Troops*. Washington, D.C.: Corps of Engineers, 1996.

McDonough, James. *Platoon Leader*. Novato, Calif.: Presidio Press, 1985.

MacGarrigle, George L. *Taking the Offensive, October 1966 to October 1967*. U.S. Army in Vietnam. Washington, D.C.: U.S. Army Center of Military History, 1998.

McMaster, H. R. *Dereliction of Duty*. New York: HarperCollins Publishers, 1997.

Meyer, Harold J. *Hanging Sam: A Military Biography of General Samuel T. Williams*. Denton, Tex.: University of North Texas Press, 1990.

Meyerson, Joel D. *Images of a Lengthy War*. U.S. Army in Vietnam. Washington, D.C.: U.S. Army Center of Military History, 1989.

Moore, Harold G., and Joseph L. Galloway. *We Were Soldiers Once, and Young: The Battle of the Ia Drang Valley*. New York: Random House, 1992.

Natkiel, Richard. *Atlas of American Wars*. Greenwich, Conn.: Bison Books, 1986.

Nolan, William K. *Sappers in the Wire: The Life and Death of Firebase Maryann*. College Park: Texas A&M Press, 1995.

Olson, James S., and Randy Roberts. *My Lai: A Brief History with Documents*. New York: St. Martin's Press, 1998.

Pagonis, William G. *Moving Mountains: Lessons in Logistics and Leadership from the Gulf War*. Boston: Harvard University Business School Press, 1992.

Palmer, Bruce R., Jr. *Intervention in the Caribbean: The Dominican Crisis of 1965*. Lexington: University Press of Kentucky, 1969.

———. *The 25-Year War: America's Military Role in Vietnam*. Lexington: University Press of Kentucky, 1984.

Peers, William R. *The My Lai Inquiry*. New York: Norton, 1979.

Pirnie, Bruce R. *Operation URGENT FURY: The United States Army in Joint Operations*. Washington, D.C.: U.S. Army Center of Military History, 1986.

Prugh, George S. *Law at War: Vietnam, 1964–1973*. Vietnam Studies. Washington, D.C.: Department of the Army, 1975.

Scales, Robert H., Jr. *Certain Victory: The U.S. Army and the Gulf War*. Washington, D.C.: Brassey's Publishing, 1994.

Schubert, Frank N., and Theresa L. Kraus, eds. *The Whirlwind War: The United States Army in Operations DESERT SHIELD and DESERT STORM*. Washington, D.C.: U.S. Army Center of Military History, 1995.

Schwarzkopf, H. Norman, and Peter Petre. *It Doesn't Take a Hero*. New York: Bantam Books, 1992.

Sheehan, Neil. *A Bright Shining Lie*. New York: Random House, 1988.

Sherrill, Robert. *Military Justice Is to Justice as Military Music Is to Music*. New York: Harper and Row, 1970.

Smith, Charles R. *Angels from the Sea: Relief Operations in Bangladesh, 1991*. Washington, D.C.: Marine Corps History and Museums Division, 1995.

Solis, Gary D. *Marines and Military Law: Trial by Fire*. Washington, D.C.: Marine Corps History and Museums Division, 1989.

———. *Son Thang: An American War Crime*. Annapolis, Md.: Naval Institute Press, 1997.

Sorley, Lewis. *Thunderbolt: General Creighton Abrams and the Army of His Times*. New York: Simon and Schuster, 1992.

Spector, Ronald H. *Advice and Support: The Early Years, 1941–1960*. U.S. Army in Vietnam. Washington, D.C.: U.S. Army Center of Military History, 1983.

———. *U.S. Marines in Grenada, 1983*. Washington, D.C.: Headquarters, U.S. Marine Corps, 1987.

Stanton, Shelby. *The Rise and Fall of an American Army: U.S. Ground Forces in Vietnam, 1965–1973*. Novato, Calif.: Presidio Press, 1985.

Stein, Jeff. *A Murder in Wartime*. New York: St. Martin's Press, 1992.

Summers, Harry G., Jr. *Historical Atlas of the Vietnam War*. Boston: Houghton Mifflin, 1995.

———. *On Strategy: A Critical Analysis of the Vietnam War*. Novato, Calif.: Presidio Press, 1982.

———. *On Strategy II: A Critical Analysis of the Gulf War*. New York: Dell, 1992.

Swain, Richard M. *"Lucky War": Third Army in DESERT STORM*. Fort Leavenworth, Kans.: U.S. Army Command and General Staff College Press, 1994.

Taylor, Telford. *Nuremburg and Vietnam: An American Tragedy*. Chicago: Quadrangle Books, 1970.

U.S. Army Center of Military History. *Resource Guide: Operation RESTORE HOPE*. Washington, D.C.: U.S. Army Center of Military History, 1994.

U.S. Army Forces Command. *The Role of FORSCOM in the Reception and Care of Refugees from Cuba in the Continental United States*. Atlanta, Ga.: Headquarters, FORSCOM, 1984.

Vien, Cao Van. *The Final Collapse*. Indochina Monographs. Washington, D.C.: Government Printing Office, 1983.

Whitlow, Robert H. *U.S. Marines in Vietnam, 1954–1964*. Washington, D.C.: Headquarters, U.S. Marine Corps, 1977.

Wiener, Frederick B. *Civilians Under Military Justice*. Chicago: University of Chicago Press, 1967.

Winthrop, William. *Military Law and Precedents*. 2d ed. Washington, D.C.: Government Printing Office, 1920.

Yates, Lawrence A. *POWER PACK: U.S. Intervention in the Dominican Republic, 1965–1966*. Leavenworth Papers no. 15. Fort Leavenworth, Kans.: Combat Studies Institute, 1988.

Articles

Adams, Thomas K. "Intervention in Haiti: Lessons Relearned." *Military Review* 76 (September–October 1996): 45–56.

Addicott, Jeffrey F., and William A. Hudson, Jr. "The Twenty-fifth Anniversary of My Lai: A Time to Inculcate the Lessons." *Military Law Review* 139 (Winter 1993): 153–85.

Althouse, James E. "Contractors on the Battlefield." *Army Logistician* 30 (November–December 1998): 14–17.

Armstrong, Edward R. "Detainee Operations in Haiti During Operation UPHOLD DEMOCRACY." *Military Police* (Winter 1995): 16–19.

Arnold, S. L., and David T. Stahl. "A Power Projection Army in Operations Other Than War." *Parameters* (Winter 1993–94): 4–26.

Ary, Vaughn A. "Accounting for Prisoners of War: A Legal Review of United States Armed Forces Identification and Reporting Procedures." *Army Lawyer* (August 1994): 16–27.

Barn, Thomas. "Fragging: A Study." *Army* (April 1977): 46.
Barnes, Rudolph C. "Legitimacy and the Lawyer in Low-Intensity Conflict (LIC): Civil Affairs Legal Support." *Army Lawyer* (October 1988): 5.
———. "Operational Law, Special Operations, and Reserve Support." *Army Lawyer* (December 1984): 1.
Berkeley, Bill. "Aftermath: Genocide, the Pursuit of Justice, and the Future of Africa." *Washington Post Magazine* (11 October 1998): 10.
Bird, Julie, and Tom Donnelly. "Friendly Fire." *Army Times* (19 September 1991): 3–4, 11.
Bond, Thomas C. "Fragging: A Study." *American Journal of Psychiatry* 133 (1976): 1328–31.
Borch, Frederic L. "Bolsheviks, Polar Bears, and Military Law: The Experiences of Army Lawyers in North Russia and Siberia in World War I." *Prologue* 30 (Fall 1998): 180–91.
———. "Law of War Training: A Practical Program." *Infantry* (November–December 1985): 33.
Borek, Ted. "Legal Services During War." *Military Law Review* 120 (Spring 1988): 1.
Bowens, Glenn. "Legal Issues in Peace Operations." *Parameters* 28 (Winter 1998–99): 51.
Burger, James A. "International Law—The Role of the Legal Adviser, and Law of War Instruction." *Army Lawyer* (September 1978): 22.
Burman, Anne L., and Teresa K. Hollingsworth. "JAGs Deployed: Environmental Law Issues." *Air Force Law Review* 42 (1997): 19–48.
"Civil Disturbance Rules of Engagement: Joint Task Force Los Angeles." *Army Lawyer* (September 1992): 32.
Clarke, Walter S. "Testing the World's Resolve in Somalia." *Parameters* (Winter 1993–94): 42–58.
Coleman, Gerald C. "Operational Law and Contingency Planning at XVIII Airborne Corps." *Army Lawyer* (March 1988): 17.
Currier, Roger M., and Irvin M. Kent. "The Boards of Review of the Armed Services." *Vanderbilt Law Review* 6 (February 1953): 241–50.
Damarest, Geoffrey. "Updating the Geneva Conventions: The 1977 Protocols." *Army Lawyer* (November 1983): 18.
Demma, Vincent. "The U.S. Army in Vietnam." In *American Military History*, rev. ed. Washington, D.C.: U.S. Army Center of Military History , 1989.

Denny, Michael C. "The Impact of Article 82 of Protocol I to the 1949 Geneva Conventions on the Organization and Operation of a Division SJA Office." *Army Lawyer* (April 1980): 14.

DeSaussure, Ariane L. "The Role of the Law of Armed Conflict During the Persian Gulf War: An Overview." *Air Force Law Review* 37 (1994): 41–68.

Doerrer, Eric A. "Civil Affairs in Haiti." *Military Review* (March–April 1996): 73–79.

Donahue, Bernard E. "Air Force Enemy Prisoner of War Operations." *Air Force Law Review* 37 (1994): 253–66.

Douglass, John J. "The Judicialization of Military Courts." *Hastings Law Journal* 22 (January 1971): 213–35.

Downing, Wayne A. "Firepower, Attrition, Maneuver—US Army Operations Doctrine: A Challenge for the 1980s and Beyond." *Military Review* 77 (January–February 1997): 144–50.

Dworken, Jonathan T. "Rules of Engagement: Lessons from RESTORE HOPE." *Military Review* (September 1994): 26–41.

Eckhardt, William. "Command Criminal Responsibility." *Military Law Review* 97 (Summer 1982): 1–34.

Erickson, Richard J. "Status of Forces Agreements: A Sharing of Sovereign Prerogative." *Air Force Law Review* 37 (1994): 137.

Ervin, Sam J., Jr. "The Military Justice Act of 1968." *Military Law Review* 45 (July 1969): 77–98.

———. "The Military Justice Act of 1968." *Wake Forest Law Review* 5 (1969): 223.

Fair, Karen V. "The Rules of Engagement in Somalia: A Judge Advocate's Primer." *Small Wars and Insurgencies* 8 (Spring 1997): 107–26.

Falk, Richard. "The Haiti Intervention: A Dangerous World Order Precedent for the United Nations." *Harvard International Law Journal* 36 (1995): 341.

Feeney, Thomas J., and Margaret L. Murphy. "The Judge Advocate General's Corps, 1982–1987." *Military Law Review* 122 (Fall 1988): 1–76.

Finnegan, Patrick. "Operational Law: Plan and Execute." *Military Review* 76 (March–April 1996): 33.

Flanagan, Edward. "100-Hour War." *Army* (April 1991): 18, 21–26.

Fontenot, Gregory. "The 'Dreadnoughts' Rip the SADDAM Line." *Army* (January 1992): 28–36.

Fratcher, William F. "History of The Judge Advocate General's Corps, United States Army." *Military Law Review* 4 (March 1959): 89–122.

Girardet, Edward. "A Forgotten Face of War: In Somalia, Where the Superpowers Once Jousted, Madness Now Reigns." *U.S. News and World Report* (4 May 1992): 38–40.

Graham, David E. "Operational Law—A Concept Comes of Age." *Army Lawyer* (July 1987): 10.

Green, Fred K. "The Cease-Fire." In *Report of the Sixth International Congress of the International Society of Military Criminal Law and the Law of War* I (May 1973): 199–249.

———. "The Concept of War and the Concept of Combatant in Modern Conflicts." *Military Law and Law of War Review* (1971): 267–309.

Grunawalt, Richard J. "The JCS Standing Rules of Engagement: A Judge Advocate's Primer." *Air Force Law Review* 42 (1997): 245–58.

Hagan, William R. "The Judge Advocate in DESERT SHIELD/STORM." *Military Advocate* (Spring 1991): 3.

Harris, Jeffrey L. "Grenada—A Claims Perspective." *Army Lawyer* (January 1986): 7.

Heinl, Robert D., Jr. "The Collapse of the Armed Forces." *Armed Forces Journal* (7 June 1971): 31.

Hodson, Kenneth J. "The Manual for Courts-Martial—1969." *Military Law Review* 57 (Summer 1972): 1–16.

———. "The Military Justice Act." *Judge Advocate Journal*, Bulletin no. 42 (May 1970): 30–38.

———. "Report of TJAG—Army." *Judge Adovcate Journal*, Bulletin no. 38 (December 1966): 9–11.

———. "Report of TJAG—Army." *Judge Advocate Journal*, Bulletin no. 41 (March 1969): 7.

———. "Report of TJAG—Army." *Judge Advocate Journal*, Bulletin no. 42 (May 1970): 4–7.

Hunt, Dennis R. "Viet Nam Hustings." *Judge Advocate Journal*, Bulletin no. 44 (July 1972): 23.

Kastenberg, Joshua E. "The Legal Regime for Protecting Cultural Property During Armed Conflict." *Air Force Law Review* 42 (1997): 277–305.

Keeva, Steven. "Lawyers in the War Room." *American Bar Association Journal* (December 1991): 52–59.

Kent, Irvin M., Jon N. Kulish, Ned E. Felder, and Herbert Green. "A Lawyer's Day in Vietnam." *American Bar Association Journal* 54 (December 1968): 1177.

Kindsvatter, Peter S. "VII Corps in the Gulf War: Deployment and Preparation for DESERT STORM." *Military Review* (January 1992): 2–16.

———. "VII Corps in the Gulf War: Ground Offensive." *Military Review* (February 1992): 16–37.

———. "VII Corps in the Gulf War: Post–Cease Fire Operations." *Military Review* (June 1992): 2–19.

King, Archibald. "Changes in the Uniform Code of Justice Necessary To Make It Workable in Time of War." *Federal Bar Journal* 22 (Winter 1962): 49–59.

Korroch, Robert E. "The Changing Practice of Military Law—Peace: The New Threat." *Federal Bar and News Journal* 37 (May 1990): 202–05.

Kuhfeld, Albert M. "Amendments to Article 15, Uniform Code of Military Justice." *Judge Advocate Journal Bulletin* 34 (1963): 69–80.

Lafferty, Patrick J. "The 511th Military Police Company in Cap Haitien, Haiti." *Military Police* (Winter 1995): 20–24.

Levie, Howard S. "Compliance by States with the 1949 Geneva Prisoner of War Convention." *Army Lawyer* (June 1974): 1.

Lewis, Paul. "A Short History of United Nations Peacekeeping." *Military History Quarterly* 5 (Autumn 1992): 33.

Linden, Eugene. "Fragging and Other Withdrawal Symptoms." *Saturday Review* (8 January 1972): 12.

Lorenz, Frederick M. "Confronting Thievery in Somalia." *Military Review* (August 1994): 46–55.

———. "Law and Anarchy in Somalia." *Parameters* (Winter 1993–94): 28.

———. "Rules of Engagement in Somalia: Were They Effective?" *Naval Law Review* 42 (1995): 62.

Martins, Mark S. "Rules of Engagement for Land Forces: A Matter of Training, Not Lawyering." *Military Law Review* 143 (Winter 1994): 3.

———. "'War Crimes' During Operations Other Than War: Military Doctrine and Law Fifty Years After Nuremberg—and Beyond." *Military Law Review* 149 (Summer 1995): 145–87.

Masterton, R. Peter. "The Persian Gulf War Crimes Trials." *Army Lawyer* (June 1991): 7.

Matthews, Ann. "Army JAG's Search for Justice." *Vietnam* (December 1998): 30–36.

Meek, Philip A. "Operation PROVIDE COMFORT: A Case Study in Humanitarian Relief and Foreign Assistance." *Air Force Law Review* 37 (1994): 225–38.

Moore, Molly. "For Female Soldiers, Different Rules." *Washington Post*, 23 Aug 90, p. D1.

Morgan, Edmund M. "The Background of the Uniform Code of Military Justice." *Vanderbilt Law Review* 6 (1953): 169–85.

———. "The Existing Court-Martial System and the Ansell Articles." *Yale Law Journal* 29 (November 1919): 52–74.

Murray, Williamson. "The Gulf War as History." *Military History Quarterly* 10 (Autumn 1997): 6–19.

Naipaul, V. S. "An Island Betrayed." *Harper's* 268, no. 1606 (March 1984): 61–72.

Nardotti, Michael J., Jr. "General Ken Hodson—A Thoroughly Remarkable Man." *Military Law Review* 151 (Winter 1996): 202–15.

Naylor, Sean D. "Desert Storm After-Action Review: Flight of Eagles." *Army Times* (22 July 1991): 8.

Nichols, Camille M. "The Logistics Civil Augmentation Program." *Military Review* (March–April 1996): 65–72.

Norsworthy, Levator, Jr. "Organization for Battle: The Judge Advocate's Responsibility Under Article 82 of Protocol I to the Geneva Conventions." *Military Law Review* 93 (Summer 1981): 9–24.

O'Brien, Edward J. "The Nuremberg Principles: Command Responsibility and the Defense of Captain Rockwood." *Military Law Review* 149 (Summer 1995): 275–91.

Palmer, Jeffrey S. "Claims Encountered During an Operational Contingency." *Air Force Law Review* 42 (1997): 201–26.

Parker, Harold E. "Report of TJAG—Army." *Judge Advocate Journal* 43 (May 1971): 7.

Parks, Hays W. "A Few Tools in the Prosecution of War Crimes." *Military Law Review* 149 (Summer 1995): 73–85.

———. "Command Responsibility for War Crimes." *Military Law Review* 62 (Fall 1973): 1–104.

———. "The Law of War Adviser." *JAG Journal* 31 (Summer 1980): 44–47.

Phillips, W. Darrell. "Fiscal Law Constraints Upon Exercise-Related Activities." *Air Force Law Review* 42 (1997): 259–76.

Prugh, George S. "The Code of Conduct for the Armed Forces." *Columbia Law Review* 56 (May 1956): 678.

———. "Denims, Pinks and Greens." Manuscript, January 1994. Historians files, OTJAG.

———. "Introduction to William Winthrop's 'Military Law and Precedents.'" *Military Law and Law of War Review* 27 (1988): 437–70.

———. "Law Practice in the Vietnam War." *Federal Bar Journal* 27 (Winter 1967): 57.

Richardson, William R. "FM 100–5: The AirLand Battle in 1986." *Military Review* 77 (January–February 1997): 174–77.

Riggs, Ronald. "The Grenada Intervention: A Legal Analysis." *Military Law Review* 109 (Summer 1985): 1–81.

Romig, Thomas J. "The Legal Basis for United States Military Action in Grenada." *Army Lawyer* (April 1985): 1.

Santere, Elyce. "From Confiscation to Contingency Contracting: Property Acquisition on or near the Battlefield." *Military Law Review* 124 (Winter 1989): 111.

Scanlon, Jerome W. "SJA Spotlight: U.S. Delegation, Four-Party Joint Military Team." *Army Lawyer* (March 1974): 14.

Schlueter, David. "The Court-Martial: A Historical Survey." *Military Law Review* 87 (Winter 1980): 129.

Schumm, Walter R., Bobby B. Polk, John Bryan, Frank Fornataro, and Jennifer Curry. "Treat Prisoners Humanely." *Military Review* 78 (January–February 1998): 83–93.

Scicchitano, J. Paul. "DESERT STORM After-Action Review: Eye of the Tiger." *Army Times* (10 June 1991): 12.

———. "What's so special about the Bryan Case?" *Army Times* (13 August 1990): 12.

Segal, D. R., T. P. Furukawa, and J. C. Lindh. "Light Infantry as Peacekeepers in the Sinai." *Armed Forces and Society* (1990): 385–403.

Selvage, Donald R. "Operation SEA ANGEL: Bangladesh Disaster Relief." *Marine Corps Gazette* 75 (November 1991): 11.

Shapiro, Bruce. "Military Resisters: Hell for Those Who Won't Go." *Nation* 118, no. 12 (18 February 1991): 194.

Sharp, Walter G., Sr. "The Effective Deterrence of Environmental Damage During Armed Conflict: A Case Analysis of the Persian Gulf War." *Military Law Review* 137 (Summer 1992): 1–66.

Silliman, Scott L. "JAG Goes to War: The DESERT SHIELD Deployment." *Air Force Law Review* 37 (1994): 85–94.

Smith, Bruce T. "Meet the Army's Newest Lieutenant Colonel: Tia Johnson." *Federal Lawyer* (February 1997): 28–31.

Stackpole, Henry C., and Eric L. Chase. "Humanitarian Intervention and Disaster Relief: Projecting Military Strength Abroad to Save Lives." *Marine Corps Gazette* 77 (February 1993): 2.

Tacitus (pseudonym). "Few Lessons Were Learned in Panama Invasion." *Armed Forces Journal* (June 1993): 54–56.

Terry, James P. "The Environment and the Laws of War: The Impact of DESERT STORM." *Naval War College Review* 45 (Winter 1992): 61.

Tomes, Jonathan P. "Primer on the AirLand Battle: What Every Judge Advocate Needs to Know About the Client's Primary Business." *Army Lawyer* (December 1983): 1.

Van Creveld, Martin. "The Gulf Crisis and the Rules of War." *Military History Quarterly* 3 (Summer 1991): 10.

Walker, Daniel, and C. George Niebank. "The Court of Military Appeals—Its History, Organization and Operation." *Vanderbilt Law Review* 6 (February 1953): 228–40.

Walsh, Gary L. "Role of the Judge Advocate in Special Operations." *Army Lawyer* (August 1989): 4.

Ward, Chester. "The UCMJ—Does It Work?" *Vanderbilt Law Review* 6 (February 1953): 186–227.

Warner, Ronald. "Planning for Foreign Claims Operations During Overseas Deployment of Military Forces." *Army Lawyer* (July 1987): 61.

Webb, James H. "The Sad Conviction of Sam Green: The Case for the Reasonable and Honest War Criminal." *Res Ipsa Loquitur; Georgetown Review of Law and Public Interest* 26 (Winter 1974): 11.

Westerman, George F. "A New Approach in Disseminating the Geneva Conventions." *Military Law Review* 45 (July 1969): 99–105.

———. "Military Justice in the Republic of Vietnam." *Military Law Review* 31 (January 1966): 137–58.

Westmoreland, William C., and George S. Prugh. "Judges in Command: The Judicialized Uniform Code of Military Justice in Combat." *Harvard Journal of Law and Public Policy* 4 (1980): 1.

Whitaker, Richard M. "Environmental Aspects of Overseas Operations." *Army Lawyer* (April 1995): 31.

Wiener, Frederick B. "Courts-Martial and the Bill of Rights: the Original Practice." *Military Law Review* Bicentennial Issue (1975): 171.

———. "Courts-Martial for Civilians Accompanying the Armed Forces in Vietnam." *American Bar Association Journal* 54 (January 1968): 24.

———. "Martial Law Today." *American Bar Association Journal* 55 (August 1969): 723.

———. "The Perils of Tinkering with Military Justice." *Army* (November 1970): 8–12.

Willis, John T. "The United States Court of Military Appeals: Its Origin, Operation and Future." *Military Law Review* 55 (Winter 1972): 39–93.

Wilson, William. "I Had Prayed to God . . ." *American Heritage: Battles and Leaders* (1994): 54–63.

Yeosock, John J. "Army Operations in the Gulf Theater." *Military Review* (September 1991): 2–15.

Zinni, Anthony C. "The SJA in Future Operations." *Marine Corps Gazette* (February 1996): 15–18.

Index

PIN : 079054-000

Made in United States
Cleveland, OH
05 November 2026

47668292R10237